BIOELECTRONICS HANDBOOK

BIOELECTRONICS HANDBOOK

MOSFETs, Biosensors, and Neurons

Massimo Grattarola

Giuseppe Massobrio

McGRAW-HILL
New York San Francisco Washington, D.C. Auckland Bogotá
Caracas Lisbon London Madrid Mexico City Milan
Montreal New Dehli San Juan Singapore
Sydney Tokyo Toronto

Library of Congress Cataloging-in-Publication Data

Bioelectronics handbook : MOSFETs, biosensors, and neurons / Massimo
   Grattarola, Giuseppe Massobrio.
       p.   cm.
   Includes bibliographical references and index.
   ISBN 0-07-003174-6
   1. Molecular electronics. 2. Bioelectronics. I. Grattarola,
Massimo. II. Massobrio, Giuseppe.
TK7874.8.B56   1998
621.3815—dc21   97-51529
                                                            CIP

McGraw-Hill
A Division of The *McGraw·Hill* Companies

Copyright © 1998 by The McGraw-Hill Companies, Inc. All rights reserved.
Printed in the United States of America. Except as permitted under the United
States Copyright Act of 1976, no part of this publication may be reproduced or
distributed in any form or by any means, or stored in a data base or retrieval
system, without the prior written permission of the publisher.

1 2 3 4 5 7 8 9 0   FGR/FGR   9 0 3 2 1 0 9 8

ISBN 0-07-003174-6

*The sponsoring editor for this book was Stephen S. Chapman, the editing
supervisor was David E. Fogarty, and the production supervisor was Pamela A.
Pelton. It was set in Times Roman by Ampersand Graphics, Ltd.*

Printed and bound by Quebecor/Fairfield.

   This book was printed on recycled, acid-free paper
containing a minimum of 50% recycled de-inked fiber.

> Information contained in this work has been obtained by The
> McGraw-Hill Companies, Inc. ("McGraw-Hill") from sources
> believed to be reliable. However, neither the McGraw-Hill nor its
> authors guarantee the accuracy or completeness of any information
> published herein, and neither McGraw-Hill nor its authors shall be
> responsible for any errors, omissions, or damages arising out of use
> of this information. This work is published with the understanding
> that McGraw-Hill and its authors are supplying information but are
> not attempting to render engineering or other professional services.
> If such services are required, the assistance of an appropriate
> professional should be sought.

To our parents

CONTENTS

Preface   xi
Acknowledgments   xiii

Part 1   Basic Properties of Silicon

Chapter 1. Semiconductor Materials   3

1.1. Chemical and Physical Bonds / 3
1.2. Electronic Orbitals / 4
1.3. Energy Bands in Metals and Semiconductors / 14
     Problems / 19
     References / 19

Chapter 2. Semiconductors and Charge Carriers   21

2.1. Optical Properties of Semiconductors / 21
2.2. Thermal Excitation of Valence Electrons / 26
2.3. Fermi-Dirac Distribution / 27
2.4. Charge Concentrations in Intrinsic Semiconductors / 32
2.5. Effective Mass / 34
2.6. Doped Semiconductors / 35
2.7. A Simplified Introduction to the Generation-Recombination Processes / 40
     Problems / 43
     References / 44

Chapter 3. Carrier Motion in Semiconductors   45

3.1. Carrier Motion / 45
3.2. Carrier Motion by Drift / 46
3.3. Carrier Motion by Diffusion / 54
3.4. Transport Equations / 58
3.5. Boltzmann Transport Equation / 60
3.6. Recombination-Generation Processes in Semiconductors / 63
3.7. Continuity Equation / 71
3.8. Hall Effect / 75
     Problems / 76
     References / 78

Part 2   Basic Properties of Biological Molecules

Chapter 4. Biological Materials   81

4.1. Physical Bonds Revisited / 81
4.2. Water and Electrolyte Solutions / 84

4.3. Optical Properties of Molecules in Solution / 89
4.4. Biological Molecules—Proteins / 92
4.5. Nucleic Acids / 96
4.6. Phospholipids Organization / 99
4.7. Cell Membrane / 99
4.8. An Overview of the Eucaryotic Cell / 105
   Problems / 106
   References / 107

Chapter 5. Motion in Solution and Chemical Reactions — 109

5.1. Diffusion in Solution / 109
5.2. Brownian Motion / 114
5.3. Electrophoresis / 117
5.4. Chemical Reactions / 121
   Problems / 130
   References / 130

Part 3 Junctions and Membranes

Chapter 6. Semiconductor Junctions — 135

6.1. pn Junction / 135
6.2. pn Junction in Equilibrium / 136
6.3. pn Junction in Nonequilibrium: Effect of the Bias Voltage / 148
6.4. Current-Voltage Characteristics of the pn Junction / 149
6.5. Charge Storage in the pn Junction / 157
6.6. Transient Behavior of the pn Junction / 160
6.7. Considerations on the Ideal pn Junction / 160
6.8. Reverse Bias: Deviations from the Ideal Diode Behavior / 161
6.9. Forward Bias: Deviations from the Ideal Diode Behavior / 163
6.10. pn Junction (Diode) Models / 167
6.11. MS Junction / 169
   Problems / 174
   References / 175

Chapter 7. Solid-Electrolyte Junctions and Membrane Transport — 177

7.1. Electrode-Electrolyte Interfaces / 177
7.2. Solution of the Poisson-Boltzmann Equation under Various Boundary Conditions / 187
7.3. Membrane Transport / 193
   Problems / 207
   References / 207

Part 4 Devices and CAD

Chapter 8. Metal-Oxide-Semiconductor (MOS) Structure — 211

8.1. MOS Structure / 211
8.2. Accumulation Operating Mode / 212
8.3. Depletion Operating Mode / 214

- 8.4. Inversion Operating Mode / *220*
- 8.5. *C-V* Plots of an MOS Structure / *225*
- 8.6. Ion Implantation for Threshold Voltage Control / *231*
- 8.7. General Analysis of the MOS Structure / *232*
  - Problems / *235*
  - References / *235*

Chapter 9. Metal-Oxide-Semiconductor Field-Effect Transistor (MOSFET) — 237

- 9.1. Enhancement-Mode MOSFET / *237*
- 9.2. Depletion-Mode MOSFET / *246*
- 9.3. MOSFET Amplifier / *251*
- 9.4. Biasing Circuits for the MOSFET / *256*
- 9.5. Small-Signal Models for the MOSFET / *260*
- 9.6. MOSFET-Based Operational Amplifier / *266*
- 9.7. Subthreshold Operation of the MOSFET / *267*
- 9.8. Contributions of Organic Chemistry to the Development of Electronic Devices / *274*
  - Problems / *276*
  - References / *280*

Chapter 10. MOSFET-Based Bioelectronic Devices: Biosensors — 281

- 10.1. Biosensor Overview / *281*
- 10.2. Ion-Sensitive Field-Effect Transistor (ISFET) / *287*
- 10.3. Enzyme Field-Effect Transistor (ENFET) / *294*
- 10.4. Cell-Based Biosensors and Sensors of Cell Metabolism / *302*
- 10.5. Light-Addressable Potentiometric Sensor (LAPS) / *304*
- 10.6. Contributions of Microfabrication Technologies to the Field of Biosensors / *307*
  - Problems / *310*
  - References / *311*

Chapter 11. Neurons and Neuronal Networks — 315

- 11.1. Short Overview of the Biology of the Neuron / *315*
- 11.2. Biophysical Description of the Action Potential / *317*
- 11.3. The Neuron as a Threshold Device / *326*
- 11.4. Synapses / *328*
- 11.5. Networks /*336*
- 11.6. Neurobioengineering Neuroelectronic Junctions / *338*
- 11.7. Silicon Neurons / *346*
  - Problems / *348*
  - References / *348*

Chapter 12. Models of Bioelectronic Devices and Computer Simulations — 351

- 12.1. SPICE Simulator / *351*
- 12.2. MOSFET Models in SPICE / *352*
- 12.3. Use of SPICE for Modeling Silicon-Based Chemical Sensors / *356*
- 12.4. Use of SPICE for Modeling Neurons (Excitable Membrane) / *380*
- 12.5. Use of SPICE for Modeling Silicon Neurons / *390*
  - References / *397*

Appendix A. Physical Constants and Material Properties — 399

A.1. Physical constants / *399*
A.2. Properties of Si, GaAs, SiO$_2$, Si$_3$N$_4$, Al$_2$O$_3$ (at 300 K) / *400*

Appendix B. Mathematical Operators — 401

B.1. Vector Differential Operator (∇) / *401*
B.2. Laplacian Operator (∇^2) / *402*
B.3. Gradient / *402*
B.4. Divergence of a Vector Field / *402*
B.5. Curl of a Vector Field / *403*
B.6. Basic Relations for the Mathematical Operators / *403*

Index 405

PREFACE

Bioelectronics is the result of the cross-fertilization between micro-/nanoelectronics and molecular biology of the cell. It deals with phenomena occuring in semiconductor materials, biological materials, aqueous solutions, and solid-liquid junctions, and represents the conceptual framework for the design of hybrid bioelectronic devices and of biologically inspired artificial devices and systems.

This book is intended to provide a contribution to the foundation of this new discipline, by describing all the aforementioned phenomena by means of a common elementary physicomathematical language; in this way MOSFETs, biosensors, and neurons are viewed under a common perspective.

The book is not an advanced treatise, but rather a basic textbook, covering in a self-consistent way topics that are usually considered in separate books. Just to give a few examples: All the tools are given to fully appreciate the biological, chemical, and electronic operational principles of silicon-based biosensors. Single neurons and neuronal networks are considered from the neurobiological, equivalent-circuit, and technological viewpoints.

The intended audience includes advanced undergraduate and beginning graduate students from several disciplines, mostly biology, electrical engineering, and physics. This book is intended to represent a self-consistent reference for teachers of courses dealing with bioengineering, biotechnology, applied biophysics, and microelectronic biosensors.

The content of the book can be divided as follows:

1. Basic notions on the properties of silicon: Chapters 1, 2, and 3.

2. Basic notions on the properties of biological molecules, aqueous solutions, and cell components: Chapters 4 and 5.

3. Junctions and fluxes of matter through membranes: Chapter 6 (*pn* junction) and Chapter 7 (solid-electrolyte junction, membrane transport).

4. Devices and CAD: Chapter 8 (MOS structure), Chapter 9 (MOSFET), Chapter 10 (biosensors), Chapter 11 (neurons and neuronal networks), and Chapter 12 (CAD applications).

Subsets of the book can be also identified, which self-consistently cover specific topics, as follows:

1. An elementary introduction to silicon devices, with emphasis on MOS technology–based devices: Chapters 1, 2, 3, 6, 8, and 9.

2. A bioengineering-biophysics–oriented introduction to molecular biology of the cell, with emphasis on interfaces, membrane transport, and neurobiology: Chapters 4, 5, 7, and 11.

3. The advanced reader, with a background both in electronics and molecular biology of the cell, should find of interest directly reading Chapters 7, 8, 9, 10, 11, and 12. These six chapters can form the basis for an advanced course in bioengineering, with emphasis on bioelectronic topics.

Massimo Grattarola
Giuseppe Massobrio

ACKNOWLEDGMENTS

The help by young coworkers, students, present and former Ph.D. students, and colleagues at the Department of Biophysical and Electronic Engineering of the University of Genova (Italy) is gratefully acknowledged. Among others, we specifically wish to mention: Sergio Martinoia, for comments and suggestions concerning Chaps. 7 and 12; Marco Bove, for comments and suggestions concerning Chaps. 11 and 12; Roberto Raiteri, for introducing us to the topic of micromechanics (Sec. 10.6); and Michele Giugliano, for his thesis work, utilized in Sec. 11.3.

A special thanks to Piet Bergveld, Professor at Twente University, the Netherlands, whose papers inspired most of Chap. 10, for his kind and very helpful comments.

The book has been in the making for about 2 years and it finally came to completion thanks to the friendly support given by Susanne and Paolo Antognetti and to the high level of patience and sympathy demonstrated by our sponsoring editor, Stephen Chapman and editing supervisor, David Fogarty.

Finallly, we gratefully acknowledge the affectionate help of our wives, daughters, and sons: Brunella, Maria, Maddalena, Laura, Lorenzo, Paolo, and Andrea. Their encouragement and patience during the many evenings and weekends that went into the writing of this book make them the unseen coworkers of this writing project.

BIOELECTRONICS HANDBOOK

P · A · R · T · 1

BASIC PROPERTIES OF SILICON

CHAPTER 1
SEMICONDUCTOR MATERIALS

Matter is held together by various kinds of forces, leading to bonds between atoms and molecules. In a very broad sense, these bonds can be divided into chemical and physical ones. Traditionally, books dealing with semiconductor materials consider chemical bonds only. Both bonds will be considered in this book. Chemical bonds and, more specifically, the covalent one will be introduced in this chapter, while physical bonds, relevant to the biological environment, will be considered in Chap. 4.

In this chapter an elementary, semiclassical description of atoms will be given, in order to introduce the concepts of discrete energy levels and of spectra of light absorption/emission. These concepts can be described at various levels, from a mere sequence of statements to a complex analysis performed with the technical instruments of quantum mechanics. We chose a very simple level and no attempts will be made to justify the results in terms of wave mechanics. The concept of energy bands in semiconductors will then be discussed, leading to the introduction of two charge carriers in semiconductors, i.e., electrons and holes.

1.1 CHEMICAL AND PHYSICAL BONDS

When two or more atoms form a molecule, as when two hydrogen atoms and one oxygen atom combine to form a water molecule, the forces that bind the atoms together within the molecule are called *covalent forces,* and the interatomic bonds formed are called *covalent bonds*. Closely related to covalent bonds are *metallic bonds,* which generate the crystalline structure of metals.

Covalent and metallic bonds are chemical bonds, which are characterized by the electrons being shared between two or more atoms, so that the discrete nature of atoms is lost. A characteristic of covalent bonds is their directionality: they are directed or oriented at well-defined angles relative to each other. Thus, for multivalent atoms (i.e., atoms with a number of covalent bonds greater than one), their covalent bonds determine the way they will arrange themselves in molecules or in crystalline solids to form an ordered, three-dimensional lattice. Relevant examples are the perfectly ordered diamond structure made by carbon atoms and the crystalline structure made by silicon atoms.

Unbonded, discrete atoms and molecules are held together by *physical forces,* which give rise to *physical bonds*. Physical bonds usually lack the specificity and strong directionality of covalent bonds. They are the most appropriate bonds for holding together molecules in liquids.

As already anticipated, semiconductor crystals (e.g., silicon) are held together by covalent bonds, that is, highly localized, short-range (i.e., less than 1 nm) bonds. The framework for their correct description is quantum mechanics. Biological structures are instead the result of a hierarchy of bonds (both chemical and physical ones), involving carbon (C) and a few other elements such as hydrogen (H), oxygen (O), and nitrogen (N), in addition to groups (mostly ions) present in the aqueous environment surrounding a biological entity.

The *primary* structure of a biological *macromolecule* (e.g., a protein) is based on covalent bonds. This type of bond holds together the backbone of the macromolecule, which, on the other hand, can also present a secondary and tertiary structure. These structures determine the three-dimensional shape of the macromolecule and are the result of interactions among regions of the macromolecule, and between the macromolecule and its surrounding environment. These interactions are not based on chemical (covalent) bonds, but rather on physical ones, such as hydrogen bonds (a particular kind of dipole-dipole bond, originally considered a quasi-covalent one), ion-dipole bonds, and dipole-dipole bonds.

Most of these physical bonds can be described within the framework of classical mechanics. They can be considered as long-range compared to the chemical bonds. Their energy relationship will be compared with the thermal energy kT in Chap. 4.

In this chapter we consider the chemical bonds of the atoms of a semiconductor; a brief description of the properties of atomic and molecular structures will lead to the concept of energy bands.

As a summary of this short section, Fig. 1.1 shows a few kinds of interaction energies.[1]

1.2 ELECTRONIC ORBITALS

The simplest way of introducing the idea of discrete energy levels in atoms and molecules is to start with the elementary description of the so-called *Bohr atom*. This description can be based on classical physics with the addition of a postulate. Let us consider an atom made by a proton and an electron (i.e., a hydrogen atom). Both particles are viewed as classical ones. Let us further postulate that the angular momentum L of the electron, which rotates around the proton in a "planetary" fashion, can have only discrete values, multiple (n) of the quantity $\hbar = h/2\pi$, h being the Planck constant (see App. A).

Then, for a circular orbit, we write

$$L_n = m_0 v r_n = n\hbar \qquad (n = 1, 2, 3, \dots) \qquad (1.1)$$

where m_0 is the electron rest mass, v is the electron velocity, and r_n is the radius of the circular orbit for a given value of n. The electron orbits are assumed to be stable, and the centripetal force on the electron is given by the Coulomb attraction between the nucleus (one positive elementary charge) and the electron (one negative elementary charge). Therefore it can be written

$$\frac{m_0 v^2}{r_n} = \frac{q^2}{4\pi\varepsilon_0 r_n^2} \qquad (1.2)$$

where q is the electronic charge and ε_0 is the permittivity of free space (see App. A). By combining Eqs. (1.1) and (1.2), we obtain

$$r_n = \frac{4\pi\varepsilon_0 (n\hbar)^2}{m_0 q^2} \qquad (1.3)$$

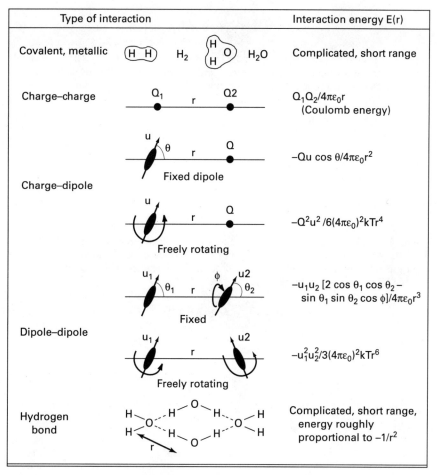

FIGURE 1.1 A few kinds of interaction energies in free space. $E(r)$ is the interaction free energy (in J); Q, electric charge (C); u, electric dipole moment (C-m); r, distance between interacting atoms or molecules (m); k, Boltzmann constant (J-K^{-1}); T, absolute temperature (K); ε_0, dielectric permittivity of free space (C^2-J^{-1}-m^{-1}). The force is obtained by differentiating the energy $E(r)$ with respect to distance r. (*Adapted from Israelachvili.*[1] *Used by permission.*)

Moreover, the total electron energy E_n, given by the sum of kinetic energy and potential energy (set at 0 at $r = \infty$), is

$$E_n = \frac{1}{2} m_0 v^2 - \frac{q^2}{4\pi\varepsilon_0 r_n} = -\frac{m_0 q^4}{2(4\pi\varepsilon_0 n \hbar)^2} \quad (1.4)$$

As will become apparent in the following, it is useful to express energies in electronvolts (eV, see App. A). By introducing numerical values, the electron energy E_n then becomes

$$E_n = -\frac{13.6}{n^2} \text{ eV} \qquad (n = 1, 2, 3, \ldots) \tag{1.5}$$

where $n = 1$ identifies an electron with the lowest admissible energy E_1. The corresponding energy state is called the *ground state*. In order to move to the first excited level ($n = 2$), an electron in the ground state needs an amount of energy ΔE_{21}, given by

$$\Delta E_{21} = E_2 - E_1 = \left(-\frac{13.6}{4} + 13.6\right) \text{eV} = 10.2 \text{ eV} \tag{1.6}$$

$\Delta E_{31} = 12.1$ eV is needed to move an electron to the second excited level ($n = 3$), and so on.

Let us assume that light is made of discrete units (*photons*), each bearing the energy E_ν given by

$$E_\nu = h\nu \tag{1.7}$$

where h is the Planck constant and ν is the frequency of the corresponding electromagnetic wave. We can predict that a collection of the above-described hydrogen atoms, all in their ground state, will absorb light at well-defined frequencies, which can be easily calculated. The first absorption frequency will be

$$\nu_1 = \frac{\Delta E_{21}}{h} \simeq 2.46 \times 10^{15} \text{ s}^{-1} \tag{1.8}$$

and the corresponding wavelength

$$\lambda_1 = \frac{c}{\nu_1} \simeq 122 \text{ nm} \tag{1.9}$$

where c is the speed of light in vacuum (see App. A).

The second absorption frequency will be

$$\nu_2 = \frac{\Delta E_{31}}{h} \simeq 2.88 \times 10^{15} \text{ s}^{-1} \tag{1.10}$$

and the corresponding wavelength will be

$$\lambda_2 \simeq 103 \text{ nm} \tag{1.11}$$

and so on (if the transitions are "allowed" ones[2]). It is worth noting that, by increasing the frequency of the impinging radiation, sooner or later the absorbing electron will become free, leaving the nucleus ionized. The corresponding radiation is then called *ionizing radiation*.

These discrete absorption frequencies form the lines of the *absorption spectrum* of the hydrogen atom H. In the same way, a collection of excited hydrogen atoms will emit photons with exactly the same frequencies. The resulting emission lines form the *emission spectrum* of H (see Fig. 1.2).

The reader should note that, in these two examples, the energy state of the collection of H atoms (i.e., the ground state in the case of the absorption spectrum and the excited states in the case of the emission spectrum) has been chosen arbitrarily. The energetic state of a particle population depends on its energy exchanges with its environment. If the

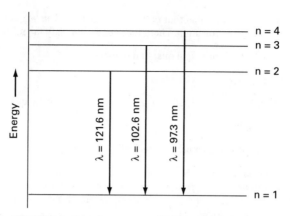

FIGURE 1.2 Emission spectrum of H. The wavelength of the emitted photons is indicated.

population is at equilibrium with its (large) surroundings at temperature T, then the occupancy of the energy state i, with energy E_i, such that

$$E_i - E_1 = \Delta E_{i1} > kT \tag{1.12}$$

where k is the Boltzmann constant, is improbable.

This qualitative statement, which will be clarified in the next chapter, implies that, at room temperature (i.e., $kT \simeq 0.026$ eV), the electrons of a population of "resting" hydrogen atoms will have an extremely high probability of being in the ground state; in other words, they should be heated to an elevated temperature in order to emit light. On the other hand, if excited by photons at room temperature, they will soon decay to the ground state, giving back to the surrounding reservoir the excess energy obtained from the photons. These considerations will be further developed when the phenomenon of *fluorescence* will be taken into account (see Chap. 2).

Historically, it was the experimental evidence of the above mentioned sharp, discrete spectra that induced Bohr to postulate the quantization of the electron angular momentum, in 1913. (This is a very simplified statement. The reader interested in the early days of quantum mechanics can consult the books listed in the Refs. 2 to 6).

Although the Bohr model was successful in explaining most of the features of the hydrogen spectra, attempts to extend this semiclassical analysis to more complex atoms (e.g., helium) were not successful. The subsequent conceptual steps were wave mechanics (proposed by Schrödinger) and matrix mechanics (proposed by Heisenberg), both around 1926.

In the wave mechanics approach, the same quantized values of energy obtained with the Bohr model are derived by solving the time-independent wave equation

$$-\frac{\hbar^2 \nabla^2 \psi}{2m_0} + U(\mathbf{r})\psi = E\psi \tag{1.13}$$

where ψ is the electron wave function, $U(\mathbf{r})$ is the potential energy, and ∇^2 is the mathematical operator described in App. B. Further analysis of this approach is beyond the scope of this book. The interested reader should consult Refs. 2 to 6.

It will suffice here to state that four *quantum numbers*, n, l, m, m_s, completely describe any electron in a multielectron atom. These four numbers satisfy the following constraints:

$n = 1; 2; 3; \ldots$ (principal quantum number)

$l = 0; 1; 2; \ldots; (n-1)$ (orbital quantum number)

$m = 0; -1, 0, 1; -2, -1, 0, 1, 2; \ldots (-l < m < l)$ (magnetic quantum number)

$m_s = \pm\frac{1}{2}$ (spin)

Moreover, the following rules must be satisfied:

1. *Pauli exclusion principle.* No two electrons can have the same four quantum numbers. As two values of spin are possible ($m_s = \pm\frac{1}{2}$), two electrons may have the same n, l, and m numbers. For example, there are two electrons only for $n = 1$. They are identified by the following four quantum numbers:

$$(1, 0, 0, +\tfrac{1}{2}); (1, 0, 0, -\tfrac{1}{2})$$

These two electrons fill up one *orbital*. Electron orbitals are named by the l value, as follows:

$l = 0 \rightarrow s$ orbital

$l = 1 \rightarrow p$ orbital

$l = 2 \rightarrow d$ orbital

$l = 3 \rightarrow f$ orbital

Therefore the orbital just considered ($n = 1$) is an s orbital. Orbitals are usually represented in cells as shown in Fig. 1.3. The arrows in the cell identify the "up" (i.e., $+\frac{1}{2}$) and "down" (i.e., $-\frac{1}{2}$) electron spins, respectively.

A maximum of 8 electrons is allowed for $n = 2$ (Fig. 1.4). The 6 electrons of the p orbitals correspond to the following quantum numbers: $(2, 1, -1, +\frac{1}{2})$; $(2, 1, -1, -\frac{1}{2})$; $(2, 1, 0, +\frac{1}{2})$; $(2, 1, 0, -\frac{1}{2})$; $(2, 1, 1, +\frac{1}{2})$; $(2, 1, 1, -\frac{1}{2})$.

If we assume that n and l characterize the energy of an orbital,[2-6] then the p orbitals (e.g., $n = 2$, $l = 1$) should have the same energy. They are said to be *degenerate*.

2. *Hund maximum multiplicity rule.* Electrons distribute themselves with the same spin value on the maximum number of degenerate orbits. So, the distribution of three electrons with $n = 2$ and $l = 1$ will be as shown in Fig. 1.5a rather than in Fig. 1.5b.

In summary, atomic orbitals have the following properties[7]:

They are three-dimensional in space and each orbital has a characteristic geometric shape. For example, s-type orbitals are spherically symmetric and p-type orbitals are similar to dumbbells, as seen in Fig. 1.6. It can be observed that the spherical symmetry of the s orbitals is somewhat reminiscent of the circular orbits of the Bohr atom, but the interpretation of the geometric shapes is profoundly different in the two cases. When describing the Bohr atom we are considering precise, deterministic orbits covered by a particle. On the contrary, the drawings of Fig. 1.6 identify regions where the probability of finding an electron (related to the module of the electron wave function) is high.

FIGURE 1.3 Cell representation of a 1s electronic orbital.

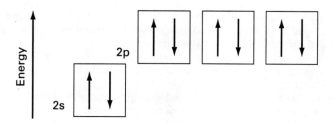

FIGURE 1.4 Energy levels and cells for $n = 2$ electronic orbitals.

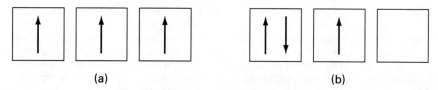

FIGURE 1.5 Maximum multiplicity rule.

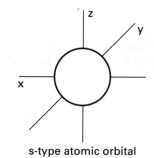

s-type atomic orbital

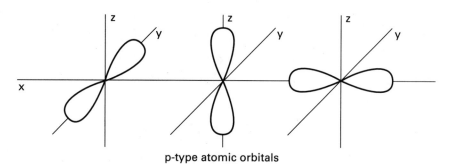

p-type atomic orbitals

FIGURE 1.6 Geometrical forms of s and p orbitals.

Only *s*-type orbitals are permitted in the lowest energy level. Both *s*- and *p*-type orbitals are allowed in the next-to-lowest energy level. Orbitals of the *s*-, *p*-, and *d*-types are permitted in the next energy level, and so forth.

All orbitals corresponding to a given energy level have roughly the same radial extent. Orbitals corresponding to higher energies have larger mean radii than orbitals corresponding to lower energy.

The set of rules and properties described so far give rise to the cell distribution shown in Fig. 1.7, where energy is increasing along the vertical direction.

The first (lowest) permitted energy level can allocate two electrons in *s*-type orbitals. These electrons are called 1*s* electrons, the 1 signifying the energy level and the *s* signifying the type of orbital. As already anticipated, the next allowed energy level, or shell, has eight possible positions for electrons, and the 2*s* configuration, or cell, is at a slightly lower energy than the 2*p* configuration. The 2*s* cell holds two electrons, while the 2*p* cell will hold six electrons (two each in the p_x, p_y, and p_z orbitals). The third shell has two 3*s* electrons, six 3*p* electrons, and ten 3*d* electrons, each at a slightly different energy level. This scheme continues through 4*s*, 4*p*, 4*d*, 4*f*, 5*s*, ... levels to infinity. As Fig. 1.7 shows, the 4*s* and 3*d* energy levels are nearly the same, as are the 5*s* and 4*d* levels, and the 6*s*, 5*d*, and 4*f* levels. These facts are important in explaining the properties of the transition elements.

A knowledge of this electronic energy-level scheme is enough to allow one to construct models from which several physical properties of atoms may be deduced. For example, the occurrence of the periodic table, a detailed explanation of chemical valence, and the theory of atomic spectroscopy (see Fig. 1.2), all follow from these simple orbital pictures.

On the basis of this scheme, we introduce the element *silicon* (Si). It has atomic number $Z = 14$ in the periodic table. It can be found in several isotopic forms, the most abundant of which has an atomic weight of 28. (We remind the reader that different isotopes of a given element have the same number of electrons and protons, but differ in the number of neutrons in their nucleus and, consequently, in their atomic weight.) The nucleus of this isotope is composed of 14 protons and 14 neutrons. There are 14 electrons surrounding the nucleus of an electrically neutral silicon atom (Fig. 1.8).

We may imagine introducing the 14 electrons one by one in the following way. The

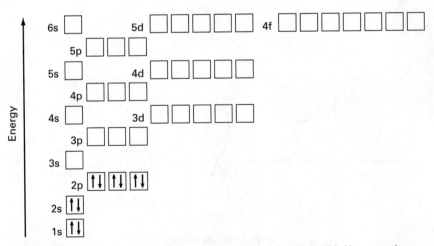

FIGURE 1.7 Energy levels and cells for the generalized atomic orbital. Each cell holds no more than two electrons. An atom with 10 electrons (i.e., Neon) is represented.

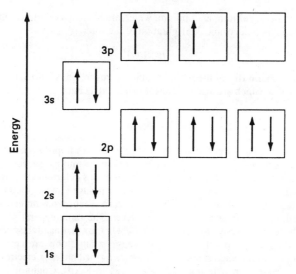

FIGURE 1.8 Cell diagram for a silicon atom: $(1s)^2(2s)^2(2p)^6(3s)^2(3p)^2$.

first two electrons occupy the lowest energy level, that is, the 1s energy level. These electrons have opposite spins, as indicated in Fig. 1.8. The next two electrons fill in the 2s level with opposite spins. The next six electrons fill in the 2p orbitals. This process goes on through the 3s level, after which there are only two electrons left. These two electrons fill in two of the six possible 3p levels, leaving four vacant levels, according to the *maximum multiplicity rule*.

This information can be summarized in the form

$$(1s)^2(2s)^2(2p)^6(3s)^2(3p)^2$$

The superscripts indicate the number of electrons in each type of orbital. This expression is called the *electronic configuration* of the atom.

The ground state cell diagram for a silicon atom in Fig. 1.8 can be simplified by noticing that the 10 electrons in the filled inner shells (1s through 2p) shield 10 of the 14 units of nuclear charge from the outer (or *valence*) electrons. For chemical purposes, we may consider a silicon atom as having only four electrons and a reduced nuclear charge of +4 electronic units. Two of the four electrons which surround the reduced nucleus are to be thought of as belonging to s-type orbitals and two to p-type orbitals.

We can then infer that any atom which has the same reduced structure as Si will be chemically similar to Si. For example, carbon ($Z = 6$) has its outer electrons in the $(2s)^2(2p)^2$ configuration and is consequently similar to Si in its chemistry. So are germanium [$Z = 32$ and outer electrons in $(4s)^2(4p)^2$ configuration] and tin [$Z = 50$, outer electrons in $(5s)^2(5p)^2$ configuration]. All of these atoms have a reduced, (or effective), electronic configuration that can be described by

$$(ns)^2(np)^2$$

As a result, they all belong to the same column of the periodic table (column IV$_A$) and tend to form similar chemical compounds and crystals.

Boron, aluminum, gallium, and indium, which are elements of interest in silicon technology, share an outer electronic configuration of the form

$$(ns)^2(np)^1$$

They belong to column III_A of the periodic table. Nitrogen, phosphorus, arsenic, and antimony (column V_A), which are also of interest in silicon technology, have the outer configuration

$$(ns)^2(np)^3$$

Table 1.1 shows a partial periodic table, with elements of importance in the technology of Si and GaAs devices. Each column is labeled by its common reduced electronic configuration. We will see that these electronic configurations determine how impurity atoms (e.g. boron and phosphorus) behave when they are introduced into a silicon crystal.

An important feature of atomic theory is that each electron has an energy level and a spatial distribution appropriate to that energy level. The same general statement can be made for each electron of any atom in a *molecule*. On the other hand the energy levels and the corresponding spatial distributions will change when atoms are combined to form molecules. In particular, when stable molecules are formed, valence electrons are normally either transferred from one atom to another (as in K^+Cl^-, Coulomb interaction), or shared between two atoms (as in H—H, covalent interaction) in the creation of a bond. The new electron orbitals that are formed when a molecule is created are called *molecular orbitals*. Like atomic orbitals, each molecular orbital corresponds to a specific value of electron energy and will hold two electrons. The orbital shapes and electron energies of molecular orbitals will be considered briefly in the following.

A powerful technique for obtaining the characteristics of molecular orbitals is to visualize them as being linear combinations of atomic orbitals. To describe this technique in an elementary way,[7] we consider the hydrogen molecule, H_2, which has the same role in molecular theory that the hydrogen atom has in atomic theory.

If two hydrogen atoms are separated by a great distance, the two electrons are in their respective $1s$ atomic orbitals. Since each of these orbitals will hold two electrons of opposite spin, there are positions for four electrons in this two-separate-atom system.

Let us now suppose that the two atoms are brought together, so that the atomic orbitals begin to overlap. When this happens, both the shape of the orbitals and the electron energy levels will change. Two new molecular orbitals will be formed as a result of this interaction. To a first approximation, the shape of these new orbitals can be obtained directly from the shape of the $1s$ orbitals that describe the ground states of the two separate hydro-

TABLE 1.1 Partial Periodic Table of the Elements of Importance for Si and GaAs Devices

II_A $(ns)^2$	III_A $(ns)^2(np)^1$	IV_A $(ns)^2(np)^2$	V_A $(ns)^2(np)^3$	VI_A $(ns)^2(np)^4$
Be	B	C	N	O
Mg	Al	Si	P	S
Zn*	Ga	Ge	As	Se
Cd*	In	Sn	Sb	Te

*Note: Zn and Cd are actually members of another column but still have the $(ns)^2$ outer electron configuration.

gen atoms. In particular, if ψ_1 and ψ_2 are the ground state orbitals for atom 1 and atom 2, respectively, then the molecular orbitals (MO) can be approximated by

$$(MO)_1 = \psi_1 + \psi_2$$

$$(MO)_2 = \psi_1 - \psi_2$$

The molecular orbital $(MO)_1$ is called a *bonding orbital*. Electrons in this kind of orbital have a high probability of being found in the space between the two nuclei where they can screen the mutual repulsion between the two positively charged protons. They have an energy level that is lower than the corresponding atomic energy level. The molecular orbital $(MO)_2$ represents an antibonding state, where the two electrons do not occupy the space between the two nuclei. Electrons in this orbital have energy levels that are higher than the electron energy levels in the original atomic orbitals. The repulsion of the two protons makes this configuration unstable. The actual energies depend on the interatomic spacing that is assumed. As shown in Fig. 1.9, the energy of the molecule (in the bonding configuration) decreases as the interatomic spacing R is decreased, up to about 0.74 Å.[8] For smaller spacing, the repulsion between the positively charged atomic cores increases rapidly and the total energy of the molecule then begins to increase too. The equilibrium spacing for the two protons is the one which gives the minimum energy for the molecule.

In the H_2 molecule, both electrons are in the bonding orbital and the antibonding orbital is empty. Figure 1.9 shows that the energy for electrons in the antibonding orbital is greater than the energy the electrons would have in isolated hydrogen atoms for all interatomic spacings.

On the basis of these premises, let us now consider the topic of crystal structure. When atoms form a solid, they frequently take up an orderly three-dimensional arrangement called a *crystal*. For each possible type of crystal, we can identify a fundamental building block, which, in principle, can be repeated in space with no limits.

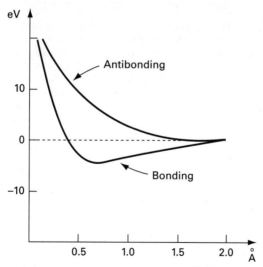

FIGURE 1.9 Total energy of H_2 molecule as a function of interatomic spacing.

Highly conductive metals, such as lithium, copper, aluminum, and silver, usually crystallize in close-packed structures where there are many nearest neighbors. For example, the fundamental building block for a lithium crystal is the body-centered cubic structure shown in Fig. 1.10. In such a structure, each lithium atom has eight nearest neighbors and six next-nearest neighbors.

A qualitative explanation of the properties of these crystals can be obtained by considering the individual atoms as being ionized and "floating" in a "gas" (actually a degenerate gas) of electrons. In a lithium crystal each atom contributes its $2s$ electron to the gas and becomes ionized [$Li^+ = (1s)^2(2s)^0$]). The ionized $2s$ electrons are free to move throughout the crystal. Therefore, they are regarded as

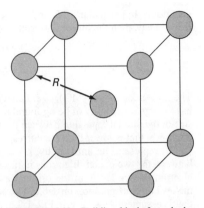

FIGURE 1.10 Building block for a body-centered cubic crystal (e.g., lithium metal).

"shared" by all the ion cores in the crystal. Such crystals are held together by a balance of two forces: the attraction between the positively charged atomic cores and the electron gas, and the repulsion of the atomic cores by each other. The resulting binding is rather weak, so that considerable movement of the positive ions is possible without spending too much energy. This qualitatively explains the general plasticity of metals. Moreover the freedom of the electrons to drift through the crystal in response to an applied electric field accounts for the high conductivity of metals.

1.3 ENERGY BANDS IN METALS AND SEMICONDUCTORS

A rigorous treatment of energy band theory is out of the scope of the book; here a qualitative description is given. The reader can find a detailed description in Refs. 8 to 12.

Let us consider the molecular orbitals and energy levels that would be formed if Li atoms were added together, one by one, to make a one-dimensional crystalline lattice.[10] We already know that changes in both the electron energy levels and the orbital patterns take place when the atomic orbitals for one atom overlap significantly with those of its neighbors. In particular, if an atom Y is added to another atom X, then two identical atomic orbitals ψ_X and ψ_Y will form two new molecular orbitals $\psi_X \pm \psi_Y$, each with its distinct energy and geometry. If we now add a third atom, we will obtain three molecular orbitals with their energies grouped around the original energy of ψ_X. Each further addition of an atom introduces one more energy value, at the same time modifying slightly those of the previous set. The process gives rise to a *band* of electron energies in the crystal. The band contains one energy level for each atom in the crystal. The described approach summarizes in a highly qualitative way the so-called linear combination of atomic orbitals (LCAO) method. It is illustrated in Fig. 1.11.

When the individual atomic energy levels (e.g., $2s$ and $2p$ of lithium) are well separated, or the distance between contiguous atoms is so large that there is no appreciable overlap of neighboring orbitals, then the separate bands of energy will remain distinct and of small width. On the other hand, if the original energies are close together, and/or neighboring atoms are sufficiently close to each other (i.e., a close-packed metallic crystal),

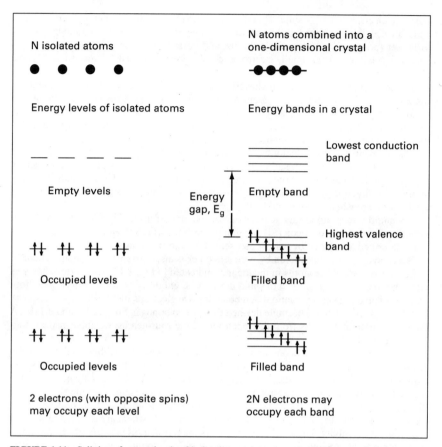

FIGURE 1.11 Splitting of energy levels of isolated atoms into energy bands in a crystal. Arrows represent electrons with different spin values. Note that in some metals, valence and conduction bands overlap and no band gap exists. (*Adapted from M. Shur.*[10] *Used by permission.*)

then the bands will merge into one another. In this case, the molecular orbital is built up from a combination of all overlapping orbitals. For example, considering again metallic lithium, the $2s$ and $2p$ bands cross and merge into one another at the equilibrium spacing. The merging of the energy bands gives 4 states per atom in the lower band and 4 states per atom in the upper band, at the equilibrium spacing. The total number of states is 8 per atom.

The conducting properties of Li crystals may be explained in terms of its energy band structure in the following way. First, the density and atomic weight of Li indicate that there are 4.6×10^{22} atoms/cm^3. As a consequence, at the equilbrium spacing, there are about 2×10^{23} energy levels/cm^3 (i.e., 4 levels/atom × 4.6×10^{22} atoms/cm^3) in an energy band having a total width of about 5 eV. The energy levels are thus spaced by an average energy of about 2.5×10^{-23} eV.

The energy band has 4 available states per atom. On the other hand, each atom has only one electron in the 2s state. Therefore the energy band is only one-quarter full. As a consequence, the electrons will be able to absorb energy from an electric field by moving to higher (previously empty) energy levels, and current will flow. The spacing between energy levels being very small, electronic conduction is assured for any small electric field.

In summary, in the Li metal the lowest energy band is partially empty. This condition makes it easy for electrons to be promoted to higher energy levels, still remaining in a bonding orbital. This leads to easy electronic conduction. In other metals (e.g., Ca), the lowest energy band is completely full, but it partially overlaps with the upper one and therefore conduction is again possible.[11]

The atoms in a semiconductor crystal have few nearest neighbors in comparison to the close packing of ions in a metallic crystal. Semiconductor crystals are held together by highly localized electronic bonds between neighboring atoms. These bonds are very strong and directional; thus, the crystals are quite hard, have a high melting point, and tend to fracture when stressed, unlike the metals.

Among the most important semiconductor materials for making transistors and integrated circuits there are silicon (Si) and gallium arsenide (GaAs). However, the properties to be described are shared by many other semiconductor materials.

As shown by x-rays scattered by pure silicon crystals, a given silicon atom is bonded to four nearest-neighbor atoms in the manner indicated in Fig. 1.12. This figure shows an atom at the center of a cubic "box" and the four neighboring atoms identifying the four vertices of an equilateral pyramid. Each bond has two electrons in it (with opposite spins), and it is a *covalent bond*. The angle defined by any two bonds in Fig. 1.12 is about 110°. A large silicon crystal results from the repetition of the four-nearest-neighbor pattern suggested in Fig. 1.12.

Since the atoms in a semiconductor crystal have few nearest neighbors and these are at relatively large distances (as compared to the metal crystals), the valence electrons will experience highly directed forces from neighboring nuclei. These forces are not enough to ionize the atoms, but they are strong enough to have a major effect on the shape of the electron valence orbitals. As a result, in a silicon crystal the 3s and 3p orbitals of the atoms are mixed to form a new set of four bonding orbitals which are directed toward the four nearest neighbors. These orbitals are called the *tetrahedral orbitals*. It can be shown that these orbitals are additive mixtures of the 3s and 3p atomic-valence orbitals, the mix-

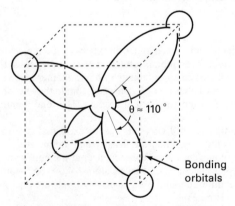

FIGURE 1.12 Bond structure of crystalline silicon: bonding orbitals and bond angles.

ing being originated by the fields of the neighboring atoms. The electrons in these bonding orbitals have energy levels which are lower than the atomic $3s$ level, which indicates that the covalent bond is a stable one. Moreover, all the valence electrons of each atom are used in these bonds, so there are none left over to produce conduction. Therefore, a perfect silicon crystal is an *insulator*.

As a direct comparison with the energy bands in Li, we can consider the energy band structure for a diamond crystal (i.e., C atoms in the lattice arrangement shown in Fig. 1.12). The difference in the atom structure here is that the electronic configuration of Li is $(2s)^1(2p)^0$, while the configuration of C is $(2s)^2(2p)^2$. Therefore the same atomic levels are involved, but the state of filling of the levels is different for the two atoms. When C atoms are brought together, the $2s$ and $2p$ atomic energy levels will originate energy bands. Here again the s and p bands merge and generate a lower band and an upper band. The lower band is formed by one $2s$ level and three $2p$ levels and corresponds to the tetrahedral bonding orbitals just described. The upper band is also generated by one $2s$ level and three $2p$ levels per atom and corresponds to a set of antibonding orbitals. Once again there are four states per atom in the lower band and four states per atom in the upper band. But in this case (as with Si and Ge), there are also *four* valence electrons per atom, so the lower band is *completely full* and the upper band is *completely empty*. As a consequence, electronic conduction is impossible, unless energy is absorbed by electrons to move from the filled band to the empty band that lies above it, by crossing an energy gap E_g, which separates the valence band from the conduction one. To fix ideas, the energy gap is 0.7 eV, 1.1 eV, and 5.5 eV for Ge, Si, and C, respectively.

The lattice structure formed by C is called the *diamond crystal* structure, since diamond is the material obtained when carbon is the atom of the structure. Silicon and germanium also crystallize in this structure since, as elements of column IV_A of the periodic table along with carbon, they also have their valence electrons in the $(ns)^2(np)^2$ configuration. However, this is not a necessary condition for the formation of this basic lattice structure. To give an example, atoms of Ga and As can also crystallize in this structure in an arrangement where atoms of one type always have atoms of the other type as nearest neighbors. Such a crystal is called a *compound semiconductor*, since Ga is an element of column III and As is an element of column V. Other III-V compound semiconductors of interest are GaP, InP, and InSb. All of these semiconductors have special properties that are useful for practical semiconductor devices, such as light-emitting diodes and infrared detectors.

In the next chapter, we will consider the particles that conduct current in a semiconductor, the existence of an ionization energy for the covalent bond and the effects of the addition of impurities on the electrical characteristics of a semiconductor. To introduce these concepts, we will use a two-dimensional representation of the silicon crystal such as the one shown in Fig. 1.13.

Some of the fundamental features of real crystals are still present in this model. For example, the model correctly shows that a Si atom has four nearest neighbors, and each valence bond is represented by two lines to indicate that there are two electrons associated with each bond. However, the reader must remember that the real structure is three-dimensional, and that the tetrahedral bonds form one continuous three-dimensional lattice throughout the solid. For this reason, we cannot assume that a given valence electron is associated to a given bond. Rather, we must suppose that it may move about through the crystal, via the system of valence orbitals, exchanging places with other valence electrons (with no net current flow). In other words, we have to consider both the system of bonding orbitals and all the bonding electrons as belonging to the whole crystal, not to the local atom. This viewpoint is fundamental for understanding many of the basic properties of semiconductors.

The band theory, qualitatively introduced so far, represents a useful starting point for

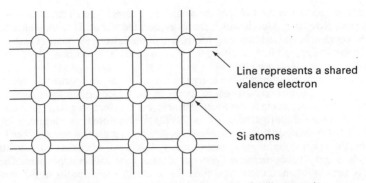

FIGURE 1.13 A two-dimensional representation of a silicon crystal.

classifying solids electrically as: conductors (no energy gap or a very small fraction of an electron-volt), insulators (very large energy gap, e.g., 8 eV) and semiconductors (energy gap in the order of 1 eV). The semiconductors have two very important properties that are not shared by conductors and insulators, namely:

1. The electrical conductivity of semiconductors can be greatly increased by adding very small amounts of appropriate impurities, which are called *dopants*. For example, by adding boron (B) to a silicon crystal in the ratio of 1 part per million (1 ppm), we can change the resistivity of the Si crystal by more than 6 orders of magnitude (i.e., the resistance between opposite faces of a 1-cm cube of crystalline Si will change from 200,000 Ω to 0.1 Ω when 1 ppm of B is added to the Si). On the contrary, the addition of 1 ppm of Al to Cu will have a negligible effect on the resistivity of the Cu.

The wide range of resistivity control available in a semiconductor is of great importance in the fabrication of transistors and integrated circuits, and, as a result, techniques for precisely introducing very small amounts of appropriate dopants into semiconductor crystals are of great relevance in the semiconductor industry.

2. The second property of semiconductors that is not shared by conductors and insulators is that the mechanism of conduction in a semiconductor is determined by the atomic structure of the dopant that is added. For example, by adding B to Si, we produce a crystal in which the majority of carriers that conduct electricity is positive. The semiconductor is then called *p* type. On the other hand, by adding phosphorus (P) to Si, we produce a crystal in which the majority of carriers that conduct current is negative, and the semiconductor is then called *n* type.

The existence of two basic types of dopant (*p* and *n* type) can be explained by the electronic configurations of the dopants themselves. Taking phosphorus (P) in silicon as an example, we observe that P has the electronic configuration $(1s)^2(2s)^2(2p)^6(3s)^2(3p)^3$, while Si has the configuration $(1s)^2(2s)^2(2p)^6(3s)^2(3p)^2$. In other words, P has its valence electrons in the same configuration as Si, except that one more *p* level is filled. Hence, we can indicate P as $(Si) + (3p)^1$.

It follows from this equivalence that, if we take out a Si atom from its position in a silicon crystal and substitute a P atom for it, the impurity atom will be able to complete all of the bonds with its four Si neighbors and have one 3*p* electron left over. This additional electron is bound to the P atom at low temperatures, but as we will see in Sec. 2.2, it becomes free at higher temperatures. Therefore, at room temperature (300 K) a P-doped Si crystal will have a free electron concentration that is in most cases equal to the number of P atoms/cm^3 that have been introduced into the crystal. The substitution of P for Si therefore produces a doped, electron-rich (or *n*-type) Si crystal. Similarly we can produce a *p*-type crystal by substituting an aluminum (Al) atom for a Si atom in a silicon crystal. Aluminum, which has the electronic configuration $(1s)^2(2s)^2(2p)^6(3s)^2(3p)^1$, is *missing* one of the electrons necessary to complete the bonds with all of its Si neighbors. It can complete three of the bonds, but will leave the fourth bond lacking one electron. At low temperatures this *defect* in the valence bond structure will be localized at the site of the Al atom. However, at higher temperatures (including room temperature) valence electrons from neighboring Si atoms will complete the bond, so that the defect will move away from the site that generated it. We will see in Chap. 2 that the defect can be considered as a free positive charge (a *hole*) in the crystal (*p*-type doping). As already indicated, this doping type is also produced in silicon by boron and other column III atoms.

PROBLEMS

1.1 *a.* On the basis of the Bohr model, calculate the wavelength of the photons absorbed in vacuum by a collection of hydrogen atoms, whose electrons are subject to the following transitions:

$$1 \to 2$$
$$1 \to 3$$
$$2 \to 3$$

  b. Is the transition $2 \to 3$ a likely one at room temperature? (*Hint:* in order to make a transition from state *a* to state *b*, state *a* should be already occupied.)

1.2 *a.* As Prob. 1.1*a*, but now the atoms have a relative permittivity $\varepsilon_r = 12$.
  b. How does the energy of the ground state compare with kT at room temperature?

1.3 Write the complete ground state electronic configuration for C ($Z = 6$) and Ge ($Z = 32$), and develop the corresponding cell diagrams for these atoms. Show that these atoms have the same reduced electronic structure of Si.

1.4 Write the complete ground state electronic configuration for B ($Z = 5$) and P ($Z = 15$), and identify the valence electrons.

REFERENCES

1. J. Israelachvili, *Intermolecular and surface forces,* 2d ed., London: Academic Press, 1992.
2. E. H. Wichmann, *Quantum physics—Berkeley physics course,* Vol. 4, New York: McGraw-Hill Education Development Center, Inc., 1967.
3. R. L. Liboff, *Introductory quantum mechanics,* Oakland, Calif.: Holden-Day, 1980.
4. A. Messiah, *Quantum mechanics,* Vol. I, Amsterdam: North Holland, 1970.
5. P. M. A. Dirac, *The principles of quantum mechanics,* 4th ed., New York: Oxford University Press, 1958.
6. L. I. Schiff, *Quantum mechanics,* 3d ed., New York: McGraw-Hill, 1968.
7. J. Gibbons, *Semiconductor electronics,* New York: McGraw-Hill, 1966.
8. C. A. Coulson, *Valence,* New York: Oxford University Press, 1952.

9. N. W. Ashwoft and N. D. Nermin, *Solid state physics,* New York: Holt, Rinehart and Winston, 1976.
10. M. Shur, *Physics of semiconductor devices,* Englewood Cliffs, N.J.: Prentice-Hall, 1990.
11. R. F. Pierret, *Advanced semiconductor fundamentals,* from *Modular series on solid state devices,* Vol. 6, Reading, Mass.: Addison-Wesley Publishing Company, 1987.
12. K. Hess, *Advanced theory of semiconductor devices,* Englewood Cliffs, N.J.: Prentice-Hall, 1988.

CHAPTER 2
SEMICONDUCTORS AND CHARGE CARRIERS

Two charge-bearing particles are the basis of the conduction properties of semiconductors: the *electron* and the *hole*. Knowledge of the properties of both types of carriers is fundamental to understanding how semiconductor devices work.

The concentration (number of particles per cubic centimeter) of electrons and that of holes in a semiconductor can be equally increased by providing energy to it, e.g., via an increase of temperature or the exposure to light of appropriate frequency. Moreover, as already anticipated, the concentration of electrons and holes in a semiconductor can be drastically modified and unbalanced by doping the semiconductor.

The basic properties of these two carriers will be discussed in this chapter. Optical properties of semiconductors and the effect of temperature on the charge carrier concentration will be considered. These two topics will be then used to introduce a short discussion on the concepts of thermal equilibrium and statistical distributions. The effective mass concept will be introduced and silicon doping will be then considered. Finally, a heuristic introduction to the generation-recombination processes will be given. In accordance with the overall level of the book, simplified theoretical methods will be provided. However, the reader should bear in mind that, for a truly appropriate description of the behavior of electrons and holes, the use of quantum mechanics is needed.

2.1 OPTICAL PROPERTIES OF SEMICONDUCTORS

Any semiconductor held at room temperature can conduct current, because the available thermal energy does generate a small amount of free charge carriers. Of course, other energy sources do produce free charges and, from a heuristic viewpoint, it is easier to consider first other energy sources such as light. The optical properties of semiconductor materials will therefore be described.

2.1.1 The Optical Absorption Band

We have already introduced the idea that the absorption of light is directly related to the energy spacing of the electronic orbitals of the absorbing material; furthermore, we have verified that the absorption spectrum of H is made up of well-separated lines. The band structure already described for Si (similar results can be obtained for GaAs and CdS) implies that these semiconductors should have a high probability of absorbing any photon with energy E_{ph} such that

$$E_{ph} = h\nu \geq E_g \quad (2.1)$$

In other words, we can define a threshold frequency ν_{th} given by

$$\nu_{th} = \frac{E_g}{h} \quad (2.2)$$

where ν_{th} is the threshold frequency for light absorption; this means that the material is transparent to light with frequency smaller than ν_{th} and it exhibits strong absorption as the frequency of the incident radiation is increased beyond ν_{th}. The corresponding photon energy $h\nu_{th}$ is, for Si, about 1.1 eV, which corresponds to infrared radiation.

The expression *strong absorption* can be experimentally quantified as follows. First of all, let us define the optical transmission ratio $T(\nu)$ for light of frequency ν as

$$T(\nu) = \frac{I_{tr}(\nu)}{I_o(\nu)} \quad (2.3)$$

where $I_o(\nu)$ is the intensity of the light entering a piece of material (e.g., a slice of Si) and $I_{tr}(\nu)$ is the intensity of the light coming out of it. Let us assume that the thickness of the silicon slice is l. The intensity $I_o(\nu)$ can be supplied by a light source equipped with a monochromator (i.e., an optical device which can select specific wavelengths). The intensity $I_o(\nu)$ and $I_{tr}(\nu)$ can be measured by a *photodetector* (i.e., a device where absorbed photons are converted into electrons). The principles of this experimental setup are sketched in Fig. 2.1.

Let us consider an infinitesimal thickness of material delimited by x and $(x + dx)$ as depicted in Fig. 2.1. It is reasonable to assume that the infinitesimal decrease in light intensity dI, occurring between x and $(x + dx)$, is proportional, with a minus sign, to the local value of I and to dx. Note that the proportionality constant α must be a function of the light frequency. These considerations are summarized by

$$I(x + dx) - I(x) = dI = -\alpha(\nu) I \, dx \quad (2.4)$$

In other words, the number of photons absorbed between x and $(x + dx)$ is proportional to the product of the number of incoming photons and the amount of absorbing material.

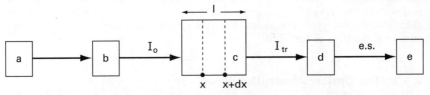

FIGURE 2.1 Sketch of an experimental setup for light absorption measurements. a, light source; b, monochromator; c, slice of material; d, photodetector [an electrical signal (e.s.) comes out of it]; e, analog/digital conversion, signal acquisition, and processing. I_o = intensity of the input light; I_{tr} = intensity of the transmitted light. Note that, in a real measurement, the light intensity I_o should be first measured by taking out the slice of material from the light path (i.e., a "blank" measurement).

These very basic concepts will be further considered in Chap. 4, where the same principles will be used to measure the concentration of an absorbing solute in a tube of solution.

By integrating Eq. (2.4), we obtain the *Beer-Lambert law:*

$$\ln \frac{I_{tr}(l)}{I_o} = -\alpha(\nu)l \tag{2.5}$$

or

$$I_{tr}(l) = I_o e^{-\alpha(\nu)l} \tag{2.6}$$

Therefore, the optical transmission ratio $T(\nu)$ is given by

$$T(\nu) = e^{-\alpha(\nu)l} \tag{2.7}$$

2.1.2 Conductivity of the Photogenerated Carriers

So far, we have been interested in the intensity of the transmitted light I_{tr}. We now shift our attention to the absorbed light, with intensity I_{abs} given by

$$I_{abs} = I_o - I_{tr} = I_o(1 - e^{-\alpha(\nu)l}) \tag{2.8}$$

Again, it is useful to consider light as made of photons. In this way we can visualize a monochromatic light shining on a silicon crystal as a bundle of incident photons, each having an energy $h\nu$. If this energy is greater than the energy gap E_g, it can ionize a covalent bond. In this case, a valence electron leaves its valence site and becomes free to wander about through the crystal. Each release of a valence electron is accompanied by the creation of a vacancy (or imperfection) in the valence structure, which is called a *hole*. In any pure semiconductor crystal, this mechanism, called *photogeneration,* creates equal numbers of holes and free electrons, and the rate at which they are created is a function of the parameters of the semiconductor material and of the incident photon flux. Threshold values for the photogeneration mechanism in various semiconductors are given in Table 2.1.

We have mentioned the fact that a minimum (*threshold*) energy is required for a photon to be absorbed. What about maximum values? If a crystal is illuminated by very highly energetic photons, some electrons will gain enough energy to leave the crystal. This process is called *photoemission*. For the sake of completeness, it should finally be mentioned that for further increases of the photon energy, or equivalently of their frequency (e.g., in the gamma-ray frequency range), the crystal becomes again transpar-

TABLE 2.1 Energy Gap and Properties of the Photon Corresponding to the Optical Threshold of Several Semiconductors

Material	E_g, eV	ν_{th}, Hz	λ_{th}, cm	Color
Ge	0.67	1.62×10^{14}	1.85×10^{-4}	Infrared
Si	1.12	2.68×10^{14}	1.12×10^{-4}	Infrared
GaAs	1.40	3.38×10^{14}	0.886×10^{-4}	Infrared
CdS	2.26	5.46×10^{14}	0.55×10^{-4}	Visible; green
C (diamond)	5.47	13.2×10^{14}	0.227×10^{-4}	Ultraviolet

ent. In this case other processes should be considered, which are beyond the aim of the book.

When light is absorbed by a semiconductor, there is an increase in conductivity: this process is called *photoconductivity* and is the basis of electronic devices known as *photoconductors,* which consist of resistors with resistance value depending on the light signal being applied to them.

To explain photoconductivity we can assume that the electrons which are released from the covalent bonds can move through the semiconductor crystal in the same way as electrons do in a metallic crystal. Thus, when an electric field is applied to the semiconductor crystal, these electrons are free to *drift* through the crystal, and they therefore contribute to an increase in conductivity.

However, the vacancy hole in the valence structure can also contribute to increasing conductivity. It is this fact which makes a semiconductor so different from a typical metal.

The vacancy left in the covalent bond by the release of a valence electron behaves as if it were a free conducting particle, with a positive electronic charge of q (1.6×10^{-19} coulomb) and a mass approximating that of the free electron. This "apparent" particle is called a *hole*.

This positive charge is actually associated with an unneutralized atom. However, when we speak of the vacancy as a "particle" (that is, hole) we associate the positive charge with the hole. Thus we say that the hole carries a positive charge of $+q$.

Moreover, we can visualize the movement of this imperfection through the lattice in the following way: When an electric field is applied to the crystal, the otherwise symmetrical paths of the valence electrons are somehow deformed in the direction opposite to the field. In this simple model the motion of the hole may be visualized as a transfer of ionization from one atom to another, carried out by the field-aided motion of the valence electrons in the tetrahedral orbitals. The direction of motion of the hole is the same as the direction of the applied electric field. Finally, the reader should note that the hole moves independently of the electron which was originally released to produce it. If we consider the hole and the free electron as two particles, then the particles are independent and give independent contributions to the photoexcited change in conductivity.

The energy band theory can help in explaining the photoconductivity process as follows: First, photons of energy E_g (or greater) promote electrons from near the top of the valence band to near the bottom of the conduction band. However, electrons at or near the bottom of the conduction band have vacant energy levels above them. These electrons can absorb energy from an applied electric field, being raised to higher energy levels. In this way, they are accelerated through the crystal. This acceleration of any electron is limited, in time, by collisions within the crystal which occur at any temperature greater than 0 K. The reader should note that without collisions, both the velocity of the electrons and the current produced by the electric field would become arbitrarily large (see Prob. 2.2). We will encounter the collision processes in Chap. 3.

Conduction within the valence structure can also take place in the same way. As the photogeneration process creates vacant energy levels at the top of the valence band, valence electrons which are at lower energy levels can be excited into these vacant levels at the top of the valence band. This corresponds to the valence electron exchange process and leads to current flow (i.e., electron motion) within the valence structure. We observe that this process can be thought of as the motion of the vacancy (hole).

In conclusion, two different and independent types of conduction can take place in a semiconductor. First, electrons which reach the conduction band, by whatever process, can generate a current. Second, the removal of electrons from the valence band, by whatever process, will create vacant energy levels. These holes can produce current flow within the valence structure by the processes just mentioned.

2.1.3 Fluorescence of Photogenerated Carriers

We already showed that a silicon crystal absorbs radiation for all frequencies $\nu > \nu_{th}$ differently from a gas of noninteracting atoms, which will absorb radiation only at those discrete frequencies that correspond to electronic transitions between allowed energy levels in the individual atoms (see Sec. 1.2).

The energy $h\nu_{th}$ is enough to break a valence bond (and therefore create an electron-hole pair). The additional energy supplied by a photon having $\nu > \nu_{th}$ could be used in several ways. What happens is that $h\nu_{th}$ is used to create the carrier pair and $h(\nu - \nu_{th})$ is converted into kinetic energy that is shared between the electron and the hole. In other words, the electron and hole move away from the point of creation with initial velocities that are related to the excess energy $h(\nu - \nu_{th})$, which is quickly lost by colliding with the atoms of the crystal as they move through the lattice. In this way, the energy $h(\nu - \nu_{th})$ is finally transformed to heat in the form of increased atomic motion. This degradation of energy in fact sets a limit on the efficiency with which a solar cell can convert the sun energy into electricity.

In summary, a photon with $\nu > \nu_{th}$ can move an electron from the top of the valence band to a level that is $h(\nu - \nu_{th})$ above the bottom of the conduction band; or it can promote an electron that is $h(\nu - \nu_{th})$ below the top of the valence band to the bottom of the conduction band; or it can produce a transition that is intermediate between these. The initial kinetic energies of any electrons and holes produced are measured from the appropriate band edges. The loss of this kinetic energy through lattice collisions transforms part of the original photon energy into heat.

If the incident photons have a frequency $\nu > \nu_{th}$, the crystal can emit radiation at the frequency ν_{th}. This phenomenon is called *fluorescence* and is analogous to the emission of radiation that takes place when an excited atom returns to its ground state (see Sec. 1.2). To follow this analogy, we can say that the crystal is in an excited state when electron-hole pairs exist within it. It returns to its ground state when the electrons and holes recombine to complete the valence structure. Every recombination event is accompanied by releasing an energy $h\nu_{th}$.

The crystal can be excited by any incident radiation in a broad absorption band, and consequently, the charge carriers will have nonzero initial velocities and will move away through the crystal. In a perfect crystal, they will lose their remaining energy by recombining. When the bond ionization energy is given up, a photon of frequency ν_{th} is produced. This gives rise to the fluorescence or emission of light at ν_{th}. As this is associated with the recombination of a hole and a electron, it is also given the name *recombination radiation*. It is the opposite of the photogeneration process by which electron-hole pairs were produced. The situation just considered is based on the assumption that the crystal is perfect. Actual crystals present imperfections (missing atoms, impurities, etc.) which frequently can *trap* electrons and holes. These traps help the recombination of a hole and an electron. The nature of most traps is such that during a recombination event the energy E_g is provided to the crystal as heat through intermediate energy levels (E_t), rather than by generating a photon. In these cases no fluorescence is observed. This is what usually happens with Ge, Si, and CdS at room temperature. A highly fluorescent material is $GaAs_{0.6}P_{0.4}$, which is used to make light-emitting diodes (LEDs).

The energy band model for the semiconductor provides a useful interpretation for the fluorescence process. This is shown in Fig. 2.2a, where an electron recombines with a hole, generating a photon of energy E_g and frequency ν_{th}. This is the *radiative recombination process*. The *nonradiative process* is schematically shown in Fig. 2.2b, where the trap arising from a crystalline defect (impurity or missing atom) is visualized as providing an energy level (E_t) in the forbidden gap, which helps the recombination process.[1,2]

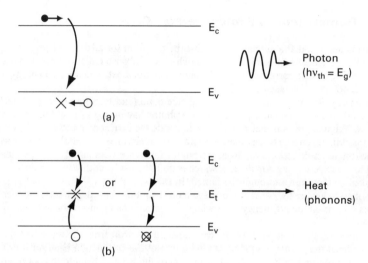

FIGURE 2.2 Energy band representation of (*a*) radiative and (*b*) nonradiative recombination processes.

The principal points of the preceding sections are illustrated in the following two examples.

Example 2.1 Suppose we illuminate a GaAs crystal with a 1-mW green light ($\lambda = 0.5 \times 10^{-4}$ cm). How thick should the crystal be to absorb 98 percent of the incident radiation?

Answer For 98 percent absorption, we require

$$\exp(-\alpha l) = 0.02 \quad \text{or} \quad \alpha l = 3.9$$

Hence, to find l we need to know α for green light.
If we assume $\alpha = 10^5$ cm^{-1} for GaAs, then $l = 3.9 \times 10^{-5}$ cm.

Example 2.2 Assume the light source of Example 2.1 is shined on a thick crystal of Si. If all of the photogenerated carriers recombine radiatively, how much light energy is converted to heat in the crystal?

Answer The incident photon energy is 2.5 eV, of which 1.1 eV will be used to create an electron-hole pair (see Table 2.1) and 1.4 eV will be distributed between the electron and hole as kinetic energy. The kinetic energy will be dissipated and thus converted to heat. Then 44 percent of the incident energy is reradiated as photons, each of which has an energy of 1.1 eV.

2.2 THERMAL EXCITATION OF VALENCE ELECTRONS

We have just observed in the previous section that, when photons are absorbed by a semiconductor, they generate pairs of free charge carriers, i.e., electrons and holes. In other words, light is a source of energy which produces free carriers by disrupting chemical bonds.

Are there free carriers in a semiconductor in the absence of this source of energy? The answer is yes, if the temperature is greater than 0 K. In a sense, thermal energy is the most obvious source of energy for breaking bonds in a semiconductor lattice, and we can predict that pairs of free charge carriers will be present in any slab of semiconductor at a temperature above 0 K. As the temperature of a crystal is increased from 0 K both the lattice atoms and the valence electrons will absorb heat. The lattice atoms will vibrate about their ideal positions, producing a continuous fluctuation of the lengths and angles of the valence bonds. The energy of electrons in these bonds will be a quantity fluctuating in time. Once in a while, these fluctuations will provide some electrons with enough energy to escape from their bond. In other words the bond has been *thermally ionized*. As in the photoionization process, the release of each valence electron is accompanied by the creation of a hole in the valence structure. Thus, the *thermal generation* mechanism (G) also creates equal numbers of holes and electrons. The rate at which holes and electrons are generated by the G process is a function of the parameters of the semiconductor material and the temperature.

Thus, there are two opposing dynamic mechanisms for the regulation of electron and hole concentrations in a semiconductor crystal: generation and recombination. These mechanisms will be described in Sec. 3.6.

Statistical mechanics methods show that the thermal-generation rate in a semiconductor with energy gap E_g is

$$G(T) = AT^a e^{-E_g/kT} \qquad (2.9)$$

where T is the absolute temperature, k is the Boltzmann constant, E_g is the energy gap, and A and a are constants which depend on the semiconductor material. Usually E_g and kT are expressed in electron-volts. In a thermal ionization event, the energy E_g, necessary for a valence electron to be raised to the conduction band, is obtained from the thermal interactions of the valence electron with its surroundings.

Now, on the basis of statistical mechanics, it can be shown that when a small system (one valence electron) is in thermodynamic contact with a large one (the rest of the crystal) which is in thermal equilibrium at a temperature T, the probability per second that the small system can absorb an energy E from the large system is proportional to

$$e^{-E/kT} \qquad (2.10)$$

Then the number of valence electrons per second per cubic centimeter that gain an energy E_g from thermal interactions with the lattice is proportional to $\exp(-E_g/kT)$. Therefore, we expect that the number of bonds per cubic centimeter which will become ionized per second, $G(T)$, should also be proportional to $\exp(-E_g/kT)$. Figure 2.3 provides an intentionally exaggerated sketch of the effect of temperature on semiconductor conduction. A more rigorous framework for the derivation of Eq. (2.10) is given in the next section.

2.3 FERMI-DIRAC DISTRIBUTION

In this section we will answer the following question: "Given a system of particles in thermal equilibrium and subject to appropriate constraints, which is the most probable way of distributing its particles into allowed energy levels?"

This is a relevant question in equilibrium statistical mechanics. By varying the constraints, the distribution can be used to describe classical particles (e.g., ions in a solution), photons, or electrons and holes. It is by answering this question that the concentration (about 10^{10} electrons/cm^3 equal to 10^{10} holes/cm^3) of electrons and holes in a slab of intrinsic silicon at room temperature can be determined.

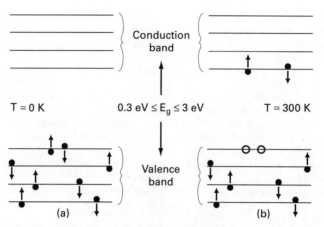

FIGURE 2.3 Schematic energy band representation of a generic semiconductor (a range of E_g is indicated). (a) At an absolute temperature in proximity of 0 K. (b) At room temperature (300 K). Note that the fraction of electron-hole couples (²⁄₈) present at room temperature is extremely exaggerated.

Let us consider a physical system (Fig. 2.4) which is characterized by discrete allowed energy levels E_i ($i = 1, 2, 3, \ldots$). Let us further assume that every level E_i contains S_i available states.[1]

The system is made of N particles and the total energy E_{tot} of the system is fixed. Let us finally assume that we are allowed to distribute any number N_i of particles in the allowed energy levels. In other words, all the sets $\{N_1, \ldots N_i, \ldots N_n\}$ are admissible which satisfy the constraints

$$N = \Sigma N_i = \text{constant} \tag{2.11}$$

$$E_{tot} = \Sigma E_i N_i = \text{constant} \tag{2.12}$$

This is a very general situation considered by statistical physics. Before commenting on it, we note that specific features of the N particles constrain the problem as follows:

- The particles can be distinguishable or undistinguishable
- Any state S_i can be available for a maximum of just one particle or for a greater number of particles.

In the following, we will consider indistinguishable particles and a maximum of one particle for any S_i.

This configuration precisely describes the distribution of electrons in a crystal if we split any level with two electrons with opposite spin into two degenerate states (i.e., states with the same energy), according to the two distinct spin values.

A point to be emphasized is that there is a given number of ways of distributing a *chosen* number N_i of particles over a *given* number S_i of states. These different ways W_i are totally equivalent and are given by

$$W_i = \frac{S_i!}{(S_i - N_i)! N_i!} \tag{2.13}$$

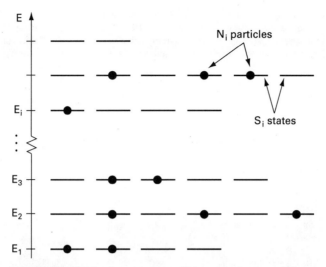

FIGURE 2.4 Multilevel energy system which contains S_i states and N_i particles at an energy E_i ($i = 1, 2, \ldots$). Each state can accomodate a maximum of one particle.

So any chosen N_i produces a configuration which can be realized in a number W_i of *equivalent* ways.

For example, if $S_i = 5$ and we choose $N_i = 3$, then we find 10 equivalent ways of producing this arrangement. If more than one level is considered, the number of different arrangements increases as the product of the individual W_i values. As Eq. (2.13) holds for any E_i level, the total number of different ways (W) in which the N electrons can be arranged in the multilevel system is

$$W = \prod_i W_i = \prod_i \frac{S_i!}{(S_i - N_i)! N_i!} \tag{2.14}$$

It should be pointed out that Eq. (2.14) is valid for any set of N_i values satisfying the restrictions given by Eqs. (2.11) and (2.12).

In a kind of "thought experiment," we can imagine repeating the procedure of distributing the electrons a number of times (or to prepare in parallel a number of identical systems). Then the frequency of occurrence of the various arrangements will be dictated by the number of ways in which they can be realized. The greater the number of ways, the most probable the arrangement. The greatest number of ways is the maximum of W. Therefore we look for the set of N_i values for which W is at its maximum. This can be obtained in the usual way by setting the total differential of W equal to zero and solving for W_{max}. The maximization procedure is greatly simplified if $dW = 0$ is replaced by $d(\ln W) = 0$ as the maximization criteria. From Eq. (2.14) we get

$$\ln W = \sum_i [\ln S_i! - \ln (S_i - N_i)! - \ln N_i!] \tag{2.15}$$

Since $d(\ln W) = dW/W$ and $W_{max} \neq 0$, then $d(\ln W) = 0$ when $dW = 0$ and the two maximization criteria are clearly equivalent.

Before proceeding it is important to note that the number of available states S_i and the number of electrons populating those states, N_i, are typically quite large for E_i values of interest in real systems, as it is for the near-band-edge part of the conduction and valence bands in semiconductors. To simplify the factorial terms in Eq. (2.15), we use Stirling's approximation, i.e.,

$$\ln x! \simeq x \ln x \quad (x \text{ large}) \tag{2.16}$$

Then Eq. (2.15) becomes

$$\ln W \simeq \sum_i [S_i \ln S_i - (S_i - N_i) \ln (S_i - N_i) - N_i \ln N_i] \tag{2.17}$$

We can perform the actual maximization. Recognizing that $dS_i = 0$ (the S_i are system constants), we obtain

$$d(\ln W) = \sum_i \frac{\partial(\ln W)}{\partial N_i} dN_i = \sum_i \ln (S_i/N_i - 1) dN_i \tag{2.18}$$

Setting $d(\ln W) = 0$ then yields

$$\sum_i \ln \left(\frac{S_i}{N_i} - 1 \right) dN_i = 0 \tag{2.19}$$

The solution of Eq. (2.19) for the most probable N_i value set is subject to the $\Sigma N_i = N$ and $\Sigma E_i N_i = E_{\text{tot}}$ restrictions. These solution constraints can be rearranged into the equivalent differential form

$$\sum_i dN_i = 0 \tag{2.20}$$

and

$$\sum_i E_i \, dN_i = 0 \tag{2.21}$$

To solve Eq. (2.19) subject to the Eqs. (2.20) and (2.21) constraints, we use the method of Lagrange multipliers. This method consists of multiplying each constraint equation by an as yet unspecified constant. Let the undetermined multipliers be $-\alpha$ and $-\beta$, respectively. By adding the resulting equations to Eq. (2.19), we obtain

$$\sum_i \left[\ln \left(\frac{S_i}{N_i} - 1 \right) - \alpha - \beta E_i \right] dN_i = 0 \tag{2.22}$$

The multipliers α and β can always be chosen such that two of the bracketed dN_i coefficients vanish, thereby eliminating two of the dN_i from Eq. (2.22). With two of the dN_i eliminated, all of the remaining dN_i in Eq. (2.22) can be varied independently, and the summation will vanish for all choices of the independent differentials only if

$$\ln \left(\frac{S_i}{N_i} - 1 \right) - \alpha - \beta E_i = 0 \quad \text{for all } i \tag{2.23}$$

Equation (2.23) is the relationship for the most probable N_i we were looking for. Solving Eq. (2.23) for N_i/S_i, we finally obtain

$$f(E_i) = \frac{N_i}{S_i} = \frac{1}{1 + e^{\alpha + \beta E_i}} \quad (2.24)$$

For densely spaced levels, as encountered in the conduction and valence bands of semiconductors, E_i may be substituted by the continuous variable E and

$$f(E_i) \rightarrow f(E) = \frac{1}{1 + e^{\alpha + \beta E}} \quad (2.25)$$

To finally derive the *Fermi-Dirac distribution function*, it is necessary to evaluate the constants α and β. Thermodynamic considerations and the analysis of real systems using the kinetic theory of gases lead to

$$\alpha = -\frac{E_F}{kT} \quad (2.26a)$$

and

$$\beta = \frac{1}{kT} \quad (2.26b)$$

where E_F is the electrochemical potential energy or *Fermi energy* of the electrons in the solid, k is the Boltzmann constant, and T is the system temperature. Thus we arrive at the final form of the *Fermi-Dirac distribution function*

$$f(E) = \frac{1}{1 + e^{(E-E_F)/kT}} \quad (2.27)$$

A plot of the Fermi-Dirac distribution function versus $(E - E_F)$ for a given number of temperatures is shown in Fig. 2.5

Let us assume that the energy level E_F is placed between the conduction and valence bands. Then, for an electron in a level of the conduction band it is $E > E_F$. If, moreover, $(E - E_F) > kT$, then Eq. (2.27) can be approximated by

$$f(E) \simeq e^{-(E-E_F)/kT} \quad (2.28)$$

From now on let us identify the Fermi-Dirac distribution function for electrons with the subscript n, i.e., $f_n(E)$. Then, since a given electronic state in a crystal must be either occupied by an electron or unfilled (i.e., occupied by a hole), we can also write the probability that a given energy level will be occupied by a hole as

$$f_p(E) = 1 - f_n(E) \quad (2.29)$$

The distribution function for holes, $f_p(E)$, is the mirror image of $f_n(E)$ around the energy level E_F. In particular, for $(E - E_F) \ll kT$, $f_p(E)$ becomes

$$f_p(E) \simeq e^{-(E_F - E)/kT} \quad (2.30)$$

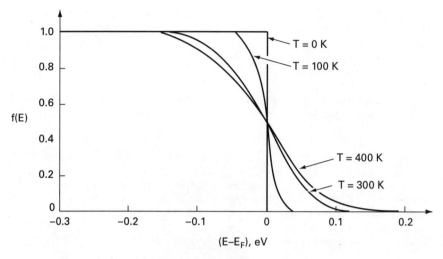

FIGURE 2.5 Plot of the Fermi function versus energy for various temperatures.

2.4 CHARGE CONCENTRATIONS IN INTRINSIC SEMICONDUCTORS

Let us now consider the procedure by which the Fermi function is used to calculate the number of electrons in the conduction band. The product of the density of available energy states times the probability that a given state is occupied (i.e., Fermi function) gives the actual distribution of electrons among the allowed energy levels. The electrons in the conduction band are concentrated near the low-energy edge of the conduction band.

The total number of electrons in the conduction band can be obtained from

$$n = \int_{E_c}^{E_{ct}} N(E) f_n(E, E_F) \, dE \tag{2.31}$$

where E_{ct} is the energy at the top of the conduction band. However, because f_n falls off exponentially with increasing E and E_{ct} is many kT away from E_c, we can extend the upper limit of the integral for n with very little effect on the result. Equation (2.31) then becomes

$$n = \int_{E_c}^{\infty} N(E) f_n(E, E_F) \, dE \tag{2.32}$$

By substituting $N(E)$ and $f_n(E, E_F)$ in Eq. (2.32) and integrating, we find

$$n = \text{const } e^{-(E_c - E_F)/kT} \tag{2.33}$$

The constant in Eq. (2.33) is called the *effective density of states*[1,2] in the conduction band and is denoted by N_c. Equation (2.33) is then written

$$n = N_c e^{-(E_c - E_F)/kT} \tag{2.34}$$

The reader should note that the exponential factor is the probability that a state at an energy level E_c will be occupied by an electron. Hence, Eq. (2.34) points out that, as far as the calculation of n is concerned, the actual density of allowed levels in the conduction band is equivalent to N_c states/cm³ concentrated at the energy level E_c.

In a similar way we can conclude that the density of holes in the valence band can be written as

$$p = N_v e^{-(E_F - E_v)/kT} \tag{2.35}$$

where E_v is the energy level at the top of the valence band and N_v is the *effective density of states* for the valence band.

The effective densities N_c and N_v are found to have the values[1–3]

$$N_c = 2.8 \times 10^{19}/\text{cm}^3 \tag{2.36a}$$

$$N_v = 1.04 \times 10^{19}/\text{cm}^3 \tag{2.36b}$$

for silicon at room temperature.

By using Eqs. (2.34) and (2.35) we can also write the *np* product in the form

$$np = N_c N_v e^{-(E_c - E_v)/kT} = N_c N_v e^{-E_g/kT} \tag{2.37}$$

where E_g is the energy gap.

In an intrinsic (i.e., undoped) semiconductor, we have

$$n = p = n_i \tag{2.38}$$

Therefore

$$np = n_i^2 = N_c N_v e^{-E_g/kT} \tag{2.39}$$

and

$$n_i = (N_c N_v)^{1/2} e^{-E_g/2kT} \tag{2.40}$$

For intrinsic silicon at room temperature Eqs. (2.38) and (2.40) give the approximate values $n_i = n = p = 10^{10}$ charge carriers/cm³. (A more precise value can be found in App. A). Let us summarize the results obtained in the last three sections:

1. In a piece of (pure) semiconductor under equilibrium conditions at temperature T greater than the absolute zero, there is a fixed concentration of charge carriers, which are electrons and holes.
2. This concentration can be estimated by putting together the information obtained from band theory with that given by the Fermi-Dirac statistics.

It must be emphasized that this fixed concentration is an *average* one, resulting from the balance between the rate of *generation* of electrons and holes and the rate of their *recombination*. Because it is an average value, *fluctuations* (i.e., "noise") should be expected around the value (the generation-recombination noise).

It is reasonable to assume that the number of recombination events per unit volume per unit time is proportional to the product of the average concentrations available for the events (i.e., free electrons and holes), or

$$R = rnp \tag{2.41}$$

where R is the recombination rate (electron-hole pairs/cm^3-s), n is the free-electron concentration (electrons/cm^3), p is the free hole concentration (holes/cm^3), and r is a proportionality constant (cm^3/s).

This statement is a first example of the *mass-action law*, which will be further considered in subsequent chapters. At any given temperature there will exist one particular concentration of holes and electrons at which the recombination rate and thermal generation are in balance. This is the thermal equilibrium density of holes and electrons for the given crystal temperature. These concentrations satisfy the equation

$$\text{Thermal-generation rate} = G(T) = rnp = \text{recombination rate} \quad (2.42)$$

Generation-recombination processes will be further discussed in this chapter (see Sec. 2.7) in a heuristic way in two useful examples: flashes of light and stationary illumination. A more general and rigorous description will be given in the next chapter, where a general *continuity equation* will be provided for the time evolution of carriers.

2.5 EFFECTIVE MASS

For an electron in free space subjected to an externally applied field E_a, we could consider Newton's laws approximately valid (at least on a not-too-detailed scale) and we could consequently assume that the electron would experience an acceleration

$$a = -\frac{qE_a}{m_0} \quad (2.43)$$

where m_0 is the *free-space mass* of the electron. However, an electron in a crystal does not move in the same way as an electron in free space. At any given time the electron must stay in one of the orbitals that correspond to an energy level in the conduction band. These orbitals are made up of appropriate linear combinations of overlapping atomic orbitals. Atoms and electrons of the crystal create an electric field E_{in} within it. This electric field results to be much larger than any external field E_a which we will apply to produce carrier motion. It is quite intuitive that the motion of an electron in a crystal cannot be calculated using Newtonian mechanics. Besides other considerations, let us underline that the interatomic spaces in which the electron moves are about the same size as the dimension of atomic orbital boundary surfaces. These extremely small dimensions imply that quantum mechanics, not Newtonian mechanics, must be used to calculate the motion of the electron. However, quantum mechanical treatment of the problem provides a very simple result; that is, as long as E_{in} is a periodic function of distance and is much greater than E_a, then the motion predicted by quantum mechanics for a real electron in a perfect crystal is the same as the motion predicted by Newton's laws for what we call an *effective-mass electron*. The effective-mass electron responds only to the externally applied field E_a (not E_{in}) according to

$$a = -\frac{qE_a}{m_n^*} \quad (2.44)$$

where m_n^* is the effective mass to be associated with the electron in the crystal.

Valence electron motion can also be described in classical terms, in this case considering a hole as a classical particle with a *positive effective mass* and a *positive charge*.

As in the free-electron case, the actual motion of the vacancy in the valence structure via the valence electrons should be calculated by quantum-mechanical methods. However, the hole, which is the classical counterpart of the valence-band vacancy, responds only to externally applied fields, in accordance with Newton's laws. In conclusion, in the effective-mass approximation, a perfect crystal behaves as a homogeneous medium in which electrons and holes move as classical particles in response to external fields only.

Measurements of effective masses for both holes and electrons show that they lie in the range from one-hundredth of the free-space mass to several times the free-space mass of an electron, depending on the material. The reader should bear in mind that the information given in this section represents an oversimplification. For example, it should be appreciated that the effective mass is not a scalar quantity (as it is the *rest mass*) but a *tensor*. Moreover, its value (which should then be expressed by nine numbers) changes according to the location of the carrier inside the energy band of a given crystal.

The interested reader can find appropriate quantum descriptions of the effective mass in Refs. 1 to 3.

Finally, it can be useful to know (see Prob. 2.3) that

$$\frac{N_v}{N_c} = \left(\frac{m_p^*}{m_n^*}\right)^{3/2} \tag{2.45}$$

2.6 DOPED SEMICONDUCTORS

So far we have been considering pure (intrinsic) semiconductor materials and their free carriers. However, in all cases of importance, selected chemical impurities (see Chap. 1) are added to the intrinsic semiconductor material to provide free carriers for conduction. The semiconductor is then said to have been *doped*.

Consider a semiconductor material uniformly doped with an impurity that can provide free electrons to the semiconductor; such an impurity is called a *donor* impurity (it comes from group V of the periodic table).

The donor atoms donate (becoming immobile positive ions) electrons to the semiconductor and increase n at the expense of p (as compared to the case of intrinsic semiconductor). Consider now a semiconductor material doped with an impurity that can capture free electrons from the semiconductor, which is equivalent to providing free holes to it; such an impurity is called an *acceptor* impurity (it comes from group III of the periodic table).

The acceptor atoms donate (becoming immobile negative ions) holes to the semiconductor, thus increasing p at the expense of n. The common donor impurities are antimony (Sb), arsenic (As), and phosphorus (P). Typical acceptor impurities are aluminum (Al), boron (B), gallium (Ga), and indium (In). The influence of these impurities on the carrier concentrations will now be considered.

2.6.1 *n*-Type Semiconductors

As already seen in Chap. 1, elements in column V of the periodic table have an atomic core with an effective charge of +5 elementary charges. This atomic core is surrounded by five valence electrons. When this kind of atom replaces a silicon atom in a silicon crystal, four of its valence electrons will form strong covalent bonds with the neighboring silicon atoms.

At absolute zero temperature, the fifth valence electron describes an orbit around its original nucleus which is very similar to the orbit of an electron around a hydrogen atom in the ground state (see Chap. 1). As a matter of fact, according to the effective-mass approximation the crystal itself can be thought of as a smooth medium, where there are now two charged particles: the partially neutralized impurity atom and the fifth valence electron contributed by the impurity atom.

The impurity atom, which is bound in the lattice, has one effective positive charge. This is because its effective nuclear charge is +5 elementary charges, four of which are neutralized by the covalent bonding electrons. The bound impurity atom thus plays the role of the proton in the hydrogen atom. The fifth valence electron moves in the field of this bound charge, though screened from it by the dielectric constant of the silicon crystal. The electron is in a Bohr orbit (an s-type hydrogen orbital; see Chap. 1).

In the real hydrogen atom, its electron moves in a vacuum, it can be characterized by its rest mass and has a ground-state binding energy E_H of -13.6 eV. In the pseudo-hydrogen atom, on the other hand, the orbiting electron moves inside the silicon crystal and is characterized by an effective mass. Hence, in the donor case, the permittivity of free space must be replaced by the permittivity of silicon and m_0 must be replaced by m_n^*. The binding energy E_B of the fifth donor electron is approximately

$$E_B \simeq -\frac{m_n^* q^4}{2(4\pi\varepsilon_r\varepsilon_0\hbar^2)^2} = \frac{m_n^*}{m_0}\frac{1}{\varepsilon_r^2} E_H \tag{2.46}$$

where ε_r is the silicon dielectric permittivity ($\varepsilon_r \simeq 12$). An exact calculation of m_n^*/m_0 is a difficult task. If we assume this ratio in the order of 0.5, then

$$E_B \simeq -0.05 \text{ eV} \tag{2.47}$$

Actual binding energies are listed in Table 2.2.

Now, at room temperature we can suppose that this electron will have an average thermal energy of $kT = 0.026$ eV. Therefore, the probability that this electron will be free from its original nucleus at room temperature is high. As an approximation,[4] we will assume that each donor atom at room temperature will donate one free electron to the crystal. This electron is free to conduct in the same way as any other electron. On the other hand, the donor impurity atom is tightly bound into the semiconductor lattice so it *cannot contribute to electronic conduction*. It does, however, provide one positive

TABLE 2.2 Properties of Acceptor- and Donor-Doped Silicon: Binding Energies

Dopants	Binding energy (absolute values, eV)
Acceptors:	
Boron	0.05
Aluminum	0.07
Gallium	0.07
Donors:	
Phosphorus	0.04
Arsenic	0.05
Antimony	0.04

elementary charge to the crystal. The crystal as a whole is still electrically neutral, of course, since there is one mobile electron in the crystal for every bound positive charge.

This doping process implies a change in the value of the semiconductor Fermi energy level, as it will be shortly discussed in Sec. 2.6.3.

The above discussion indicates that the introduction of donor impurities increases the electron concentration n in the silicon crystal which is called *n-type* silicon. Let us then calculate what the electron and hole concentrations will be when a semiconductor crystal is doped with N_D donor impurities/cm³. The reasoning used to calculate the carrier concentrations in any doped crystal has the two following features:

1. Recombination and generation must always be in balance in thermal equilibrium.
2. The crystal must always be space-charge neutral.

To maintain electron and hole concentrations in a steady-state condition, electrons and holes should recombine at the same rate as they are generated [see Eq. (2.42)]. Of course the values of n and p in a donor-doped crystal cannot be any longer n_i and p_i, since the electron concentration has been increased by the donor doping. Let us indicate the new electron and hole concentrations with n_n and p_n, respectively. To calculate n_n and p_n, we need to know the new generation rate $G(T)$.

Let us consider a typical donor doping density of 10^{17} phosphorus atoms/cm³ in a silicon crystal. Since there are 5×10^{22} Si atoms/cm³, a donor doping level $N_D = 10^{17}$/cm³ amounts to only one doping atom for every 5×10^5 Si atoms. This means that the strength of a given valence bond in the crystal is practically unaffected by doping. It follows that the energy E_g required to break a given valence bond has also not changed as a result of doping. Then the thermal generation rate, which depends only on E_g and T, is the same in doped and intrinsic crystals, at least as long as the doping density $N_D \ll N_{Si}$. We may therefore write

$$G(T)_{\text{doped}} = G(T)_{\text{intrinsic}} \quad (2.48)$$

We already know that $G(T)$ in an intrinsic crystal can be expressed in terms of n_i as

$$G(T)_{\text{intrinsic}} = rn_i^2 \quad (2.49)$$

Hence we can write

$$G(T)_{\text{doped}} = rn_i^2 \quad (2.50)$$

In conclusion,

$$n_n p_n = n_i^2 \quad (2.51)$$

Equation (2.51) means that if we increase the electron concentration from n_i to n_n by adding donor impurities, the probability for recombination of a hole will increase, with the result that the equilibrium hole concentration will be reduced from p_i to p_n.

Equation (2.51) is valid for any doped semiconductor under equilibrium conditions as long as the appropriate value of n_i is used and the doping concentration is not too high. Equation (2.51) provides a very important relation which n_n and p_n must satisfy in thermal equilibrium. However, another independent relation between n_n and p_n is needed to determine their values. This second relation arises from the space-charge neutrality condition. Let us enumerate the types of charges that exist in a silicon crystal that has been doped with N_D donors/cm³: electrons (mobile negative charge) at a concentration of n_n/cm³, ionized donor atoms (fixed positive charge) at a concentration of N_D/cm^3, and holes (mobile positive charge) at a concentration of p_n/cm^3. To have electrical neutrality in the crystal, we require

$$q n_n = q(N_D + p_n) \tag{2.52a}$$

or

$$n_n = N_D + p_n \tag{2.52b}$$

Let us assume that we know the donor doping density N_D. Then we have two independent equations in the two unknowns n_n and p_n:

$$n_n p_n = n_i^2 \tag{2.53}$$

$$n_n - p_n - N_D = 0 \tag{2.54}$$

Combining Eqs. (2.53) and (2.54) we get

$$n_n^2 - N_D n_n - n_i^2 = 0 \tag{2.55}$$

Solving the quadratic equation for n_n yields:

$$n_n = \frac{N_D}{2} + \left(\frac{N_D^2}{4} + n_i^2\right)^{1/2} \tag{2.56}$$

Example 2.3 Let us calculate at room temperature and at 100°C the carrier concentrations in a phosphorus-doped silicon crystal where the concentration of phosphorus atoms is 10^{18} atoms/cm^3.

Answer From the generation-recombination balance we have, for any temperature,

$$n_n p_n = n_i^2 \tag{E2.1}$$

Space-charge neutrality implies (assuming all donors to be ionized),

$$n_n = N_D + p_n \tag{E2.2}$$

To calculate n_n and p_n we need to know N_D and n_i. Let us assume that $n_i(25°C) = 10^{10}/\text{cm}^3$. Then

$$n_n p_n = 10^{20}/\text{cm}^6 \tag{E2.3}$$

From Eq. (2.56) we deduce that n_n is practically $10^{18}/\text{cm}^3$. Then

$$p_n = \frac{n_i^2}{n_n} = 10^2/\text{cm}^3 \tag{E2.4}$$

In conclusion,

$$n_n(T=25°C) = 10^{18}/\text{cm}^3 \qquad p_n(T=25°C) = 10^2/\text{cm}^3$$

At 100°C, we can assume $n_i = 1.5 \times 10^{12}/\text{cm}^3$. Then,

$$n_n p_n = 2.25 \times 10^{24}/\text{cm}^6$$

We still have $n_n = N_D = 10^{18}/\text{cm}^3$, leading to

$$n_n(100°C) = 10^{18}/\text{cm}^3 \qquad p_n(T=100°C) = 2.25 \times 10^6/\text{cm}^3$$

It is worth noting that n_n did not change with temperature, but p_n increased by 4 orders of magnitude. Devices which depend on p_n for their operation can then be expected to be quite sensitive to temperature.

2.6.2 *p*-Type Semiconductors

A *p-type silicon* crystal is obtained by substituting silicon atoms with an element from column III of the periodic table. This element has, in its neutral state, an effective nuclear charge of +3 elementary charges. It is surrounded by three valence electrons in an $(ns)^2(np)^1$ configuration. When such an atom is used to substitute a silicon atom in the lattice, the valence electrons of the impurity atom do not complete the normal valence structure. On the contrary, they form only three covalent bonds with the neighboring silicon atoms. This creates a vacancy in the valence bond structure. The motion of valence electrons within the valence-bonding orbitals is such that the hole is not confined. Rather, it can move away to some new position. When it happens, a net negative charge is created near the acceptor atom (and a net positive charge in proximity of the hole). The attraction between these charges limits the extent to which the hole can move away. As in the case of donor atoms, the acceptor atom is tightly bound in the lattice. It contributes one elementary negative charge, since in the crystal it is surrounded by one more electron than it has in the neutral state. The bound negative charge is neutralized macroscopically by the presence of the hole. From the point of view of the effective mass, the hole is a mobile positive charge. At low temperatures in an acceptor-doped crystal, it describes a hydrogenlike orbit around a negative charge which is bound in the lattice. At room temperature, however, the hole has a high probability of being ionized. Therefore, at room temperature we can assume that every acceptor atom in the crystal gives a free hole for conduction. Actual binding energies for acceptor-atom hole ground states are listed in Table 2.2.

By a reasoning similar to that used for an *n*-type semiconductor, we can arrive at the following two equations which n and p must satisfy in an acceptor-doped (*p*-type) crystal:

$$n_p p_p = n_i^2 \qquad (2.57)$$

$$p_p = N_A + n_p \qquad (2.58)$$

where N_A is the acceptor doping concentration and the symbols p_p and n_p indicate the hole and electron concentrations in *p*-type material, respectively. The solution of Eqs. (2.57) and (2.58) gives a quadratic equation, as in the case of *n*-type semiconductors [Eq. (2.55)]. For practical levels of acceptor doping ($N_A \gg n_i$), it is

$$p_p = N_A \qquad (2.59)$$

$$n_p = \frac{n_i^2}{N_A} \qquad (2.60)$$

In summary, a given slice of semiconductor material always has two types of carriers in it. By chemically doping it, we can choose one of these carriers as the *majority carrier*. We can therefore make two basically different types of material, *n*-type and *p*-type.

The reader should notice that any doped semiconductor becomes practically intrinsic at sufficiently high temperatures where the intrinsic carrier concentration, which increases with increasing temperature, exceeds the fixed dopant density ("intrinsic temperature region").

Donors and acceptors tend to negate each other. Indeed, it is possible to produce intrinsiclike material by doping it but making $(N_D - N_A) = 0$ ("compensated material").

2.6.3 Fermi Energy Level in Doped Semiconductors

The arguments developed in Sec. 2.4 can be used to express the equilibrium carrier concentration for any (nondegenerate) doping condition.[1] By recalling that

$$n = N_c e^{(E_F-E_c)/kT} \tag{2.61}$$
and
$$p = N_v e^{(E_v-E_F)/kT} \tag{2.62}$$
we can also write for intrinsic silicon
$$n = n_i = N_c e^{(E_i-E_c)/kT} \tag{2.63}$$
and
$$p = n_i = N_v e^{(E_v-E_i)/kT} \tag{2.64}$$
where we indicate with E_i the Fermi energy level for intrinsic silicon. Combining Eqs. (2.61) to (2.64), we obtain
$$n = n_i e^{(E_F-E_i)/kT} \tag{2.65}$$
and
$$p = n_i e^{(E_i-E_F)/kT} \tag{2.66}$$
where n and p represent the actual carrier concentrations, E_F is the actual Fermi level, and n_i and E_i are the corresponding values for the intrinsic semiconductor.

Equations (2.65) and (2.66) can be rearranged as
$$E_F - E_i = kT \ln (n/n_i) \tag{2.67}$$
and
$$E_i - E_F = kT \ln (p/n_i) \tag{2.68}$$
Of course, for intrinsic silicon
$$n = p = n_i \tag{2.69}$$
and
$$E_F = E_i \tag{2.70}$$
On the other hand, for a typical p-doping situation, if we assume $p = N_A$ then
$$E_F = E_i - kT \ln (N_A/n_i) \tag{2.71}$$
Conversely, for a typical n-doping situation, by assuming $n = N_D$ we obtain
$$E_F = E_i + kT \ln (N_D/n_i) \tag{2.72}$$

In conclusion, Eqs. (2.71) and (2.72) show that the Fermi energy level, which has the value E_i for an intrinsic semiconductor, moves down in energy from E_i with increasing N_A (acceptor doping) and upward in energy with increasing N_D (donor doping).

2.7 A SIMPLIFIED INTRODUCTION TO THE GENERATION-RECOMBINATION PROCESSES

Let us summarize the features of generation and recombination processes:

A recombination event takes place when a free electron combines with a hole (or vacant valence bond) to complete the valence structure on a local basis. Recombination thus reduces the number of free electrons and holes. Processes such as thermal- or photogeneration produce hole-electron pairs, and in a steady-state condition, the average generation and recombination rates balance. In a simplified way, the recombination rate per unit volume (that is, the number of hole-electron pairs which recombine per second per unit volume) is given by Eq. (2.42).

Let us now assume that a short flash of light homogeneously falls on an n-type semiconductor slice. The light will generate holes and electrons in equal concentrations. Immediately after the flash, the total electron and hole concentrations will be

$$n = n_n + n_\nu \tag{2.73a}$$

$$p = p_n + p_\nu \tag{2.73b}$$

where n_ν and p_ν are the excess concentrations generated by the flash everywhere through the slice.

The recombination rate immediately after the light flash is

$$R = r(n_n + n_\nu)(p_n + p_\nu) \tag{2.74}$$

while the thermal-generation rate is still

$$G = rn_i^2 \tag{2.75}$$

Recombination is then greater than generation and therefore there will be a net recombination rate U. The general form for U is

$$U = R - G = r(np - n_i^2) \tag{2.76}$$

Immediately after the flash, it will be

$$U = r[(n_n + n_\nu)(p_n + p_\nu) - n_i^2] \tag{2.77}$$

Because of recombination, the values of n and p will decay with time until they have been reduced to n_n and p_n, respectively, and then recombination and thermal generation will again be in balance. The rate of loss of holes by recombination is the same as the rate of loss of electrons, so that

$$U = -\frac{dp}{dt} = -\frac{dn}{dt} \tag{2.78}$$

Using Eq. (2.76) with Eq. (2.78), we find

$$-\frac{dp}{dt} = -\frac{dn}{dt} = r(np - n_i^2) \tag{2.79}$$

If we are dealing with steady illumination instead of a flash, photogeneration could also be included, by expanding the equation to read

$$-\frac{dp}{dt} = -\frac{dn}{dt} = r(np - n_i^2) - G_\nu \tag{2.80}$$

where G_ν represents the photogeneration rate, i.e., number of electron-hole couples generated by light per second.

Before proceeding further, we wish to note that Eqs. (2.79) and (2.80) describe one possible process by which charge concentration changes with time inside a semiconductor. Of course, processes other than photoinduction can be present, and here we are referring to just one among several processes. These considerations will be further developed in the next chapter.

As already noticed, the flash of light creates holes and electrons in equal numbers ($p_\nu = n_\nu$), and recombination removes holes and electrons in equal numbers. Therefore, space-charge neutrality implies at any instant:

$$n(t) - n_n = p(t) - p_n \qquad (2.81)$$

We can use Eq. (2.81) to rewrite the expression for the time evolution of the minority carrier as

$$-\frac{dp}{dt} = r[(p - p_n + n_n)p - n_n p_n] \qquad (2.82)$$

Equation (2.82) is a nonlinear, ordinary differential equation describing how p varies with time after the light flash. We look for a solution under the conditions that $p(t)$ has an initial value

$$p(0) = p_n + p_\nu \qquad (2.83)$$

It can be easily verified that the final value $p(\infty)$ will be p_n, by equating to zero the left-side term of Eq. (2.82).

The solution of Eq. (2.82) is found by assuming that the light flash creates only a small number of carrier pairs compared to n_n, and then $(p - p_n)$ can be neglected in comparison with n_n. This condition, which is known as the *low injection condition*[4,5] justifies our choice of considering the minority carriers' time evolution, the concentration of the majority carriers being virtually unaffected by the flash. Then Eq. (2.82) can be simplified to

$$-\frac{dp}{dt} = rn_n(p - p_n) \qquad (2.84)$$

We then define a minority carrier lifetime by the formula

$$\frac{1}{\tau_p} = rn_n \qquad (2.85)$$

so that Eq. (2.84) is rewritten as

$$\frac{dp}{dt} = -\frac{p - p_n}{\tau_p} \qquad (2.86)$$

The solution of the linear differential Eq. (2.86) is

$$p(t) = p_n + Ae^{-t/\tau_p} \qquad (2.87)$$

The constant A can be evaluated from the known value of p at $t = 0$, giving

$$A = p_\nu \qquad (2.88)$$

and
$$p(t) = p_n + p_\nu e^{-t/\tau_p} \qquad (2.89)$$

Equation (2.89) can be also written as

$$p(t) - p_n = p_\nu e^{-t/\tau_p} \qquad (2.90)$$

According to Eq. (2.90), the excess carrier population disappears in an exponential fashion. The decay process is characterized by the parameter τ_p, the *hole lifetime*.

The recombination of electrons (minority carriers) in a p-type material can be studied in exactly the same way. The linearized recombination law becomes

$$\frac{dn}{dt} = -\frac{n - n_p}{\tau_n} \tag{2.91}$$

so long as $n \ll p_p$. The *electron lifetime* in p-type material, τ_n, is defined by

$$\tau_n = \frac{1}{rp_p} \tag{2.92}$$

Then the excess concentration of electrons created by a light flash will decay according to

$$n(t) = n_n + n_r e^{-t/\tau_n} \tag{2.93}$$

Under the same hypotheses of low injection level, the solution of Eq. (2.80) gives for the case of steady illumination for n-type silicon

$$p(t) = p_n + G_\nu \tau_p (1 - e^{-t/\tau_p}) \tag{2.94}$$

and

$$n(t) = n_p + G_\nu \tau_n (1 - e^{-t/\tau_n}) \tag{2.95}$$

for p-type silicon.

The just-described process of photogeneration can be viewed as a band-to-band process. On the contrary, the thermal creation and annihilation of carriers, which is going on at all times in all semiconductors, is typically dominated by indirect thermal recombination-generation through recombination-generation centers (see Fig. 2.2). The whole topic will be considered in much greater detail in Chap. 3.

PROBLEMS

2.1 Let us use the experimental apparatus sketched in Fig. 2.1 with an hypothetical material displaying the following optical properties:

Illumination with blue light ($\lambda = 400$ nm), 0.5 percent of the incident light is transmitted.
Illumination with red light ($\lambda = 600$ nm), 100 percent of the incident light is transmitted.

  a. What is the color of the material when illuminated with white light (of course a qualitative answer is expected)?
  b. What is the value of α for $\lambda = 600$ nm?
  c. Assuming that, for $\lambda = 400$ nm, $\alpha = 10^5$ cm^{-1}, calculate the thickness of the material.

2.2 Suppose that an electron with resting mass m_0 is accelerated in vacuum by an electric field of 1 V/cm for 1 μm.
  a. What will be its final kinetic energy?
  b. Compare this energy with kT at room temperature and with the silicon energy gap. If this energy is transferred to the silicon crystal, is it enough to ionize it?

2.3 Let us assume that the effective masses of electrons and holes m_n^*, m_p^* in a silicon crystal are known. Then, by making use of the relationships utilized in the development of the Fermi-Dirac statistics, express the Fermi level as a function of m_n^*, m_p^*. (Hint: $n = p$, therefore ...)

2.4 A beam of monochromatic photons of energy 2 eV is shined on a silicon crystal. The power of the beam is 1 mW. Assuming that all the photons are absorbed in the crystal and produce electron-hole pairs, what is the photogeneration rate (that is, number of hole-electron pairs produced per second)? Assume that these hole-electron pairs all recombine radiatively, each recombination event giving off a photon of energy E_g. How much light energy is converted into heat within the crystal?

2.5 A silicon bar is doped with 10^{17} atoms/cm^3 of phosphorus. It is illuminated with 10^{20} photons/s-cm^3. Suppose that each photon generates an electron-hole pair. Assume the minority lifetime = 10 μs. Calculate the concentration of the minority carriers after 5 μs. After 5 μs the source is switched off. Calculate the concentration of the minority carriers 5 μs later (i.e., 10 μs after $t = 0$).

REFERENCES

1. R. F. Pierret, *Advanced semiconductor fundamentals,* from *Modular series on solid state devices,* Vol. VI, Reading, Mass.: Addison-Wesley, 1989.
2. S. M. Sze, *Physics of semiconductor devices,* 2d ed., New York: John Wiley & Sons, 1981.
3. M. Shur, *Physics of semiconductor devices,* Englewood Cliffs, N.J.: Prentice-Hall, 1990.
4. J. F. Gibbons, *Semiconductor electronics,* New York: McGraw-Hill, 1966.
5. A. S. Grove, *Physics and technology of semiconductor devices,* New York: John Wiley & Sons, 1967.

CHAPTER 3
CARRIER MOTION IN SEMICONDUCTORS

In Chap. 2 it was shown that there are two carriers of electricity in a semiconductor crystal: electrons and holes. In this chapter we describe the three basic processes by which electrons and holes move in a semiconductor; these processes are *drift, diffusion,* and *generation-recombination*. The discussion of electron and hole motion will be then extended in Chap. 4 to ion motion in aqueous solutions.

Even if introduced individually, the various types of processes must be understood to occur simultaneously inside any given semiconductor or bioelectronic system. Together with the photogeneration and thermal-generation processes described in Secs. 2.1 and 2.2, these processes form the physical basis for all the semiconductors described in the book; moreover, they lead to the basic set of equations used in solving bioelectronic device problems.

3.1 CARRIER MOTION

The most important property about semiconductors is their capability to carry electric current. As current is defined as the time rate at which charge is carried across a given surface in a direction normal to it, the current will depend on both the number of charges free to move and the speeds at which they move.

Electrons and holes in semiconductors are almost "free particles," as they are not associated with any particular lattice site. The influences of crystal forces are buried in an effective mass m^* (see Sec. 2.5) that differs somewhat from the free-carrier mass. From the laws of statistical mechanics, we can say that electrons and holes will have the thermal energy associated with classical free particles, i.e., $kT/2$ units of energy per degree of freedom (x, y, z dimensions), where k is the Boltzmann constant and T is the absolute temperature. This means that carriers in a crystal are not stationary, but they move with random velocities. From the principle of equipartition of energy (since $v_{th}^2 = v_x^2 + v_y^2 + v_z^2$), the mean-square *thermal velocity* v_{th} of the carriers is related to the temperature by the equations

$$\tfrac{1}{2} m^* v_{th}^2 = \tfrac{3}{2} kT \tag{3.1a}$$

or

$$v_{th} = \sqrt{\frac{3kT}{m^*}} \tag{3.1b}$$

The *thermal velocity* v_{th} is then the velocity of the free carriers due to heat absorbed from the environment. Using the values of m_n^* and m_p^* given in App. A.2, we find that at room temperature ($T = 300$ K), v_{th} is 2.3×10^7 cm/s for electrons and 1.88×10^7 cm/s for holes.

As in thermal equilibrium, the carriers in a semiconductor move about in a random way with a mean velocity v_{th} (a mechanism that implies that, on the average, as many carriers of each type flow from a side through a plane in the semiconductor as flow through the same plane from the opposite side), therefore, in thermal equilibrium there can be no electric current. The direction of motion of a charged particle is randomized by its interaction with atoms and ions in the semiconductor crystal: the charged particle either collides with the atoms, or is deflected by the ions. Both types of event are called *collisions* (or *scattering*), and in each case the direction of motion of the carriers is changed.

The frequency of collisions is given by the sum of the frequencies of each of the individual scattering mechanisms, i.e.,

$$\frac{1}{\tau_c} = \frac{1}{\tau_I} + \frac{1}{\tau_L} + \cdots \qquad (3.2)$$

where τ_c is the mean scattering time for carriers, and τ_I and τ_L represent the components of τ_c due to the ionized impurity scattering mechanism and to the lattice scattering mechanism, respectively. The dots in Eq. (3.2) state that other mechanisms due to different scattering processes (such as acoustic, optical, and piezoelectric scattering) can be taken into account, even if typically only the two indicated components are dominant for a given value of temperature and impurity concentration.

If some order can be given to the random motion of the carriers by modifying the thermal equilibrium, a net motion of charge in a particular direction, and then a current flow, can result. *Electric fields* and *gradients of carrier concentrations* are two such ordering factors, and they result in *drift* and *diffusion* currents, respectively. These currents will be described in the following sections.

In order to properly understand the carrier motion, a knowledge of quantum mechanics is essential; however, as discussed in Sec. 2.5, we can use for the carriers the concept of *effective mass,* which incorporates the major quantum features of the electronic motion in the semiconductor material. In this way we can discuss semiconductor electronics in terms of the familiar language of classical physics by using electrons with charge $-q$ and effective mass m_n^* and holes with charge $+q$ and effective mass m_p^*. The subscripts n and p are used to refer the properties of electrons and holes, respectively.

3.2 CARRIER MOTION BY DRIFT

In this section we describe the first of the three basic processes by which electrons and holes move in a semiconductor: the *drift process.*

3.2.1 Drift Velocity and Mobility

If thermal equilibrium is disturbed by the presence of an electric field, a net displacement of carriers can result, establishing a current flow. Between collisions, the carriers move with a velocity given by the vector sum of the random postcollision thermal velocity v_{th} and the velocity due to the electric field, in the field direction for holes and opposite to it for electrons. This latter velocity is called the *drift velocity* and it is usually indicated as v_d.

Let us consider the case of an electric field applied to a semiconductor sample along the positive x direction. To get an expression for the drift velocity, we notice that between collisions the electric field E causes the carriers to gain momentum. The carriers accelerate until they suffer another collision. The time between collisions, τ_c, will not be the same for all carriers, nor for any carrier all of the time; thus the velocities reached between collisions will vary. We define the drift velocity as the mean of the drift velocities of all carriers at a given time. Similarly, τ_c is the mean of all times between collisions. If all the momentum gained by a carrier is lost to the lattice in collisions, we can write,[1-3] for electrons,

$$-q\tau_{cn} E(x) = m_n^* v_{dn}(x) \qquad (3.3a)$$

Hence,
$$v_{dn}(x) = -\left(\frac{q\tau_{cn}}{m_n^*}\right) E(x) = -\mu_n E(x) \qquad (3.3b)$$

For holes,

$$+q\tau_{cp} E(x) = m_p^* v_{dp}(x) \qquad (3.4a)$$

hence,
$$v_{dp}(x) = \left(\frac{q\tau_{cp}}{m_p^*}\right) E(x) = \mu_p E(x) \qquad (3.4b)$$

The proportionality constants μ_n and μ_p are defined as the *mobility* of electrons and holes, respectively. The quantity μ is by convention always defined to be a positive number and its units are cm^2/(V-s).

An intuitive (but approximate) derivation of Eqs. (3.3) and (3.4) arises from the following considerations. The average velocity with which a particle (hole or electron) drifts can be calculated by finding the average x displacement produced by the field $E(x)$ per collision, and then dividing the result by the mean free time. We first observe that the effective mass approximation holds between collision points, so the x acceleration $a(x)$ along each path is, from *Newton's law*,

$$m^* a(x) = q|E(x)| \qquad (3.5)$$

The net displacement after N collisions is then

$$S_x = \frac{a(x)}{2} \sum_{i=1}^{N} \tau_{c,i}^2 = \frac{a(x) N \overline{\tau_c^2}}{2} \qquad (3.6)$$

Hence the average net displacement per collision is

$$\overline{x} = \frac{S_x}{N} = \frac{a(x)\overline{\tau_c^2}}{2} \qquad (3.7)$$

where $\overline{\tau_c^2}$ stands for the average of the square of the free times. Thus, we can write

$$v_d(x) = \frac{\overline{x}}{\tau_c} = a(x)\tau_c = \left(\frac{q\tau_c}{2m^*}\right)|E(x)| \qquad (3.8)$$

Equation (3.8) is approximate. A rigorous analysis, solving the Boltzmann transport equation (see Sec. 3.5), leads to equations like Eq. (3.8), but without the factor 2 in denominator, as pointed out by Eqs. (3.3b) and (3.4b).

The values of μ_n and μ_p, as well as of m_n^* and m_p^*, for intrinsic Si and GaAs are given in App. A.2.

From the definition of μ, and from Eq. (3.2), we can write

$$\frac{1}{\mu} = \frac{1}{\mu_I} + \frac{1}{\mu_L} + \cdots \qquad (3.9)$$

where μ_I and μ_L represent the components of μ due to ionized impurity scattering mechanisms and lattice scattering mechanisms, respectively. These are the two most important scattering mechanisms.

Impurity scattering occurs when an electron travels past a fixed charged particle (e.g., an ionized acceptor or donor) and its path is deflected by the charge on the fixed particle.

Lattice scattering results from the thermal vibration of the atoms of the crystal lattice, which breaks the periodicity of the lattice and then hinders the motion of electrons.

Equations (3.3b) and (3.4b) point out that the drift velocity is linearly dependent on the electric field, provided that τ_c and m^* are independent of E (low-field conditions). The experimental data for drift velocity, as shown in Fig. 3.1, point out that the linear relationship between v_d and E holds up to quite high values of the electric field. However, when E is above about 3×10^3 V/cm in n-type silicon, for example, μ can no longer be considered to be independent of E, so the linearity does not hold up. At fields about 10^5 V/cm, the drift velocity saturates at its *scattering-limited* value, which is close to that of the thermal velocity v_{th}.

Mobility is an important parameter for devices in which the current is due to drift. The higher the mobility, the faster the carriers will move. Figure 3.1 shows that n-type GaAs de-

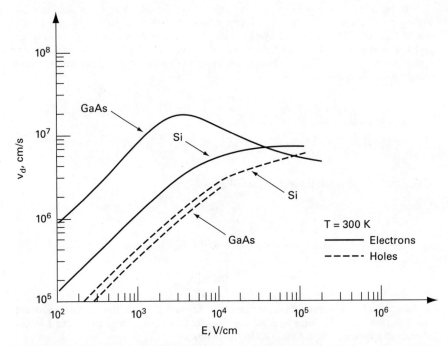

FIGURE 3.1 Dependence of drift velocity v_d on the electric field E in silicon (Si) and gallium arsenide (GaAs). (*After S. M. Sze.*[3])

vices are very suitable for high-speed applications, provided that they are designed to operate at fields in the linear region. Figure 3.1 also shows that mobility is *material-dependent*: this arises through the effective mass term in Eqs. (3.3b) and (3.4b). Mobility also depends on *temperature* and on the total *concentration* of charged doping impurity $N_T = N_D + N_A$.

Experimental measurements of electron and hole mobilities in silicon at room temperature versus the total ionized impurity concentration N_T are shown in Fig. 3.2. We can see that the mobility reaches a maximum value at low impurity concentrations, corresponding to the lattice-scattering mechanism, and that both electron and hole mobilities decrease

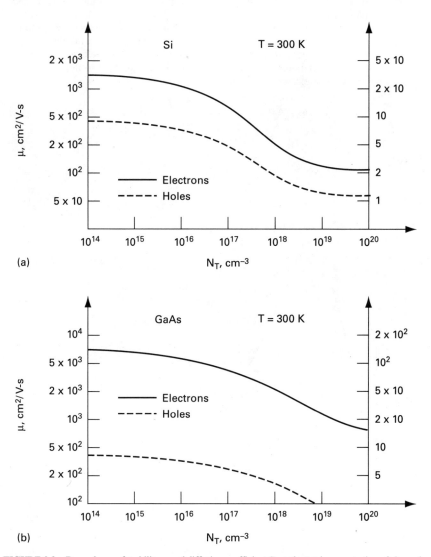

FIGURE 3.2 Dependence of mobility μ and diffusion coefficient D on the total concentration of charged doping N_T for (*a*) Si and (*b*) GaAs. (*After S. M. Sze.*[3])

with increasing impurity concentration, approaching a minimum value at high concentrations. We can also see that the mobility of electrons is larger than the mobility of holes, as it happens in many semiconductors.[3]

Experimental measurements of the influence of temperature on the mobility of holes in silicon for two different impurity concentrations are shown in Fig. 3.3. We can distinguish two regions: at low temperatures, impurity scattering dominates and separate curves are observed for the different doping concentrations. At high temperatures, lattice scattering dominates and the impurity concentration has little effect on the mobility, as indicated by the merging of the curves. The mobility is seen to decrease with increasing temperature in this range.[4]

In the analysis of semiconductor devices it is often necessary to know the magnitude of the low-field mobility for different doping concentrations and temperatures. Although this information is available graphically (Figs. 3.2 and 3.3), it is convenient to have the data enclosed in an equation. Such equations, which give a good fit to the experimental data for the mobility of electrons and holes in silicon, are[5]

$$\mu_n = 88 T_n^{-0.57} + \frac{1252 T_n^{-2.33}}{1 + \left(\dfrac{N_T}{1.26 \times 10^{17}\, T_n^{2.4}}\right) 0.88 T_n^{-0.146}} \qquad (3.10)$$

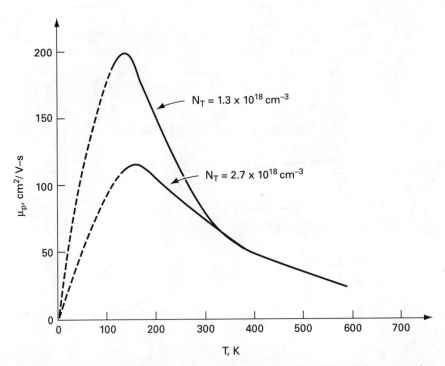

FIGURE 3.3 Dependence of hole mobility μ_p on absolute temperature T and donor doping concentration N_T in n-type silicon. (*From A. S. Grove.*[4] *Used by permission.*)

$$\mu_p = 54.3 T_n^{-0.57} + \frac{407 T_n^{-2.33}}{1 + \left(\dfrac{N_T}{2.35 \times 10^{17} \, T_n^{2.4}}\right) 0.88 T_n^{-0.146}} \quad (3.11)$$

where N_T is the total impurity concentration in cm^{-3}, T is in kelvins, and $T_n = T/300$. At 300 K the equations reduce to

$$\mu_n = 88 + \frac{1252}{1 + 6.984 \times 10^{-18} \, N_T} \quad (3.12)$$

$$\mu_p = 54.3 + \frac{407}{1 + 3.745 \times 10^{-18} \, N_T} \quad (3.13)$$

3.2.2 Drift Current

As an application of the drift mobility concept, we calculate the drift current in a sample of semiconductor material of cross section A, as shown in Fig. 3.4. The electric field E due to the applied voltage causes carriers to move in opposite directions with their drift velocities, but, because of the negative sign on the charge of the electron, the hole and electron contributions to the total current add.

The electron current density flowing in the direction of the applied electric field across a plane (dashed in Fig. 3.4) perpendicular to the x direction can be found by summing the product of the charge on each electron times its drift velocity over all electrons per unit volume n. Thus, the *electron drift current density* is given by

$$J_{n,\text{drift}}(x) = \frac{I_{n,\text{drift}}(x)}{A} = \sum_{i=0}^{n}(-q)v_i = -qnv_{dn}(x) = qn\mu_n E(x) \quad (3.14)$$

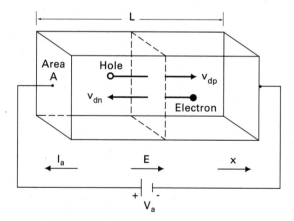

FIGURE 3.4 Drift velocity v_{dn} of electrons and drift velocity v_{dp} of holes in an electric field E due to an applied voltage V_a. (*Kindly provided by Laura Massobrio.*)

where Eq. (3.3b) has been used for the drift velocity. Similarly, the *hole drift current density* is given by

$$J_{p,\text{drift}}(x) = \frac{I_{p,\text{drift}}(x)}{A} = \sum_{i=0}^{p}(+q)v_i = qpv_{dp}(x) = qp\mu_p E(x) \quad (3.15)$$

The *total drift current density* in the x direction can now be written as the sum of the electron and hole components:

$$J_{\text{drift}}(x) = J_{n,\text{drift}}(x) + J_{p,\text{drift}}(x) = q(n\mu_n + p\mu_p)E(x) \quad (3.16a)$$

Ion motion in a solid can conduct electricity, in addition to electron and hole motion. The relationship for field-induced motion of *ions* is given by an expanded version of Eq. (3.16a), i.e.,

$$J_{\text{drift}}(x) = \left[\sum_i c_i z_i q \mu_i + qn\mu_n + qp\mu_p\right] E(x) \quad (3.16b)$$

where μ_i is the mobility for each mobile ion i, c_i is the ion concentration, and z_i is the ion valence.

3.2.3 Conductivity and Resistivity

The drift current and the electric field can be related through the vector form of *Ohm's law*:

$$\mathbf{J}_{\text{drift}} = \sigma \mathbf{E} \quad (3.17)$$

where σ is the *conductivity* of the material. Now considering the one-dimensional (x direction) form of Eq. (3.16a) and comparing it to Eq. (3.17), we have

$$\sigma = q(n\mu_n + p\mu_p) \quad (3.18)$$

The parameter σ is then a material's property because of its dependence on the carrier concentrations and on the mobilities. These terms depend on the doping concentrations and on the temperature. Semiconductor material specifications often refer to the *resistivity* of the material. The resistivity ρ is just the reciprocal of the conductivity and is commonly expressed in units of ohm-centimeters (Ω-cm). Thus

$$\rho = \frac{1}{\sigma} = \frac{1}{q(n\mu_n + p\mu_p)} \quad (3.19)$$

Example 3.1 Calculate σ and ρ at room temperature for both intrinsic silicon material and silicon material doped with 10^{17} donor atoms/cm^3.

Answer Referring to App. A.2, and taking into account Eqs. (3.18) and (3.19), we can write for the intrinsic silicon

$$\sigma = (1.6 \times 10^{-19})(1.45 \times 10^{10})(480 + 1350) = 4.25 \times 10^{-6}(\Omega\text{-cm})^{-1} \quad (E3.1a)$$

$$\rho = \frac{1}{\sigma} = 2.35 \times 10^5 \ \Omega\text{-cm} \quad (E3.1b)$$

When the silicon is doped, n and p will change, as indicated in Sec. 2.6. For $N_D = 10^{17}$/cm^3, $n_n = 10^{17}$/cm^3 and p_n is negligible. Also using Eq. (3.12), we find $\mu_n = 777.44$ cm^2/(V-s). The conductivity and resistivity of the semiconductor are then

$$\sigma \simeq qn\mu_n = (1.6 \times 10^{-19})(10^{17})(777.44) = 12.45 \ (\Omega\text{-cm})^{-1} \quad \text{(E3.2)}$$

and from Eq. (3.19), we obtain

$$\rho = \frac{1}{\sigma} = 0.080 \ \Omega\text{-cm} \quad \text{(E3.3)}$$

It is worth noting that the resistivity of the doped semiconductor material is *more than one million times less* than the resistivity calculated for intrinsic material. This comparison shows that the resistivity of a semiconductor can be controlled over a wide range by doping. We can also control the resistivity *type* by doping with either donor or acceptor impurities. These are characteristic properties of the most used semiconductors, and distinguish them from insulators and conductors.

3.2.4 Temperature Coefficient of Semiconductor Resistivity

If we go back to the calculation of σ for intrinsic silicon, we can rewrite Eq. (3.18) as

$$\sigma = q(\mu_n + \mu_p)n_i \quad (3.20)$$

Using the expression for n_i given by Eq. (2.40), we obtain

$$\sigma = q(\mu_n + \mu_p)BT^{3/2}e^{-qE_g/2kT} \quad (3.21)$$

where B is a material- and temperature-dependent parameter and μ_n and μ_p are functions of temperature. As the mobility of an intrinsic semiconductor is determined by lattice scattering (see Sec. 3.2.1), the temperature dependence of mobility due to this mechanism can be approximated by T^{-a} ($1.5 < a < 2.5$). Therefore, the temperature variation of mobility tends to cancel the $T^{3/2}$ term in Eq. (3.21); thus we can approximate the conductivity of the intrinsic material as follows:

$$\sigma \simeq (\text{constant}) \ e^{-qE_g/2kT} \quad (3.22)$$

From Eq. (3.22), the change in σ for a small change in T is

$$\frac{d\sigma}{\sigma} = \frac{qE_g}{2kT}\frac{dT}{T} \quad (3.23)$$

which indicates that the conductivity of the intrinsic material increases rapidly as T increases.

3.2.5 Charge Control Analysis of Current Flow

In the previous sections we have seen how the current density **J** can be related to the electric field **E** by calculating the number of carriers that flow per second through a cross section. In the following we present an alternative formulation of this problem, which takes into account the total charge and the transit time between the contacts.

For this purpose let us reconsider Fig. 3.4, assuming that it represents an *n*-type semiconductor sample. The sample has cross section A, length L, and majority carrier concentration n_n. The time required for an electron on the right-hand end of the semiconductor sample to drift to the left-hand end is

$$\tau_t = \frac{L}{v_{dn}} = \frac{L}{\mu_n E} \quad (3.24)$$

If the electric field is created by applying a voltage V_a as shown, then

$$E = \frac{V_a}{L} \qquad (3.25)$$

Combining Eqs. (3.24) and (3.25), we obtain

$$\tau_t = \frac{L^2}{\mu_n V_a} \qquad (3.26)$$

where τ_t is called the *transit time* between the contacts.

The *total charge* of electrons in the semiconductor sample at any given time is

$$Q_n = q n_n A L \qquad (3.27)$$

Hence the current is

$$I_a = \frac{Q_n}{\tau_t} \qquad (3.28)$$

Substituting the value of τ_t obtained in Eq. (3.26), we find

$$I_a = \frac{\mu_n Q_n}{L^2} V_a \qquad (3.29)$$

We can now calculate the conductance of the semiconductor sample as

$$g = \frac{\mu_n Q_n}{L^2} = \frac{q \mu_n n_n A}{L} = \frac{\sigma_n A}{L} \qquad (3.30)$$

This method of analysis is called *charge control analysis*.

3.3 CARRIER MOTION BY DIFFUSION

In this section we describe the second of the three basic processes by which electrons and holes move in a semiconductor: the *diffusion process*.

3.3.1 Diffusion Process

In Sec. 3.2 we have considered the drift current which flows in a semiconductor material when an electric field is applied to the semiconductor, giving rise to Ohm's law behavior. However, another important component of the current exists in a semiconductor if there is a spatial variation (*gradient*) of carrier concentration within the semiconductor. This component of the current is called *diffusion current*.

Let us assume, for example, that electrons can be introduced into a *p*-type semiconductor material by injection from an adjacent *n*-type semiconductor material, as will be described in Chap. 6. Assuming that there is no electric field in the *p*-type semiconductor material to move the carriers by drift, they move only by *diffusion*. In this example, the concentration of electrons injected into the *p*-type semiconductor will decrease as they diffuse deeper into the semiconductor due to recombination (see Sec. 3.6) with majority carrier holes. Diffusion is a process whereby particles, as a result of their random thermal motion, tend to spread out or redistribute, migrating on a macroscopic scale from regions of high particle concentration toward and into regions of low particle concentration. The diffusion process operates so as to produce a uniform distribution of particles throughout the material. The diffusing particle need not be charged; random, thermal motion, not interparticle repulsion, is the enabling action behind the diffusion process.

The general features of the diffusion process can be explained readily by means of a simple example. If some molecules of a gas are introduced into one end of an empty container, they soon reach, by diffusion motion, a uniform concentration throughout the container. The driving force that makes the carriers diffuse is their *concentration gradient*.

To understand the origin of the diffusion current, we consider the situation of a semiconductor with carrier concentrations, generically indicated by C_c, that vary only in one dimension, as shown in Fig. 3.5.

Thus the number of carriers flowing (*diffusing*) per unit time across a unit area A of the plane at $x = 0$ in the hypothetical one-dimensional semiconductor of Fig. 3.5 is given by

$$F_c = -D_c \frac{dC_c}{dx} \qquad (3.31)$$

where D_c is a proportionality constant called the *diffusion coefficient* for the carriers; it describes the ease with which the carriers diffuse in the lattice. The usual units for D_c are cm^2/s. Equation (3.31) is also known as *Fick's first law* in one dimension.

The minus sign in Eq. (3.31) appears because carriers flow down the concentration slope, from regions of high concentration to regions of lower concentration, along the $+x$ direction; therefore the derivative has a negative sign.

Net transport of carriers in the x-direction will cease when $dC_c/dx = 0$ everywhere, i.e., when the carriers are uniformly distributed in the material.

It is important to note that the particle flux density resulting from diffusion depends on the carrier concentration gradient and *not* on the concentration itself; it is the concentration imbalance that matters, not the value of the concentration. Moreover, diffusion has nothing to do with the fact that diffusing carriers are charged.

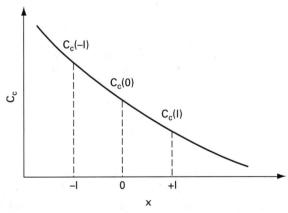

FIGURE 3.5 Shape of the concentration of carriers diffusing away from their source.

3.3.2 Diffusion Current

To understand the origin of the diffusion current, we consider the case of a sample of semiconductor material with a carrier concentration C_c that varies only in one dimension, e.g., the x direction. We assume: (1) that the semiconductor is at uniform temperature, so that the average energy of the diffusing carriers does not vary with x and (2) that no electric fields are applied to the semiconductor—only the concentration $C_c(x)$ is variable.

Referring to Fig. 3.5 (where the diffusing species consists of electrons) we consider the number of carriers crossing the plane at $x = 0$ per unit collision time per unit area. On the average, the carriers crossing the plane at $x = 0$ from left, at each collision time, started at approximately $x = -l$, where $l = v_{th} \tau_c$ is the mean-free path of a carrier.

The average rate per unit area of carriers crossing the plane at $x = 0$ from the left depends on the concentration of carriers that started at $x = -l$; therefore

$$F_1(x) = \tfrac{1}{2} v_{th} C_c(-l) \tag{3.32}$$

The factor ½ appears, since half of the carriers travel to the right and half travel to the left after a collision at $x = -l$. Similarly, the concentration of carriers crossing the plane at $x = 0$ from the right is given by

$$F_2(x) = \tfrac{1}{2} v_{th} C_c(l) \tag{3.33}$$

Thus the net rate F_c of carrier flow (diffusion) from the left is

$$F_c(x) = F_1 - F_2 = \tfrac{1}{2} v_{th} [C_c(-l) - C_c(l)] \tag{3.34}$$

Approximating the concentrations at $x = \pm l$ by the first two terms of a Taylor series expansion, we obtain[1]

$$F_c(x) = \tfrac{1}{2} v_{th} \left\{ \left[C_c(0) - \frac{dC_c}{dx} l \right] - \left[C_c(0) + \frac{dC_c}{dx} l \right] \right\} = -v_{th} l \frac{dC_c}{dx} \tag{3.35}$$

We can now apply Eq. (3.35) in the calculation of the diffusion current in the situation where the carrier concentration C_c represents an electron concentration n. As each electron carries a negative charge $-q$, the diffusive flow to the right in Fig. 3.5 produces an electron diffusion current density in the negative x-direction, given by

$$J_{n,\text{diff}}(x) = (-q) F_c = q v_{th} l \frac{dn}{dx} \tag{3.36}$$

Equation (3.36) can be written in a more useful (and commonly used) form by applying the theorem for the equipartition of energy (see Eq. 3.1) to this one dimensional case. This implies

$$\tfrac{1}{2} m_n^* v_{th}^2 = \tfrac{1}{2} kT \tag{3.37}$$

Taking into account the relation $l = v_{th} \tau_{cn}$, together with Eq. (3.3b), Eq. (3.36) can be written in the form

$$J_{n,\text{diff}}(x) = q \left(\frac{kT}{q} \mu_n \right) \frac{dn}{dx} \tag{3.38}$$

The quantity in parentheses on the right-hand side of Eq. (3.38) is defined as the *diffusion coefficient* D_n for electrons, i.e.,

CARRIER MOTION IN SEMICONDUCTORS

$$D_n = \left(\frac{kT}{q}\right)\mu_n \qquad (3.39)$$

Equation (3.39) is known as *Einstein's relation* for electrons. It relates the two important constants that characterize free-carrier transport by drift and by diffusion in a semiconductor.

The values of diffusion constant of electrons and holes at room temperature can be obtained from Fig. 3.2, where the right-hand axis is labeled in terms of diffusivities. An analysis leading to the rigorous expression of the diffusion coefficient arises from the solution of the Boltzmann transport equation, as described in Sec. 3.5. The *electron diffusion current density* in one dimension is usually written as

$$J_{n,\text{diff}}(x) = qD_n \frac{dn}{dx} \qquad (3.40)$$

If in Fig. 3.5, the situation involves holes, the hole diffusion current density along the x direction is given by

$$J_{p,\text{diff}}(x) = (+q)F_c = -qv_{\text{th}}l\frac{dp}{dx} \qquad (3.41)$$

An expression analogous to Eq. (3.38) can then be written for the *hole diffusion current density*. Thus

$$J_{p,\text{diff}}(x) = -q\left(\frac{kT}{q}\mu_p\right)\frac{dp}{dx} = -qD_p\frac{dp}{dx} \qquad (3.42)$$

where the *diffusion coefficient* for holes, D_p, is related to hole mobility through *Einstein's relation* for holes

$$D_p = \left(\frac{kT}{q}\right)\mu_p \qquad (3.43)$$

Note that Eqs. (3.39) and (3.43) state that D and μ have the same dependence on doping impurity concentration (see Fig. 3.2) and on temperature (see Fig. 3.3).

The quantity kT/q is called the *thermal voltage*, and its magnitude is equal to 25.86 mV at room temperature. As we will see in Chap. 7, electrochemists and biologists write $RT/\mathscr{F}$ in place of kT/q.

The difference in signs in Eqs. (3.40) and (3.42) arises as a matter of definition. If we define current as positive when flowing in the $+x$ direction, then a positive value for (dn/dx) will result in electron flow in the negative x direction (i.e., down the concentration gradient). However, the electron carries a negative charge, so conventional current flows in the $+x$ direction; hence the $+$ sign in Eq. (3.40). Likewise, a positive value of (dp/dx) will also result in a hole flow in the $-x$ direction, and therefore in a negative current density, since the charge on the hole is positive.

It is worth pointing out that Eqs. (3.40) and (3.42) are special forms of Eq. (3.31). In fact, the flux F_c (generic particles crossing the unit area per second) leads directly to the carrier diffusion current density J, and dC_c/dx can be regarded as the carrier concentration gradients, dn/dx or dp/dx. The values of the diffusion coefficients for electrons and holes in intrinsic Si and GaAs are given in App. A.2, and will be used in the analysis of semiconductor devices.

The *total electron and hole diffusion current density* in the x direction can now be written as the sum of the electron and hole components, i.e.,

$$J_{\text{diff}}(x) = J_{n,\text{diff}}(x) + J_{p,\text{diff}}(x) = q\left(D_n \frac{dn}{dx} - D_p \frac{dp}{dx}\right) \tag{3.44a}$$

In addition to electron and hole motion, ion motion in a solid can conduct electricity. The relationship for the concentration gradient–induced motion (diffusion) of ions is given by an expanded version of Eq. (3.44a), that is,

$$J_{\text{diff}}(x) = \sum_i (-qz_i)D_i \frac{dc_i}{dx} + q\left(D_n \frac{dn}{dx} - D_p \frac{dp}{dx}\right) \tag{3.44b}$$

where D_i is the diffusion coefficient for each mobile ion i, c_i is the ion concentration and z_i is the ion valence.

The general, three-dimensional form of Eq. (3.44a) is

$$\mathbf{J}_{\text{diff}} = q(D_n \nabla n - D_p \nabla p) \tag{3.45}$$

where the mathematical operator ∇ is used for the vector $(\partial/\partial x, \partial/\partial y, \partial/\partial z)$. The properties of ∇ are described in App. B.

3.4 TRANSPORT EQUATIONS

In semiconductor devices, drift and diffusion currents are usually present at the same time. The *pn* junction considered in Chap. 6 is an example of such a situation. The *total hole current density* in the x direction is the sum of the drift and diffusion components, that is, by Eqs. (3.15) and (3.42),

$$J_p(x) = q\left[\mu_p p E(x) - D_p \frac{dp}{dx}\right] = q\mu_p\left[pE(x) - \frac{kT}{q}\frac{dp}{dx}\right] \tag{3.46}$$

Similarly, taking into account Eqs. (3.14) and (3.40), the *total electron current density* in the x direction is given by

$$J_n(x) = q\left[\mu_n n E(x) + D_n \frac{dn}{dx}\right] = q\mu_n\left[nE(x) + \frac{kT}{q}\frac{dn}{dx}\right] \tag{3.47}$$

The general, three-dimensional forms of Eqs. (3.46) and (3.47) are

$$\mathbf{J}_p = q(\mu_p p \mathbf{E} - D_p \nabla p) \tag{3.48}$$

$$\mathbf{J}_n = q(\mu_n n \mathbf{E} + D_n \nabla n) \tag{3.49}$$

Equations (3.48) and (3.49) are known as the *transport equations* for holes and electrons, respectively. The first and second terms in the parentheses on the right-hand side of Eqs. (3.48) and (3.49) represent the drift and diffusion current densities, respectively. The signs of these current components follow from their derivation as pointed out in Sects. 3.2.2 and 3.3.2. A useful picture summarizing the relation between carrier flow and current for electron and hole motion by diffusion and drift is shown in Fig. 3.6

The *total current density* is given by the sum of transport Eqs. (3.48) and (3.49), i.e.,

$$\mathbf{J}_{\text{tot}} = \mathbf{J}_p + \mathbf{J}_n \tag{3.50}$$

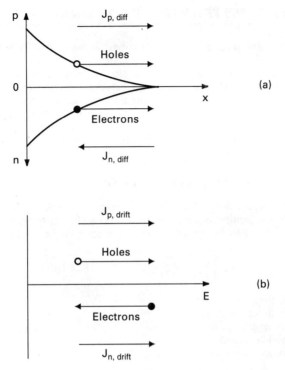

FIGURE 3.6 Carrier flow and current relationship for carrier motion by (a) diffusion and (b) drift.

Equation (3.50) can be evaluated at any plane perpendicular to the direction of the carrier flow, provided that the same plane is used for both electron and hole currents. This note is important, for example, in situations where the hole and electron currents vary with distance, e.g. in a long base diode (see Sec. 6.4.3).

Equations (3.48) and (3.49) do not include the effects caused by an externally applied magnetic field **H**, effects which should require two other current components, $(\mathbf{J}_{p\perp} \tan \theta_p)$ and $(\mathbf{J}_{n\perp} \tan \theta_n)$ to be added to these equations. The electron component $\mathbf{J}_{n\perp}$ represents the current component of $\mathbf{J}_n$ perpendicular to the magnetic field, being $\tan \theta_n = q\mu_n n R_H |\mathbf{H}|$ (which has negative value because the Hall coefficient R_H is negative for electrons). Similar results are obtained for the hole current (see Sec. 3.8).

Equations (3.48) and (3.49) are valid in *low electric field* conditions, i.e., as far as the diffusion coefficients are related to the carrier mobilities by the Einstein relations given by Eqs. (3.39) and (3.43). These equations, moreover, are valid for nondegenerate semiconductors; at high carrier concentrations, these relations have to be modified.

In *high electric field* conditions, carriers are heated by the field and the carrier energy becomes larger than the average thermal energy $3kT/2$. This modifies the scattering mechanisms, changing the transport conditions. As a consequence, the electron and hole velocities are no longer proportional to the electric field when the electric field is high (see Fig. 3.1). The diffusion coefficients also become dependent on the electric field.

3.5 BOLTZMANN TRANSPORT EQUATION

The *Boltzmann transport equation* (BTE) describes the distribution $f(\mathbf{v}, \mathbf{r}, t)$ of carriers as a function of velocity $\mathbf{v}$, space coordinate $\mathbf{r}$, and time t. The quantity $f(\mathbf{v}, \mathbf{r}, t)\, d\mathbf{v}\, d\mathbf{r}$ defines the probability of finding a particle (hole, electron, molecule) in a velocity range between $\mathbf{v}$ and $(\mathbf{v} + d\mathbf{v})$ having a space coordinate between $\mathbf{r}$ and $(\mathbf{r} + d\mathbf{r})$.

Therefore, the BTE is not a quantum mechanical equation, as it specifies $\mathbf{v}$ and $\mathbf{r}$ simultaneously, ignoring the Heisenberg uncertainty relation. This function is often referred to as *occupancy function*, and it is also indicated as $f(\mathbf{p})$, or f for simplicity, $\mathbf{p} = m\mathbf{v}$ being the momentum vector. The BTE is useful in many instances of transport problems.

3.5.1 Boltzmann Transport Equation Formulation

In a semiconductor, if the electron or hole concentrations are less than about $10^{18}/cm^3$, they have random velocities corresponding to the maxwellian distribution.

If we refer to electrons, for example, their number occupying a momentum range dV_p is

$$dn = \left[\frac{n}{(2m^* \pi kT)^{3/2}} e^{-|\mathbf{p}|^2/2m^*kT}\right] dV_p = f_0\, dV_p \quad (3.51)$$

where m^* = electron effective mass
n = electron concentration
$\mathbf{p}$ = momentum vector
T = absolute temperature
k = Boltzmann constant

If a force (e.g., an electric field) is applied to disturb the equilibrium condition, the occupancy function will no longer be spherical. The nonequilibrium distribution can be written in spherical harmonics as

$$f(\mathbf{p}) = f_0(p) + f_1(p) \cos\theta + f_2(p) \sin\theta \cos\phi + f_3(p) \sin\theta \sin\phi + \cdots \quad (3.52)$$

where θ is the angle between the momentum vector and the x axis, and ϕ is the angle between the projection of the momentum vector on the yz plane and the y axis. The coefficients f_1, f_2, f_3 give the current density in the x, y, and z directions respectively, i.e.,

$$J(x) = \frac{4\pi}{3} \frac{q}{m^*} \int_0^\infty p^3 f_1(p)\, dp \quad (3.53a)$$

$$J(y) = \frac{4\pi}{3} \frac{q}{m^*} \int_0^\infty p^3 f_2(p)\, dp \quad (3.53b)$$

$$J(z) = \frac{4\pi}{3} \frac{q}{m^*} \int_0^\infty p^3 f_3(p)\, dp \quad (3.53c)$$

If an electric field is chosen along one dimension (e.g., the x axis), $f_2(p)$ and $f_3(p)$ vanish.

The distribution function $f(\mathbf{p})$ varies by acceleration associated with applied fields as well as by collisions that tend to restore thermal equilibrium. Thus we can write[6,7]

$$\frac{df}{dt} = \frac{\partial f}{\partial t}\bigg|_{fields} + \frac{\partial f}{\partial t}\bigg|_{diffusion} + \frac{\partial f}{\partial t}\bigg|_{collisions} \quad (3.54)$$

The rate of change of $f(\mathbf{p})$ associated with the *electric field* is obtained from a consideration of momentum space. A uniform force makes the distribution to move with a "velocity" $\dot{\mathbf{p}}$(i.e., a force). In a time Δt, the distribution moves a "distance" $\Delta p = \dot{p}\Delta t$ in the direction of the force. Such a motion of the distribution produces a rate of change[6]

$$\left.\frac{\partial f}{\partial t}\right|_{\text{fields}} = -\nabla_{\mathbf{p}} f \cdot \dot{\mathbf{p}} \quad (3.55)$$

The mathematical operator $\nabla_{\mathbf{p}} f$ is used for the vector $(\partial f/\partial v_x, \partial f/\partial v_y, \partial f/\partial v_z)$, v_x, v_y, v_z being the components of the velocity along the x, y, and z axes, respectively.

The rate of change of $f(\mathbf{p})$ associated with a *gradient of carriers* in real space is obtained by consideration of a small volume element in $\mathbf{p}$ space. The carriers in this element move with velocity $\mathbf{p}/m^*$ and in a time Δt move a distance $\Delta \mathbf{r} = \mathbf{p}\Delta t/m^*$. The rate of change of $f(\mathbf{p})$ associated with the motion of the carriers in real space and with no force applied, is[6]

$$\left.\frac{\partial f}{\partial t}\right|_{\text{diffusion}} = \frac{-\mathbf{p} \cdot \nabla_{\mathbf{r}} f(\mathbf{p})}{m^*} \quad (3.56)$$

The mathematical operator $\nabla_{\mathbf{r}} f$ is used for the vector $(\partial f/\partial x, \partial f/\partial y, \partial f/\partial z)$.

The terms involving collisions are more complicated, mainly for high applied electric field conditions, where departure from equilibrium is significant. Here, we refer to low electric field conditions, and assume that there is no exchange (gain or loss) of energy between the carriers and the crystal. Thus we can write[6]

$$\left.\frac{\partial f}{\partial t}\right|_{\text{collisions}} = -\frac{f - f_0}{\tau} \quad (3.57)$$

where τ is the rate of decay of the current carried by carriers in the element dV_p, due to collisions. If the scattering is isotropic (i.e., the behavior of the material is the same regardless of the direction of any of the field vectors) as is the case with collisions on lattice vibrations, then the decay time and collision time are identical. It should be noted that τ is a function of momentum. Equation (3.54) can then be written as

$$\frac{\partial f}{\partial t} = -\nabla_{\mathbf{p}} f(\mathbf{p}) \cdot \dot{\mathbf{p}} - \frac{\mathbf{p}}{m^*} \cdot \nabla_{\mathbf{r}} f(\mathbf{p}) - \frac{f - f_0}{\tau} \quad (3.58)$$

or equivalently,

$$\frac{\partial f}{\partial t} = -\nabla_{\mathbf{p}} f(\mathbf{p}) \cdot \mathbf{F} - \nabla_{\mathbf{r}} f(\mathbf{p}) \cdot \mathbf{v} - \frac{f - f_0}{\tau} \quad (3.59)$$

Equations (3.58) or (3.59) are two forms of the *Boltzmann transport equation*.

3.5.2 Application of BTE to Carrier Mobility and Diffusion

When an electric field is applied along the x direction of a crystal which also has a carrier gradient along the x direction, the f_2 and f_3 components in Eq. (3.52) vanish. Then, if a force $\mathbf{F} = (\pm q)\mathbf{E}$ produced by an electric field is applied to holes or electrons, Eq. (3.59), at the steady-state condition, becomes

$$0 = -\frac{\partial f}{\partial p_x} qE - \frac{p_x}{m^*}\frac{\partial f}{\partial x} - \frac{f-f_0}{\tau}$$

$$= -qE\left(\frac{\partial f_0}{\partial p}\cos\theta + \frac{\partial f_1}{\partial p}\cos^2\theta + \frac{f_1}{p}\sin^2\theta\right)$$

$$-\frac{p}{m^*}\cos\theta\left(\frac{\partial f_0}{\partial x} + \frac{\partial f_1}{\partial x}\cos\theta\right) - \frac{f_1\cos\theta}{\tau} \tag{3.60}$$

In Eq. (3.60), the relation $p(x) = p\cos\theta$ and Eq. (3.52) have been used. The orthogonality properties of the trigonometric functions allow us to separate relations for $f_0(p)$ and $f_1(p)$; then multiplying by $\cos\theta\sin\theta\, d\theta$ and integrating over the sphere, Eq. (3.60) gives

$$\frac{f_1}{\tau} + qE\frac{\partial f_0}{\partial p} + \frac{p}{m^*}\frac{\partial f_0}{\partial x} = 0 \tag{3.61}$$

If we know $f_0(p)$, we obtain $f_1(p)$ directly. The assumption of an equilibrium maxwellian distribution for $f_0(p)$ as expressed by Eq. (3.51), together with Eqs. (3.53a) and (3.61), gives

$$J(x) = -\frac{4\pi}{3}\left[\frac{q^2E}{m^*}\int_{p=0}^{\infty} p^3\tau(p)\frac{\partial f_0}{\partial p}dp + \frac{q}{m^{*2}}\int_0^{\infty} p^4\tau(p)\frac{\partial f_0}{\partial x}dp\right]$$

$$= \frac{q\left(qEn - kT\frac{\partial n}{\partial x}\right)\int_0^{\infty} p^4\tau f_0\, dp}{3m^{*2}kT\int_0^{\infty} p^2 f_0\, dp} = \frac{q^2En\langle v^2\tau\rangle}{m^*\langle v^2\rangle} - \frac{q}{3}\langle v^2\tau\rangle\frac{\partial n}{\partial x} \tag{3.62}$$

The first term of Eq. (3.62) represents the *drift* component of the total current density $J(x)$; the term proportional to the electric field $E(x)$ and the carrier concentration n, is called the *carrier mobility* μ, i.e.,

$$\mu = \frac{q\langle v^2\tau\rangle}{m^*\langle v^2\rangle} \tag{3.63}$$

The second term of Eq. (3.62) represents the *diffusion* component of the total current density $J(x)$: the term proportional to the gradient of carriers without the electronic charge is called the *diffusion coefficient D*, i.e.,

$$D = \frac{\langle v^2\tau\rangle}{3} \tag{3.64}$$

The average of $v^2\tau$ appears in both diffusion coefficient and mobility, so that a relation between these two parameters can be written as

$$D = \frac{kT}{q}\mu \tag{3.65}$$

which is known as the *Einstein relationship* (see Sec. 3.3.2).

Taking into account that electrons bear a charge $-q$, and holes carry a charge $+q$, Eq. (3.62) can be written for electrons and holes along the x-direction, as

$$J_n(x) = \frac{q^2 E(x) n}{m_n^*} \frac{\langle v^2 \tau \rangle}{\langle v^2 \rangle} + \frac{q}{3} \langle v^2 \tau \rangle \frac{dn}{dx} = q\mu_n n E(x) + q D_n \frac{dn}{dx} \quad (3.66)$$

$$J_p(x) = \frac{q^2 E(x) p}{m_p^*} \frac{\langle v^2 \tau \rangle}{\langle v^2 \rangle} - \frac{q}{3} \langle v^2 \tau \rangle \frac{dp}{dx} = q\mu_p p E(x) - q D_p \frac{dp}{dx} \quad (3.67)$$

The subscripts n and p are used to refer the properties of electrons and holes respectively. The last form of Eqs. (3.66) and (3.67) is the same encountered in Eqs. (3.47) and (3.46), respectively.

3.6 RECOMBINATION-GENERATION PROCESSES IN SEMICONDUCTORS

Perturbation of a semiconductor equilibrium condition causes a modification in the carrier concentrations inside the semiconductor. *Recombination-generation (R-G)* is the mechanism by which the carrier excess or deficit in the semiconductor is maintained (if the perturbation is maintained) or eliminated (if the perturbation is taken off). As nonequilibrium conditions are dominant during device operation, the recombination-generation mechanism plays an important role in defining the characteristics of a device.

Recombination can be then regarded as a balancing process for carrier generation. It is thus not directly responsible for carrier motion. However, the recombination process always affects the transport of carriers from one place to another, and it is therefore appropriate to consider it in this chapter.

Formally, recombination-generation mechanisms can be defined as follows:

Generation is a process by which electrons and holes (carriers) are created.

Recombination is a process by which electrons and holes are annihilated or destroyed.

Unlike drift and diffusion, generation and recombination are not single processes: There are a number of different ways in which carriers can be created and destroyed within a semiconductor, though one process will often be the dominant one.

Band-to-band and *R-G intermediate center* recombination-generation are the most common processes involved in semiconductors. The *band-to-band* process involves the direct annihilation of a conduction band electron and a valence band hole. This process is typically radiative (see Chap. 2): that is, the excess energy released during the process is converted into the production of a photon. The *R-G intermediate center* process is due to the presence of allowed energy levels E_t in the midgap region of a semiconductor, introduced by certain impurity atoms. This process is typically nonradiative; thermal energy (heat) is released during the process, or equivalently lattice vibrations (phonons) are produced. Other recombination processes can occur in semiconductors, but are not relevant in the devices considered in this book; the reader can find a wide description of these mechanisms in Ref. 8. Any of the above recombination processes can be reversed to generate carriers.

3.6.1 Characterization of Recombination-Generation Processes

Let us consider a semiconductor with only one type of *R-G* center, which introduces allowed states at an energy E_t into the middle of the bandgap. Although actual semiconductors can contain a number of deep-level centers, the process is dominated by a single center.

It is convenient to remind here some definitions that will be useful in the analysis of the *R-G* processes. In particular:

n_0, p_0	Carrier concentrations in the semiconductor under equilibrium conditions. The subscript zero is used to denote explicitly the condition of thermal equilibrium for the involved quantity.
n, p	Carrier concentrations in the semiconductor under arbitrary conditions.
$\begin{cases} \Delta n = n - n_0 = n' \\ \Delta p = p - p_0 = p' \end{cases}$	Deviations in the carrier concentrations from their equilibrium values. Δn and Δp can be either positive or negative, where a positive deviation corresponds to a carrier excess and a negative deviation corresponds to a carrier deficit.
$\left.\dfrac{\partial n}{\partial t}\right\|_{R\text{-}G}$	Time rate of change in the electron concentration due to both *R-G* center recombination and generation.
$\left.\dfrac{\partial p}{\partial t}\right\|_{R\text{-}G}$	Time rate of change in the hole concentration due to *R-G* center recombination-generation.
n_t	Number of *R-G* centers per cubic centimeter that are filled with electrons.
p_t	Number of empty *R-G* centers per cubic centimeter.
N_t	Total number of *R-G* centers per cubic centimeter, $N_t = (n_t + p_t)$.

The quantities $\partial n/\partial t|_{R\text{-}G}$ and $\partial p/\partial t|_{R\text{-}G}$ indicate net rates, considering the effects of both recombination and generation. In particular, for example, $\partial n/\partial t|_{R\text{-}G}$ will be negative if there is a net loss of electrons ($R > G$) or positive if there is a net gain of electrons ($G > R$). The symbolism $|_{R\text{-}G}$ points out that the carrier concentration change is caused by recombination-generation via *R-G* centers, excluding non–*R-G* processes.

3.6.2 Recombination-Generation Processes through Intermediate Centers

The theory of the recombination-generation process taking place through the action of intermediate energy-level or recombination-generation centers has been worked on by Hall, and later by Shockley and Read.[9] The various steps that take place in the recombination-generation processes through intermediate-level centers are shown in Fig. 3.7. The arrows indicate the transition of the electron during the particular process. The figure shows the case of a center of energy E_t placed in the middle of the energy gap ($E_t = E_i$) with a single-level center which can present two charge states: negative and neutral. This implies that the states considered are acceptor-type (i.e., neutral when empty, and negative when full), but a similar statement also applies to donor-type states. The trap is then able to communicate with both the conduction and valence bands through the four processes we are going to consider in some detail.

The possible transitions are then (*a*) electron capture at an *R-G* center, (*b*) electron emission from an *R-G* center, (*c*) hole capture at an *R-G* center, and (*d*) hole emission from an *R-G* center. The transitions *c* and *d* can also be thought of as (*c*) an electron trapped at an *R-G* center falling into a vacant valence band state and (*d*) a valence band electron being excited to the *R-G* level. As only transitions *a* and *b* affect the electron concentration, and only transitions *c* and *d* affect the hole concentration, we can write

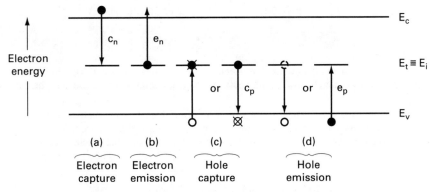

FIGURE 3.7 Possible transitions between a single R-G center and the energy bands. The arrows indicate the direction of electron transition. (*Adapted from R. F. Pierret.*[8] *Used by permission.*)

$$\left.\frac{\partial n}{\partial t}\right|_{R\text{-}G} = \left.\frac{\partial n}{\partial t}\right|_{(a)} + \left.\frac{\partial n}{\partial t}\right|_{(b)} \quad (3.68a)$$

$$\left.\frac{\partial p}{\partial t}\right|_{R\text{-}G} = \left.\frac{\partial p}{\partial t}\right|_{(c)} + \left.\frac{\partial p}{\partial t}\right|_{(d)} \quad (3.68b)$$

In the following we will then examine each process, assuming that the semiconductor is nondegenerate.

    a. Electron capture. An electron falls from the conduction band into an empty *R-G* center. The rate at which this process takes place is proportional to the concentration of electrons n in the conduction band, the concentration of empty (of electrons) *R-G* centers, and the probability that an electron flows near an *R-G* center and is captured by it. The concentration of empty *R-G* centers is given by their total concentration N_t times 1 minus the probability $f(E_t)$ that they are occupied. When *thermal equilibrium* applies, f is the Fermi function given by Eq. (2.27).

    The probability per unit time that an electron is captured by an *R-G* center is given by the product of the electron thermal velocity v_{th} and a parameter σ_n called the *electron capture cross section*. The capture cross section describes the ability of the center to capture an electron. The product $v_{\text{th}}\sigma_n$ can be considered as the volume swept out per unit time by a particle with cross section σ_n. If the center is within this volume, the electron is captured by it. The capture cross section is generally determined experimentally for a given type of center. The above discussion can be mathematically translated to write the *total rate of capture of electrons* by the intermediate centers as

$$\left.\frac{\partial n}{\partial t}\right|_{(a)} = -(v_{\text{th}}\sigma_n)n\{N_t[1-f(E_t)]\} = -c_n n\{N_t[1-f(E_t)]\} = -c_n n p_t \quad (3.69)$$

where

$$p_f = [1 - f(E_t)] \quad (3.70)$$

is the *probability that the center is empty.* Thus,

$$p_t = N_t[1 - f(E_t)] \quad (3.71)$$

represents the *number of empty R-G centers* per cubic centimeter, and

$$c_n = v_{th}\sigma_n \tag{3.72}$$

is the *electron capture coefficient* (cm³/s), defined as a positive constant. Since c_n is taken to be a positive quantity, a minus sign is added to the right-hand side of Eq. (3.69) to account for the fact that electron capture acts to reduce the number of electrons in the conduction band.

b. *Electron emission.* The emission of an electron from the intermediate center into the conduction band takes place at a rate given by the product of the concentration of centers occupied by electrons $N_t f(E_t)$ times the probability e_n that the electron makes this jump. Thus,

$$\left.\frac{\partial n}{\partial t}\right|_{(b)} = e_n[N_t f(E_t)] = e_n n_t \tag{3.73}$$

where

$$n_t = [N_t f(E_t)] \tag{3.74}$$

is the number of *R-G* centers per cubic centimeter that are *filled with* electrons, and e_n (s⁻¹), the *emission probability of electrons*, is again a positive-defined constant. The proportionality constant acts to increase the number of electrons in the conduction band.

The definition of the electron emission probability e_n implies that it depends on the centers not occupied within the conduction band, and also on their position within the forbidden bandgap. Similar arguments can be applied to processes *c* and *d*.

c. *Hole capture.* Holes can be captured from the valence band at a rate that is proportional to the concentration of centers occupied by electrons $N_t f(E_t)$ the concentration of holes, and a transition probability. This probability can then be described by the product of the hole thermal velocity v_{th} and the *hole capture cross section* σ_p of the center. Thus,

$$\left.\frac{\partial p}{\partial t}\right|_{(c)} = -(v_{th}\sigma_p)p[N_t f(E_t)] = -c_p p[N_t f(E_t)] = -c_p p n_t \tag{3.75}$$

where

$$c_p = v_{th}\sigma_p \tag{3.76}$$

is called the *hole capture coefficient*. The dependence on the concentration of centers occupied by electrons $[N_t f(E_t)]$ arises from the fact that the capture of holes from valence band through a center corresponds to the transition of an electron from the center to the valence band.

d. *Hole emission.* This process describes the excitation of an electron from the valence band into the empty center. Similarly to electron emission, hole emission probability is given by

$$\left.\frac{\partial p}{\partial t}\right|_{(d)} = e_p\{N_t[1 - f(E_t)]\} = e_p p_t \tag{3.77}$$

where e_p is the *hole emission probability* and depends on factors similar to those that affect e_n.

An interesting analogy may be noted: the above results can be obtained from chemical-reaction-type arguments. In fact, each of the fundamental processes can be compared to a chemical reaction. If, for example, we consider process a, this in chemical terms becomes

$$\text{Electron} + \text{empty } R\text{-}G \text{ center} \rightarrow \text{filled } R\text{-}G \text{ center} \tag{3.78}$$

From chemistry, we know that an irreducible chemical reaction of the form $A + B \rightarrow C$ proceeds at a rate given by (see Chap. 5)

$$\text{Rate} = (\text{constant}) [A] [B] \tag{3.79}$$

where [A] and [B] are the concentrations of the reacting components. For the process a chosen as an example, the reacting components are electrons and empty R-G centers; the corresponding concentrations are n and p_t, respectively.

Going back to the generation-recombination processes in semiconductors, substituting the basic process-rate equations into Eqs. (3.68), we obtain

$$\left. \frac{\partial n}{\partial t} \right|_{R\text{-}G} = e_n n_t - c_n n p_t = -r_n \tag{3.80a}$$

$$\left. \frac{\partial p}{\partial t} \right|_{R\text{-}G} = e_p p_t - c_p p n_t = -r_p \tag{3.80b}$$

where r_n and r_p are the net electron and hole recombination rates respectively; they are positive if recombination prevails and negative if generation prevails.

3.6.3 Net Recombination Rates under Equilibrium and Steady-State Conditions

Equations (3.80) simplify if equilibrium conditions are considered. Under equilibrium conditions, each fundamental process and its inverse must self-balance independently of any other process that may take place in the semiconductor. This statement is also known as the *principle of detailed balance*.[9]

When this principle is applied to the R-G center interaction, it requires process a to self-balance with its inverse process b, and process c to self-balance with its inverse process d. Thus, under equilibrium conditions

$$r_n = 0 \tag{3.81a}$$

$$r_p = 0 \tag{3.81b}$$

Equations (3.81) provide the mathematical basis to correlate the emission and capture coefficients, when introduced in Eq. (3.80).

It can be demonstrated[4,8] that the electron emission coefficient e_n can be written as

$$e_n = c_n N_c e^{(E_t - E_c)/kT} = c_n n_i e^{(E_t - E_i)/kT} = c_n n_i \frac{1 - f(E_t)}{f(E_t)} \tag{3.82}$$

Equation (3.82) states that the emission probability e_n of electrons increases exponentially as the center energy level E_t approaches the conduction band edge E_c. As with Eq. (3.82), we can calculate the hole emission coefficient, which is then given by

$$e_p = c_p N_v e^{-(E_t - E_v)/kT} = c_p n_i e^{-(E_t - E_i)/kT} = c_p n_i \frac{1 - f(E_t)}{f(E_t)} \tag{3.83}$$

Again we note that the emission probability e_p increases exponentially as the center level E_t approaches the edge of the valence band E_v.

Before analyzing the net recombination under steady-state conditions, it is convenient to remember that steady-state or quasi-steady-state conditions imply that the rate of change of system variables, such as n, p, and E, is slow compared to the rates of the dominant fundamental process taking place inside the semiconductor. In *both* the *equilibrium* and *steady-state* conditions, the average values of all macroscopic variables within a system are constant with time; that is, dn/dt, dp/dt, and dE/dt are all zero.

The difference between *equilibrium* and *steady-state* conditions within a small dx section of semiconductor is shown in Fig. 3.8 as suggested by Ref. 8.

As n_t does not vary with time, and assuming that n_t varies only by the *R-G* center interaction, we can write under *steady-state* conditions

$$\frac{dn_t}{dt} = \left.\frac{\partial p}{\partial t}\right|_{R-G} - \left.\frac{\partial n}{\partial t}\right|_{R-G} = r_n - r_p = 0 \quad (3.84)$$

or
$$r_n = r_p \quad (3.85)$$

After some mathematical calculations, we can write[8]

$$R = r_n = r_p = \frac{n_p - n_i^2}{\dfrac{1}{c_p N_t}(n + n_k) + \dfrac{1}{c_n N_t}(p + p_k)} \quad (3.86)$$

where R is the *net steady-state recombination rate* and

$$n_k = \frac{p_{t0} n_0}{n_{t0}} = \frac{e_n}{c_n} \quad (3.87a)$$

$$p_k = \frac{n_{t0} p_0}{p_{t0}} = \frac{e_p}{c_p} \quad (3.87b)$$

are computable constants. The subscript zero indicates that all the involved quantities have to be evaluated under equilibrium conditions. Since $1/(c_n N_t)$ and $1/(c_p N_t)$ have units of time, we can define them as time constants, i.e.,

$$\tau_n = \frac{1}{c_n N_t} \quad (3.88a)$$

$$\tau_p = \frac{1}{c_p N_t} \quad (3.88b)$$

Equations (3.88) define two important material parameters: the *electron minority carrier lifetime* τ_n and the *hole minority carrier lifetime* τ_p. These two parameters can be considered as the average time an excess minority carrier will live in a "sea" of majority carriers. Moreover the carrier lifetimes vary inversely with the *R-G* center concentration N_t, but they are independent of the doping concentration. When Eqs. (3.88) are introduced into Eq. (3.86), we obtain

$$R = \frac{np - n_i^2}{\tau_p\left(n + \dfrac{e_n}{c_n}\right) + \tau_n\left(p + \dfrac{e_p}{c_p}\right)} \quad (3.89)$$

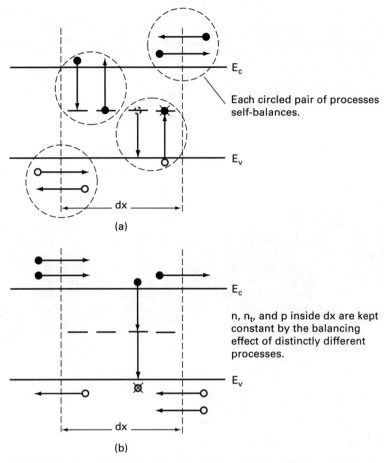

FIGURE 3.8 A small dx section of a semiconductor under (a) equilibrium and (b) steady-state conditions. (*From R. F. Pierret.[8] Used by permission.*)

Equation (3.89) applies to any steady-state situation and gives the net recombination rate for both electrons and holes. This equation is usually known as the *Shockley-Hall-Read (SHR) function*. To understand the meaning of Eq. (3.89), we consider the special situation of equal electron and hole capture cross section, i.e., $\sigma_n = \sigma_p$, and equal minority carrier lifetimes, i.e., $\tau_n = \tau_p \equiv \tau$. Under these assumptions, Eq. (3.89) becomes

$$R = \frac{np - n_i^2}{\tau\left[(n+p) + 2n_i \cosh\left(\dfrac{E_t - E_i}{kT}\right)\right]} \tag{3.90}$$

where Eqs. (3.82) and (3.83) have been used.

Equation (3.90) shows that R is positive, and therefore there is net recombination if the pn product exceeds n_i^2. The sign changes and there is net generation if the $p\,n$ product is less than n_i^2. The term $(pn - n_i^2)$ represents the restoring "force" for free carriers in a nonequilibrium condition; it is also the driving force for recombination, as it represents the deviation from the equilibrium condition. The "resistance" to this recombination process represented by the term $(n + p)$, increases with n and with p, and it reaches its minimum when the sum $(n + p)$ is at its minimum value. The third term in the denominator increases as E_t moves away from the middle of the energy gap E_i and approaches either the conduction-band or the valence-band edge. In this case, one of the emission processes becomes more probable and this reduces the effectiveness of the recombination center. This is because, after an electron is captured by the center, a hole must be captured by it next, in order to complete the recombination process. If, however, the energy level of the center is very near the conduction-band edge, it will be more probable to re-emit the captured electron into the conduction band, by preventing the completion of the recombination process. (Similar considerations hold for centers near the valence-band edge.)

Thus a recombination center is most effective if the two emission probabilities are about the same, i.e., when its energy level is near the middle of the energy bandgap.

Equation (3.89) can be simplified under certain operating conditions. In particular, in what follows we consider the SHR function under low-level injection conditions.

The level of injection specifies the magnitude of changes in the carrier concentrations resulting from a perturbation. Consider now the case when excess carriers of both types are somehow introduced into the semiconductor in equal concentrations in order to preserve space-charge neutrality. Two situations can then take place. First, the excess minority-carrier concentration approaches the equilibrium majority-carrier concentration; this condition is referred to as *high-level injection*. Second, the excess minority-carrier concentration is much less than the equilibrium majority-carrier concentration; this condition is referred to as *low-level injection* (see Sec. 2.7). Low-level injection is here considered. It leads to simplified expressions of R which will be used in many practical situations. Low-level injection characteristics can be described mathematically as

$$\Delta n, \Delta p \ll n_0 \quad (n \simeq n_0) \text{ for an } n\text{-type semiconductor} \quad (3.91a)$$

$$\Delta n, \Delta p \ll p_0 \quad (p \simeq p_0) \text{ for a } p\text{-type semiconductor} \quad (3.91b)$$

To point out quantitatively the meaning of low-level injection, let us consider, for example, an $N_D = 10^{15}/\text{cm}^3$ doped n-type Si at room temperature, subject to a perturbation where $\Delta p = \Delta n = 10^9/\text{cm}^3$. For the given material $n_0 \simeq N_D = 10^{15}/\text{cm}^3$ and $p_0 \simeq n_i^2/N_D \simeq 10^5/\text{cm}^3$. Then, $n = n_0 + \Delta n \simeq n_0$ and $\Delta p = 10^9/\text{cm}^3 \ll n_0 = 10^{15}/\text{cm}^3$. The situation is evidently one of low-level injection. It should be noted, however, that $\Delta p \gg p_0$.

After some mathematical calculations, we can write at low-level injection

$$R = \frac{\Delta p}{\tau_p} \quad \text{for } n\text{-type semiconductor} \quad (3.92a)$$

$$R = \frac{\Delta n}{\tau_n} \quad \text{for } p\text{-type semiconductor} \quad (3.92b)$$

Equation (3.89) simplifies to

$$R = \frac{\Delta p}{\tau_n + \tau_p} \quad (3.93)$$

under *high-level injection* conditions where $\Delta n = \Delta p \gg n_0$ or p_0.

3.6.4 Surface Recombination-Generation

In many semiconductor devices under particular conditions, *surface recombination-generation* becomes as important as the bulk recombination-generation considered in the preceding sections. The same processes which take place in the semiconductor bulk also take place at the semiconductor surface: Electrons and holes can be captured at surface centers; electrons and holes can be emitted from surface centers.

The physical similarity between surface and bulk recombination-generation leads to a parallel mathematical description of the processes, allowing one to formulate the surface relationships by direct deduction from the corresponding bulk results.

A difference, however, has to be taken into account: whereas a single level usually dominates bulk recombination-generation, the surface-center interaction involves centers distributed in energy throughout the bandgap. Thus, it is necessary to integrate the single-level surface rates over the energy bandgap. An exhaustive description of this process can be found in Refs. 4 and 8. We need not specify it further for the aim of the book.

3.7 CONTINUITY EQUATION

Drift, diffusion, thermal recombination-generation, or other processes result in a change in the carrier concentrations with time. The combined effect of these carrier actions can be accounted for by equating the change in the carrier concentrations per unit time (dn/dt or dp/dt) to the sum of the $\partial n/\partial t$ or $\partial p/\partial t$ due to each process. This is mathematically expressed by

$$\frac{\partial n}{\partial t} = \frac{dn}{dt}\bigg|_{\text{drift}} + \frac{\partial n}{\partial t}\bigg|_{\text{diff.}} + \frac{\partial n}{\partial t}\bigg|_{\substack{\text{thermal} \\ R\text{-}G}} + \frac{\partial n}{\partial t}\bigg|_{\substack{\text{other} \\ \text{processes}}} \quad (3.94a)$$

$$\frac{\partial p}{\partial t} = \frac{dp}{dt}\bigg|_{\text{drift}} + \frac{\partial p}{\partial t}\bigg|_{\text{diff.}} + \frac{\partial p}{\partial t}\bigg|_{\substack{\text{thermal} \\ R\text{-}G}} + \frac{\partial p}{\partial t}\bigg|_{\substack{\text{other} \\ \text{processes}}} \quad (3.94b)$$

The effect of each process is obtained by using the requirement of conservation of carriers. Electrons and holes cannot appear and vanish at a given point, but must be transported to or created at the given point via some type of carrier action. Spatial and time continuity in the carrier concentrations must be preserved. For this reason Eqs. (3.94) are known as the *continuity equations*.

3.7.1 Derivation of the Continuity Equation

To derive the continuity equation, let us consider the electron concentration within an infinitesimal volume of semiconductor material of area A and thickness dx located at x. The electron concentration can change in time if there is a difference between the electron currents at opposite sides of the infinitesimal volume. This situation can happen, for example, if the current is due to diffusion and dn/dx varies along the distance dx, or if there is a difference between the rates at which electrons are generated and recombine within the infinitesimal volume. This can be described in an equationlike form:

$$\text{rate of change of number of electrons} = \text{change due to spatial difference in current} + \text{change due to difference in rates of generation and recombination} \quad (3.95)$$

72   BASIC PROPERTIES OF SILICON

Referring to Eq. (3.95) and expressing it mathematically, we can derive the one-dimensional (x direction) continuity equation for electrons. The first component on the right-hand side in Eq. (3.95) can be found by dividing the currents at each side of the infinitesimal volume by the electron charge; then we indicate the last component by the formalism used in Eqs. (3.94). The rate of change in the number of electrons within the infinitesimal volume is then

$$\frac{\partial n(x,t)}{\partial t}(A\,dx) = \left[\frac{J_n(x)}{-q} - \frac{J_n(x+dx)}{-q}\right]A + \left[\frac{\partial n}{\partial t}\bigg|_{\text{thermal} \atop R\text{-}G} + \frac{\partial n}{\partial t}\bigg|_{\text{other} \atop \text{processes}}\right](A\,dx) \quad (3.96)$$

Dividing Eq. (3.96) by the volume ($A\,dx$), we obtain

$$\frac{\partial n(x,t)}{\partial t} = -\frac{1}{q}[J_n(x) - J_n(x+dx)]\frac{A}{A\,dx} + \left[\frac{\partial n}{\partial t}\bigg|_{\text{thermal} \atop R\text{-}G} + \frac{\partial n}{\partial t}\bigg|_{\text{other} \atop \text{processes}}\right] \quad (3.97)$$

Expanding $J_n(x + dx)$ in a Taylor series yields

$$J_n(x + dx) = J_n(x) + \frac{\partial J_n(x)}{\partial x}dx + \cdots \quad (3.98)$$

and substituting into Eq. (3.97), we derive the *one-dimensional continuity equation for electrons*:

$$\frac{\partial n(x,t)}{\partial t} = \frac{1}{q}\frac{\partial J_n(x)}{\partial x} + \left[\frac{\partial n}{\partial t}\bigg|_{\text{thermal} \atop R\text{-}G} + \frac{\partial n}{\partial t}\bigg|_{\text{other} \atop \text{processes}}\right] \quad (3.99)$$

A similar continuity equation applies to holes, except that the sign of the first term on the right-hand side of Eq. (3.99) is changed because of the charge associated with a hole. Thus the *one-dimensional continuity equation for holes* is

$$\frac{\partial p(x,t)}{\partial t} = -\frac{1}{q}\frac{\partial J_p(x)}{\partial x} + \left[\frac{\partial p}{\partial t}\bigg|_{\text{thermal} \atop R\text{-}G} + \frac{\partial p}{\partial t}\bigg|_{\text{other} \atop \text{processes}}\right] \quad (3.100)$$

To obtain equations that can be solved, we must relate the quantities on the right-hand side of Eqs. (3.99) and (3.100) to the carrier concentrations n and p. This is easy to do for the current terms, since J_n and J_p have been written in terms of carrier concentrations in Eqs. (3.47) and (3.46). When these equations are used, we obtain for the one-dimensional case

$$\frac{\partial n(x,t)}{\partial t} = \mu_n n(x)\frac{\partial E(x)}{\partial x} + \mu_n E(x)\frac{\partial n(x)}{\partial x} + D_n\frac{\partial^2 n(x)}{\partial x^2} + \left[\frac{\partial n}{\partial t}\bigg|_{\text{thermal} \atop R\text{-}G} + \frac{\partial n}{\partial t}\bigg|_{\text{other} \atop \text{processes}}\right]$$

$$(3.101a)$$

$$\frac{\partial p(x,t)}{\partial t} = -\mu_p p(x)\frac{\partial E(x)}{\partial x} - \mu_p E(x)\frac{\partial p(x)}{\partial x} + D_p\frac{\partial^2 p(x)}{\partial x^2} + \left[\frac{\partial p}{\partial t}\bigg|_{\text{thermal} \atop R\text{-}G} + \frac{\partial p}{\partial t}\bigg|_{\text{other} \atop \text{processes}}\right]$$

$$(3.101b)$$

In the above equations we have assumed that mobility μ and diffusion constant D are not functions of x. Although this assumption is not valid in some cases, the major physical effects are included in Eqs. (3.101) and the more exact formulations are seldom considered.

The general three-dimensional forms of Eqs. (3.99) and (3.100) are

$$\frac{\partial n}{\partial t} = \frac{1}{q}\nabla\cdot\mathbf{J}_n + \left[\left.\frac{\partial n}{\partial t}\right|_{\text{thermal R-G}} + \left.\frac{\partial n}{\partial t}\right|_{\text{other processes}}\right] \quad (3.102a)$$

$$\frac{\partial p}{\partial t} = -\frac{1}{q}\nabla\cdot\mathbf{J}_p + \left[\left.\frac{\partial p}{\partial t}\right|_{\text{thermal R-G}} + \left.\frac{\partial p}{\partial t}\right|_{\text{other processes}}\right] \quad (3.102b)$$

where the mathematical operator ∇ is used for the vector $(\partial/\partial x, \partial/\partial y, \partial/\partial z)$. Equations (3.102) are general, as no assumptions (other than the validity of a Taylor series truncation) have been made in deriving them. This generality often makes the continuity equations rather difficult to solve exactly (see Sec. 3.7.2 for simplified forms). In many semiconductor devices, for instance, the doping concentration changes with distance (often in three dimensions), which results in the diffusion current, and sometimes the generation-recombination rates, being position-dependent. Moreover, the current can contain a drift component due to a field that depends on the charge concentration through *Poisson's equation,* that, in one dimension, is

$$-\frac{d^2\phi}{dx^2} = \frac{dE}{dx} = \frac{\rho}{\varepsilon_s} = \frac{q}{\varepsilon_s}(N_D - N_A + p - n) \quad (3.103)$$

where ϕ is the electrostatic potential, ρ the charge concentration, $\varepsilon_s = \varepsilon_0\varepsilon_r$ the dielectric permittivity of the semiconductor, and N_D and N_A the concentrations of ionized donors and acceptors, respectively. The general three-dimensional form of Eq. (3.103) is

$$\nabla\cdot\mathbf{E} = \frac{\rho}{\varepsilon_s} \quad (3.104)$$

A useful semiconductor equation can be derived from the expressions of the continuity equations. Subtracting Eq. (3.102a) from Eq. (3.102b), we obtain

$$q\frac{\partial}{\partial t}(p-n) + \nabla\cdot(\mathbf{J}_n + \mathbf{J}_p) = 0 \quad (3.105)$$

From Eq. (3.103), we find

$$q\frac{\partial}{\partial t}(p-n) = \frac{\partial\rho}{\partial t} \quad (3.106)$$

Differentiating Poisson's equation [Eq. (3.104)], with respect to time and substituting $\partial\rho/\partial t$ derived from Eqs. (3.105) and (3.106), we obtain

$$\varepsilon_s\frac{\partial}{\partial t}\nabla\cdot\mathbf{E} + \nabla\cdot(\mathbf{J}_n + \mathbf{J}_p) = 0 \quad (3.107)$$

Integration of Eq. (3.107) over the space coordinates gives

$$\mathbf{J}(t) = \mathbf{J}_n + \mathbf{J}_p + \varepsilon_s\frac{\partial}{\partial t}\mathbf{E} \quad (3.108)$$

where $\mathbf{J}(t)$ is the total current density, including the *displacement current* $\varepsilon_s\,\partial\mathbf{E}/\partial t$.

3.7.2 Simplified Forms of Continuity Equations

As already mentioned, it is seldom necessary to deal with the full form of Eqs. (3.102), which are partial differential equations being functions of time and position. Thus, they have an infinity of solutions, one of which applies to a given problem because it matches boundary and initial conditions. In the following, we present situations that allow us to consider simplified forms of the continuity equations.

Steady State. Under steady-state conditions, the time dependence on the left-hand side of Eqs. (3.101) or (3.102) vanishes and only position-dependent derivatives remain. The continuity equations then simplify to ordinary differential equations. Thus, in steady-state conditions, we can write

$$\frac{\partial n}{\partial t} = 0 \tag{3.109a}$$

$$\frac{\partial p}{\partial t} = 0 \tag{3.109b}$$

Dark Operation. The dark operation condition applied to Eqs. (3.101) or (3.102) simply implies setting

$$\left.\frac{\partial n}{\partial t}\right|_{\text{other processes}} = 0 \tag{3.110a}$$

$$\left.\frac{\partial p}{\partial t}\right|_{\text{other processes}} = 0 \tag{3.110b}$$

No Electric Field. If the electric field is zero or negligible in the region under consideration, the drift component of J in Eqs. (3.101) or (3.102) can be neglected. This situation leads us to consider only the diffusion components, i.e., for the one-dimensional case [Eqs. (3.101)],

$$-\frac{1}{q}\frac{\partial J_p}{\partial x} = D_p \frac{\partial^2 p}{\partial x^2} \tag{3.111a}$$

$$\frac{1}{q}\frac{\partial J_n}{\partial x} = D_n \frac{\partial^2 n}{\partial x^2} \tag{3.111b}$$

Constant Electric Field. Even if the electric field is constant, some terms in Eqs. (3.101) or (3.102) can be neglected. For example, the first term in each of the one-dimensional continuity equations [(Eqs. 3.101)] drops out.

Uniform Carrier Currents. The condition in which the carrier currents are constant throughout the region under consideration also simplifies the continuity Eqs. (3.101) or (3.102). Thus, referring for example to the one-dimensional form of the continuity equations [Eqs. (3.101)], this condition gives

$$\frac{\partial J_n}{\partial x} = 0 \tag{3.112a}$$

$$\frac{\partial J_p}{\partial x} = 0 \tag{3.112b}$$

Uniform Doping Concentration, Field-Free Region, No Light, Steady State. This situation is quite commonly a good approximation to the conditions in some regions of semiconductor devices (see Chap. 6). The uniform doping concentration condition implies that there is no positional variation of the thermal equilibrium carrier concentrations, so the diffusion current, which is the only component present in a field-free region, can be expressed in terms of the excess carrier concentration. Combining this result with the other conditions specified for this situation gives, for the one-dimensional case assumed as example,

$$D_p \frac{d^2 p'_n}{dx^2} = \frac{p'_n}{\tau_p} \tag{3.113a}$$

$$D_n \frac{d^2 n'_p}{dx^2} = \frac{n'_p}{\tau_n} \tag{3.113b}$$

3.8 HALL EFFECT

If a sample of semiconductor carrying a current **I** is placed in a transverse magnetic field **B**, an electric field **E** is induced in the direction perpendicular to both **I** and **B**. This phenomenon, known as the *Hall effect*, is used to determine whether a semiconductor is *n*-type or *p*-type and to find the carrier concentration. The mobility μ can be calculated by measuring the conductivity σ.

The Hall effect is then based on the vector force that exists because particles are in motion with a velocity **v** across a magnetic field **B**. The resulting *Lorentz force* is given by

$$\mathbf{F} = q(\mathbf{v} \times \mathbf{B}) \tag{3.114}$$

where q is the charge of the particle that is in motion. If we let the current I be in the positive x direction and B be in the positive z direction, a force will be exerted in the negative y direction on the current carriers. The current I can be holes moving from left to right or electrons moving from right to left in the semiconductor sample of Fig. 3.9. Note that the current I is associated to a drift velocity $v(x)$ caused by an electric field $E(x)$ applied in the x direction.

Therefore, independently of whether the carriers are holes or electrons, they will be forced downward toward surface S_2 in Fig. 3.9. If the sample is an *n*-type semiconductor,

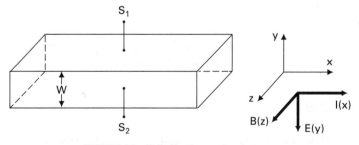

FIGURE 3.9 Hall effect in a semiconductor.

so that the current is carried by electrons, these carriers will accumulate on surface S_2 which becomes negatively charged with respect to surface S_1. Thus a potential V_H, called the Hall voltage, appears between surfaces S_2 and S_1. If the polarity of V_H is positive at surface S_1, then, as explained above, the carriers must be electrons. Conversely, if surface S_2 is positively charged with respect to surface S_1, the semiconductor must be p-type. These results have been verified experimentally, justifying the bipolar (electrons and holes) nature of the current in a semiconductor, and confirm the quantum-mechanical fact that the hole acts like a classical free positive-charge carrier.

The explanation above can be expressed in a mathematical way. In the equilibrium condition, the electric field intensity $E(y)$ caused by the Hall effect must exert a force on the carrier which balances the magnetic (Lorentz) force, i.e.,

$$qE(y) = qv_d(x)B(z) \tag{3.115}$$

where $v_d(x)$ is the carrier drift velocity in the x direction. The *Hall voltage* is given by

$$V_H = E(y)W = v_d(x)B(z)W \tag{3.116}$$

where W is the distance between the two surfaces of the semiconductor.

Since, for a p-type semiconductor, we know from (3.15) that

$$J_p(x) = qp\mu_p E(x) = qpv_{dp}(x) \tag{3.117}$$

the Hall electric field can be written as

$$E(y) = \frac{J_p(x)}{qp}B(z) = R_H J_p(x)B(z) \tag{3.118}$$

The term

$$R_H = -\frac{1}{qp} = -\frac{E(y)}{J_p(x)B(z)} \tag{3.119}$$

is called the *Hall coefficient*. A similar argument is valid for the electrons, where we find that

$$R_H = \frac{1}{qn} = \frac{E(y)}{J_n(x)B(z)} \tag{3.120}$$

Therefore, by measuring the Hall voltage V_H for a known current and magnetic field, we can compute the hole density from

$$p = \frac{1}{qR_H} = \frac{J_p(x)B(z)}{qE(y)} = \frac{I(x)B(z)W}{qV_H A} \tag{3.121}$$

where A is the cross-sectional area of the semiconductor.

All the quantities on the right side can be measured, and therefore p is determined.

PROBLEMS

3.1 A small concentration of minority carriers is injected into a homogeneous Si semiconductor sample at one point. An electric field of 10 V/cm is applied across the semiconductor sample, and this electric field moves the minority carriers a distance of 1 cm in a time 250 μs. Calculate the drift velocity, the mobility, and the diffusion coefficient of the minority carriers.

3.2 Calculate the electron and hole concentrations, the resistivity, and the position of the Fermi level of a Si semiconductor sample doped with 1.1×10^{16} B/cm^3 and 9×10^{15} P/cm^3, at 300 K. For the values of the parameters not specified, you can refer to App. A.

3.3 A Si semiconductor sample doped with 10^{15} P/cm^3 is exposed to 10-mW light. The sample is 1 cm long and 10^{-2} cm^2 in area. Previous measurements show that the minority carrier lifetime is 20 μs. Calculate the resistance of the sample of semiconductor with and without light source. For the values of the parameters not specified, you can refer to App. A.

3.4 Consider a p-type semiconductor, and derive the energy level of the centers from which a trapped electron is as likely to be re-emitted into the conduction band as it is to recombine with holes. Will centers with energy above or below this level be efficient recombination centers?

3.5 A uniformly donor-doped Si semiconductor sample maintained at room temperature is suddenly illuminated with light at time $t = 0$. Assuming $N_D = 10^{15}$/cm^3, $\tau_p = 10^{-6}$ s, and a light-induced creation of 10^{17} electrons and holes per cm^{-3}-s throughout the semiconductor, determine $\Delta p_n(t)$ for $t > 0$. For the values of the parameters not specified, you can refer to App. A.

3.6 A uniformly donor-doped Si semiconductor cylindrical sample is 1 cm long and 1 mm in diameter. A voltage source of 1 V is applied to the semiconductor sample through a resistor $R_g = 50\ \Omega$. At room temperature the current flowing through the circuit is 10 mA. Calculate:
  a. The doping concentration when the contact potentials between the two metal-semiconductor junctions can be neglected.
  b. The current that will flow in the circuit when the room temperature is increased by 150°C, assuming the carrier mobility is independent of temperature, and taking into account the relation

$$n_i(T) = n_i(T_0)2.5^{(T-T_0)/10} \tag{P3.1}$$

For the values of the parameters not specified, you can refer to App. A.

3.7 A uniformly doped square sample of Ge has a cross-section of 1 cm^2 and a thickness of 200 μm, and it is doped with 10^{17} Ga/cm^3. Assume $\tau_n = 1.1 \times 10^{-4}$ s and $\mu_n = 3600$ cm^2/V-s. If 10^{15} electrons/cm^3 were injected into the semiconductor sample from $x = d$, find the minority carrier current as a function of x, assuming the following boundary conditions: On one side (call it $x = 0$), the excess carrier concentration is 0; on the other side ($x = d$), the electron concentration is 10^{15}/cm^3.
If you evaluate this current at $x = 0$, you will find a current of a certain percentage of its maximum value. What happened to the percentage that vanished?

3.8 The intrinsic carrier concentration n_i is given by the relation

$$n_i = f(m)T^{3/2}e^{-E_g/2kT} \tag{P3.2}$$

where $f(m)$ is a function of the masses of the electron and of the hole and is assumed to be independent of temperature. Determine the value of $f(m)$ for silicon by plotting $\log n_i$ versus $1/T$ using the range $200 < T < 1000$, and assuming that $n_i \simeq 10^{10}$ carriers/cm^3 at 300 K. This result can be used to calculate the product of the effective density of states in the valence band N_v and in the conduction band N_c.

3.9 When holes are injected into an n-type semiconductor under steady-state conditions, the hole concentration p_n obeys the differential equation

$$D_p \frac{d^2 p_n}{dx^2} - \frac{p_n - p_{n0}}{\tau_p} = 0 \tag{P3.3}$$

Solve Eq. (P3.3) for a semi-infinite semiconductor sample, using the boundary conditions $p_n(x = 0)$ = constant and $p_n(x = \infty) = p_{n0}$. Note that $x = 0$ represents the surface of the semiconductor sample at which holes are injected. Solve Eq. (P3.3) also for the case where the semiconductor sample has a finite length W, for which $p_n(W) = p_{n0}$. This condition assumes that all excess carriers are extracted from the sample at $x = W$. Take the diffusion length $L_p^2 = D_p \tau_p$.

REFERENCES

1. R. S. Muller and T. I. Kamins, *Device electronics for integrated circuits,* New York: John Wiley & Sons, 1977.
2. D. L. Pulfrey, and N. G. Tarr, *Introduction to microelectronic devices.* Englewood Cliffs, N.J.: Prentice-Hall, 1989.
3. S. M. Sze, *Physics of semiconductor devices,* New York: John Wiley & Sons, 1969.
4. A. S. Grove, *Physics and technology of semiconductor devices,* New York: John Wiley & Sons, 1967.
5. N. D. Arora, J. R. Hauser, and D. J. Roulston, "Electron and hole mobilities in silicon as a function of concentration and temperature," *IEEE Trans. Electron Devices,* ED-29(2): 292–295, 1982.
6. J. L. Moll, *Physics of semiconductor devices,* New York: McGraw-Hill, 1964.
7. K. Hess, *Advanced theory of semiconductor devices,* Englewood Cliffs, N.J.: Prentice-Hall, 1988.
8. R. F. Pierret, *Advanced semiconductor fundamentals,* from *Modular series on solid state devices,* Vol. VI, Reading, Mass.: Addison-Wesley, 1989.
9. W. Shockley and W. T. Read, "Statistics of the recombination of holes and electrons," *Phys. Rev.,* 87: 835–849, 1952.

PART 2

BASIC PROPERTIES OF BIOLOGICAL MOLECULES

CHAPTER 4
BIOLOGICAL MATERIALS

The typical "habitat" of biological systems is made of electrolyte solutions, i.e., water solutions containing ions as solutes. We begin this chapter by characterizing the basic properties of these solutions in an elementary way. The reader will appreciate that, at this elementary level of description, several main points of this chapter are the same as those considered for describing the basic properties of semiconductors, e.g., thermal ionization, electroneutrality, diffusion and drift, and the mass action law. Of course, there are obvious differences, as we move from the realm of solid-state matter to the realm of liquid matter. In liquid matter *physical bonds* play a relevant role and therefore they will be considered again (see Chap. 1) and in more detail. Once the elementary properties of electrolyte solutions are summarized, the main biological actors will be taken into account, i.e., nucleic acids, proteins, and lipids. The cell membrane will be also described and finally a short overview of the eucaryotic animal cell will be given.

In accordance with the purpose of the book, we will not probe deeply in our description of these fundamental macromolecules. Advanced books are quoted in the proper sections of the chapter.

4.1 PHYSICAL BONDS REVISITED

As anticipated in Chap. 1, physical bonds usually lack the strong directionality and specificity of the covalent bonds considered in describing interactions inside a silicon crystal. Physical bonds are perfectly appropriate for holding molecules together in liquids, where the molecules can move about and rotate, while still remaining bonded to each other. In contrast to covalent binding, in physical binding the electrons of the different atoms do not merge, but are only "perturbed" by the bond. Nevertheless, physical binding forces, such as the Coulomb force, can be as strong as the covalent forces.

Physical binding forces display a vast range of strength, from "strong" charge-charge interactions, down to "weak" dipole-induced dipole interactions. A few relevant examples taken from this range will be considered in the following sections.

4.1.1 Ion-Ion Interactions

The interaction energy (or free energy) for the Coulomb interaction between two ions is given by

$$E(R) = \frac{z_1 z_2 q^2}{4\pi\varepsilon_0 \varepsilon_r R} \tag{4.1}$$

where ε_0 is the permettivity of vacuum, ε_r is the relative permittivity of the medium, R is the distance between the two charges, and z_1, z_2 are integer positive or negative numbers, named *valence*, indicating how many elementary charges q are carried by the two ions. For example, the valence is $z = +1$ for monovalent *cations* such as K$^+$, $z = -1$ for monovalent *anions*, such as Cl$^-$, $z = +2$ for divalent cations such as Mg^{2+}, and so on.

The intensity of the *Coulomb force* is given by

$$F(R) = -\frac{dE(R)}{dR} = \frac{z_1 z_2 q^2}{4\pi\varepsilon_0 \varepsilon_r R^2} \qquad (4.2)$$

Let us calculate the strength of the interaction between two isolated ions (e.g., Na$^+$ and Cl$^-$) in contact in vacuum ($\varepsilon_r = 1$). In this simple case R is the sum of the two ionic radii (about 0.28 nm) and the binding energy is

$$E(R) \simeq -8.4 \times 10^{-19} \, J \simeq -5.2 \, \text{eV} \qquad (4.3)$$

The negative sign means that the binding is spontaneous, i.e., that positive work $W = -E$ must be exerted to break the bond. The numerical value of E is about 200 times kT at room temperature. This value of the binding energy is in the same range of the covalent ones.

Two isolated ions represent a simplified approximation for estimating the mean energy of an ionic bond in a salt lattice. Nevertheless, this approximation gives the right order of magnitude. A more appropriate calculation, made by considering the 12 next-nearest neighbors in a cubic NaCl lattice, is less than twice the approximate value.

By assuming $\varepsilon_r = 80$ (water relative permittivity) and letting the distance R between charges be equal to the sum of the radii, then Eq. (4.1) can be utilized to give an approximate estimate of the positive free energy ΔE necessary to separate Na$^+$ and Cl$^-$ in water.

This energy can be used to estimate the solubility (mole fraction) X_s of ions in water forming a saturated solution in equilibrium with the solid can be then approximated by

$$X_s \simeq e^{-\Delta E/kT} \qquad (4.4)$$

A derivation of Eq. (4.4) and further considerations on the ionic bond can be found in Ref. 1.

4.1.2 Ion-Dipole Interactions

The electrostatic interaction between an ion ($Q = zq$) and a dipole is depicted schematically in Fig. 4.1. The total interaction energy will be the sum of the Coulomb energy of Q with $-Q_d$ at B and of Q with $+Q_d$ at A:

$$E(R) = -\frac{QQ_d}{4\pi\varepsilon_0 \varepsilon_r}\left[\frac{1}{OB} - \frac{1}{OA}\right] \qquad (4.5)$$

where

$$OA = [(R + \tfrac{1}{2} l \cos \theta)^2 + (\tfrac{1}{2} l \sin \theta)^2]^{1/2} \qquad (4.6)$$

$$OB = [(R - \tfrac{1}{2} l \cos \theta)^2 + (\tfrac{1}{2} l \sin \theta)^2]^{1/2} \qquad (4.7)$$

l and R being the dipole length and the length of the segment OM, respectively.

At separations $R > l$, we can approximate $OA \simeq R + \tfrac{1}{2} l \cos \theta$ and $OB \simeq R - \tfrac{1}{2} l \cos \theta$. Moreover, by neglecting $(l^2/4) \cos^2 \theta$ in comparison with R^2, we obtain

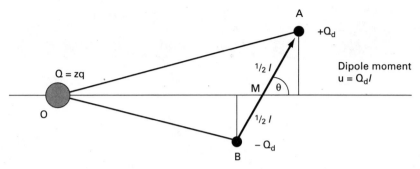

FIGURE 4.1 Ion-dipole configuration resulting in Eq. (4.5).

$$E(R) \simeq -\frac{zqu \cos \theta}{4\pi\varepsilon_0\varepsilon_r R^2} \qquad (4.8)$$

We can use Eq. (4.8) to calculate the interaction between an Na$^+$ ion and a water dipole in *vacuum* (i.e., in a cloud). A drastic simplification for the water dipole is to consider it as a spherical molecule of radius 0.14 nm and with a point dipole of moment 1.85 D (1 D = 1 debye = 3.336 × 10^{-30} C-m). Thus, for an Na$^+$ ion (valence $z = 1$, radius = 0.09 nm) near a water molecule, the maximum interaction energy will be given by Eq. (4.8) for $\theta = 0°$.

$$E(R, \theta = 0°) \simeq -1.6 \times 10^{-19} \text{ J} \qquad (4.9)$$

The energy value in Eq. (4.9) is about $40kT$ at room temperature. For the small divalent cation Mg^{2+} ($z = 2$, radius = 0.06 nm) the value rises to about $100kT$.

When ion-water interactions take place in *bulk water*, the above energies will be reduced by a factor of about 80 (the bulk water relative permittivity value); even then the strength of interaction will exceed kT at room temperature for small divalent (or multivalent) ions and it is not negligible for small monovalent ions. We can conclude that small or multivalent ions in water will tend to orient the water dipoles around them, with $\theta = 0°$ near cations, and $\theta = 180°$ near anions. As a result cations and anions in solution have a number of water molecules orientationally bound to them. The ions are said to be *hydrated* and the number of bound water molecules (typically 4 to 6) is known as the *hydration number*. The bound dipoles form a *hydration shell*. The reader should realize that there is a continuous exchange of bound dipoles with bulk water. The lifetime of a bound water dipole can range, at room temperature, from about 10^{-9} s for small monovalent cations such as Li$^+$, to 10^{-8} s for Ca^{2+}, and to 10^{-6} to 10^{-5} s for Mg^{2+}.

Large monovalent anions (such as Cl$^-$, $z = -1$, radius = 0.18 nm) are generally less hydrated than cations, and the lifetimes are comparable to those of two bulk water molecules (about 10^{-11} s).[1] Hydration modifies the dielectric properties of water and, consequently, the short-range Coulomb interaction among ions.

We will see in subsequent sections that water bonds also *polyelectrolytes,* such as DNA and proteins.

4.1.3 Dipole-Dipole Interaction and Hydrogen Bonding

When two polar molecules are near each other, there is a dipole-dipole interaction. It can be shown[1] that for two point dipoles of moment u_1 and u_2 at a distance R (see Fig. 4.2) the interaction energy is

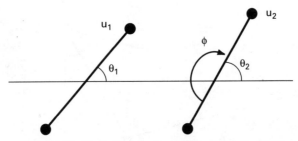

FIGURE 4.2 Dipole-dipole configuration resulting in Eq. (4.10).

$$E(R, \theta_1, \theta_2, \phi) = -\frac{u_1 u_2}{4\pi\varepsilon_0 \varepsilon_r R^3}[2\cos\theta_1 \cos\theta_2 - \sin\theta_1 \sin\theta_2 \cos\phi] \quad (4.10)$$

For dipole moments in the order of 1 D, typical energies are weaker than kT at room temperature. An *exception* is given by the bond F$^-$—H$^+$, O$^-$—H$^+$, N$^-$—H$^+$.

F, O, and N are the most *electronegative* elements, that is, they tend to strongly attract the molecular electronic orbitals when making a covalent bond with other elements. This fact results in a very strong dipole when the very small, electron-depleted H atom (almost a bare proton) is bonded.

To fix the ideas, let us consider a collection of water molecules: in each of them the presence of O$^-$—H$^+$ bonds generates a very strong dipole and therefore each dipole (i.e., each H$_2$O molecule) strongly interacts with other dipoles (i.e., other H$_2$O molecules). Because of this strong dipole-dipole interaction, water is said to be an *associate liquid*. A similarly strong dipolar interaction could be also made by H$^+$ if other electronegative groups, such as —N or C=O, were present.

In all of these situations, a very strong type of directional dipole-dipole interaction is generated, which can be in the order of 10 to 40 kJmol^{-1}. Such a dipole-dipole interaction is named *hydrogen bond*. This is the bond which allows the organization of the secondary structures of proteins and which stabilizes the DNA double helix (see Sec. 4.5).

4.2 WATER AND ELECTROLYTE SOLUTIONS

In a very rudimentary sense, a 150-mM NaCl water solution at 37°C (310 K) is the basic environment of a mammalian cell. Under these conditions water is very poorly ionized (as a silicon crystal is) and NaCl is totally dissociated in hydrated Na$^+$ cations and hydrated Cl$^-$ anions. Other chemical species can then be added, thus making the situation more complex.

4.2.1 Water as a Collection of Partially Ionized Molecules

Under standard conditions, 1 liter of water is equal to 1 kg of water and it contains $(1000/18) = 55.6$ moles $= 335 \times 10^{23}$ water molecules. At room temperature (300 K), a very small fraction of these dipolar molecules is thermally ionized, according to the *generation-recombination scheme:*

$$[H_2O] \underset{k_2}{\overset{k_1}{\rightleftarrows}} [H^+] + [OH^-] \quad (4.11)$$

The square brackets indicate concentrations (*molarity,* i.e., moles/liter, M). According to the *mass action law,* Eq. (4.11) corresponds to the differential equations.

$$\frac{d}{dt}[H_2O] = -k_1[H_2O] + k_2[H^+][OH^-] \qquad (4.12a)$$

$$\frac{d}{dt}[H^+] = k_1[H_2O] - k_2[H^+][OH^-] \qquad (4.12b)$$

$$\frac{d}{dt}[OH^-] = \frac{d}{dt}[H^+] \qquad (4.12c)$$

with the constraint that the sum of the number of un-ionized molecules and of the number of H^+—OH^- pairs is, at any time, a fixed constant (see also Sec. 5.4.1).

Under equilibrium conditions ($d/dt = 0$), Eqs. (4.12a) to (4.12c) reduce to the equality

$$k_1[H_2O] = k_2[H^+][OH^-] \qquad (4.13)$$

or

$$\frac{[H^+][OH^-]}{[H_2O]} = \frac{k_1}{k_2} = K_{eq} \qquad (4.14)$$

If we assume that the fraction of ionized molecules is very small, we can approximate $[H_2O]$ with the total concentration of water and incorporate it into the equilibrium constant, i.e.,

$$[H^+][OH^-] = K_w \qquad (4.15)$$

Under standard conditions

$$K_w = 10^{-14} \, M^2 \qquad (4.16)$$

That is,

$$[H^+] = [OH^-] = 10^{-7} \, M \qquad (4.17)$$

or

$$\frac{10^{-7}}{10^3} 6.02 \times 10^{23} = 6.02 \times 10^{13} \text{ ions/cm}^3 \qquad (4.18)$$

We wish to remind the reader that, under the same standard conditions (see Chap. 2),

$$n_i = p_i = 1.45 \times 10^{10} \text{ carriers/cm}^3 \qquad (4.19)$$

It is customary to translate Eq. (4.17) into the expression

$$pH = 7 \qquad (4.20)$$

where, of course

$$pH = -\log [H^+] \qquad (4.21)$$

The notion of pH can be extended to other water solutions where the presence of other chemicals unbalances Eqs. (4.16) and (4.17). Chemicals which are essential to understand the properties of biological molecules are *acids* and *bases*. A great many of the low-molecular-weight metabolites and macromolecular components of living cells are acids and bases, whose main feature is the potential to ionize. According to the Brönsted concept of acids and bases, an *acid is a substance that donates protons* (hydrogen ions), and a *base is a substance that accepts protons*.[2]

Acids and bases can be *strong* or *weak*. A strong acid is a substance that ionizes almost 100 percent in aqueous solutions. For example HCl in solution is essentially 100 percent ionized to H_3O^+ and Cl^-; i.e.,

$$HCl + H_2O \rightarrow H_3O^+ + Cl^- \tag{4.22}$$

where H_3O^+ (the *hydronium ion*, or conjugate acid of water) is the actual form of the hydrogen ion (i.e., a proton) in solution. The ionization of HCl can be represented as the simple dissociation

$$HCl \rightarrow H^+ + Cl^- \tag{4.23}$$

and, for all practical purposes, H_3O^+ and H^+ can be used interchangeably. A strong base is a substance that ionizes almost 100 percent in aqueous solutions yielding OH^- ions.

Potassium hydroxide is an example of a strong inorganic base that ionizes according to

$$KOH \rightarrow K^+ + OH^- \tag{4.24}$$

Strong acids and strong bases affect the H^+ and OH^- concentration of an aqueous solution in exactly the same way as acceptors and donors unbalance the concentration of electrons and holes in a silicon crystal.

Under the same approximations utilized in Chap. 2, the reader can easily verify that a 1-mM concentration of HCl will shift the H^+ concentration from 10^{-7} M to 10^{-3} M. Assuming that the water H^+ and OH^- generation process is not affected (i.e., $[H] \times [OH^-] = 10^{-14}$ M^2), the new OH^- concentration is

$$[OH^-] = 10^{-11} M \tag{4.25}$$

The analogy with the doping of a semiconductor is apparent. Of course, if the strong acid/base concentration is equal or even smaller than 10^{-7} M, then the final result is obtained by solving a quadratic equation. The reader should remember that this is also the case for silicon doping comparable to the intrinsic carrier concentration. This analogy is illustrated in Example 4.1.

Example 4.1   Find the pH of a 10^{-8}-M solution of HCl.

Answer   By indicating $[H^+] = X$, the second order equation

$$X^2 - 10^{-8} X - 10^{-14} = 0 \tag{E4.1}$$

should be solved, which gives

$$[H^+] = 10.51 \times 10^{-8} M \tag{E4.2}$$

or

$$pH = 6.98 \tag{E4.3}$$

In contrast to strong acids and bases, *weak* acids and *weak* bases ionize to a limited extent in aqueous solutions. The effects of their partial ionization on the solution pH is analyzed in the next section.

4.2.2 Partial Ionization and Buffer Solutions

In an aqueous solution a weak acid HA (e.g., acetic acid, CH_3—COOH) ionizes to a limited extent as follows:

$$\text{HA} + \text{H}_2\text{O} \rightleftharpoons \text{A}^- + \text{H}_3\text{O}^+ \quad (4.26)$$

[Conjugate acid]$_1$ [conjugate base]$_2$ [conjugate base]$_1$ [conjugate acid]$_2$

The proton released from the weak acid HA is accepted by water to form H_3O^+. It should be noted that, according to the Brönsted definition, HA is a conjugated acid and A^- is the corresponding conjugated base. Every time a Brönsted acid loses a proton, then a Brönsted base is produced. The substance that accepts the proton is a different Brönsted base; by accepting the proton, another Brönsted acid is produced. Thus, in every ionization of an acid or base, two conjugate acid–conjugate base pairs are involved [see Eq. (4.26)].

Equation (4.26) is a reversible reaction described by the equilibrium constant

$$K = \frac{[\text{H}_3\text{O}^+][\text{A}^-]}{[\text{HA}][\text{H}_2\text{O}]} \quad (4.27)$$

If the "practically" constant value of $[\text{H}_2\text{O}]$ is incorporated into K and $[\text{H}^+]$ is substituted for $[\text{H}_3\text{O}^+]$, Eq. (4.27) becomes

$$K_a = \frac{[\text{H}^+][\text{A}^-]}{[\text{HA}]} \quad (4.28)$$

which corresponds to the dissociation reaction:

$$\text{HA} \rightleftharpoons \text{H}^+ + \text{A}^- \quad (4.29)$$

One of the few simple examples of a weak inorganic base is given by ammonia:

$$\text{NH}_4\text{OH} \rightleftharpoons \text{NH}_4^+ + \text{OH}^- \quad (4.30)$$

Among organic bases, amines (R—NH_2, where R is a generic radical) play an important role. The reader is referred to Ref. 2 for more on this topic.

Considering again a weak acid, the dissociation of HA yields H^+ and A^- in equal concentrations. Therefore we can write

$$K_a = \frac{[\text{H}^+][\text{A}^-]}{[\text{HA}]} = \frac{[\text{H}^+]^2}{[\text{HA}]} \quad (4.31)$$

If we assume that the degree of ionization is small so that [HA] is practically equal to the total concentration of the acid $[\text{HA}]_T$, and if we further assume that the contribution of water to $[\text{H}^+]$ is negligible, then the final $[\text{H}^+]_F$ concentration is simply given by

$$[\text{H}^+]_F = (K_a[\text{HA}]_T)^{1/2} \quad (4.32)$$

or

$$\text{pH} = \tfrac{1}{2}(\text{p}K_a + \text{p}[\text{HA}]) \quad (4.33)$$

where $\text{p}K_a = -\log K_a$ and $\text{p}[\text{HA}]_T = -\log [\text{HA}]_T$.

Equation (4.31) can be rearranged into

$$\log[\text{H}^+]_F = \log K_a + \log \frac{[\text{HA}]_T}{[\text{A}^-]} \quad (4.34)$$

or

$$\text{pH} = \text{p}K_a + \log \frac{[\text{A}^-]}{[\text{HA}]_T} \quad (4.35)$$

Equation (4.35) is a useful starting point for the calculation of the *capacity* of a *buffer*.

88 BASIC PROPERTIES OF BIOLOGICAL MOLECULES

Conjugate acids and conjugate bases can form *pH-buffer* solutions, which are of great relevance for any biological system. A pH buffer is a substance (or a mixture of substances) that allows a solution to maintain a near constant pH upon the addition of small amounts of H$^+$ or OH$^-$ ions to it. Amino acids and proteins, which will be considered in Sec. 4.4.1, are examples of biological buffers. The capacity β of a buffer can be defined as the number of moles per liter of H$^+$ and OH$^-$ required to cause a given change in pH.

It can be shown[2] that

$$\beta = \frac{2.3[A^-][HA]}{[A^-] + [HA]} \quad (4.36)$$

or

$$\beta = \frac{2.3\, K_a[H^+][C]}{(K_a + [H^+])^2} \quad (4.37)$$

where C is the total concentration of buffer components (i.e., $[C] = [A^-] + [HA]$).

We conclude this section with a definition quite important for subsequent descriptions of amino acids and proteins, i.e., the definition of *polyprotic acids*.

A polyprotic acid ionizes in successive steps, according to the following scheme

$$\begin{array}{c} H_2A \xrightleftharpoons{k_{a1}} H^+ + HA^- \\ \Big\updownarrow k_{a2} \\ H^+ \\ + \\ A^{2-} \end{array} \quad (4.38)$$

where

$$K_{a1} = \frac{[H^+][HA^-]}{[H_2A]} \quad (4.39)$$

$$K_{a2} = \frac{[H^+][A^{2-}]}{[HA^-]} \quad (4.40)$$

When an intermediate ion of a polyprotic acid is dissolved in water, it undergoes both ionization as an acid and ionization as a base or hydrolysis. There will be a specific pH value at which species with no net charge are present. This is the *isoelectric point* (pI) of that specific polyprotic acid. This definition will be further considered in Sec. 4.4.1 in a discussion of amino acids.

4.2.3 Ionic Strength

The *ionic strength* of an electrolyte solution is defined as

$$I = \tfrac{1}{2}\sum_{i=1}^{n} c_i z_i^2 \quad (4.41)$$

where c_i is the concentration of the ith ion, z_i is its valence, and the summation extends over all the ions in the solution.

It is easy to verify that for a single 1:1 electrolyte (e.g., NaCl) the ionic strength is equal to its concentration. A 2:2 totally ionized salt (such as Mg_2SO_4) has an ionic strength which is 4 times that of a 1:1 salt with the same concentration.

It is important to keep in mind that ionic strength is a property of the solution and is not a property of any particular ion in the solution. Only the *net* charge of an ion is used in calculating ionic strength. Therefore, an un-ionized compound (e.g., a weak acid) or a dipolar compound does not contribute toward the ionic strength of a solution. An increase in the ionic strength of a solution implies an increase in the electrostatic screening by ions of the macromolecules present in the solution. Biological complexes held together by electrostatic interactions can be separated into their single macromolecular components simply by increasing the ionic strength of the solution. Moreover, because of this screening effect, proteins and DNA tend to assume a folded conformation when the solution ionic strength is increased, as will be discussed in the following sections.

4.3 OPTICAL PROPERTIES OF MOLECULES IN SOLUTION

The Beer-Lambert law [see Eq. (2.6)] can be usefully utilized for estimating the concentration of molecular species in solution. For the reader's convenience we introduce again here the equation utilized to deduce the intensity I_{tr} of the light of frequency $\nu > \nu_{th}$, transmitted by a slab of semiconductor of thickness l, exposed to light of intensity I_o

$$I_{tr}(l) = I_o e^{-\alpha(\nu)l} \tag{4.42}$$

We can adapt Eq. (4.42) to a solution inside a tube of thickness l, and containing a solute of concentration c, by writing

$$I_{tr}(l) = I_o e^{-\varepsilon(\nu)cl} \tag{4.43}$$

where ε, known as the *molar extinction coefficient* of the solute, is a function of the frequency ν of the incident light. Equation (4.43) deserves two comments:

- Equation (4.43) is correct only if the tube containing the solution and the solvent does not absorb any of the incident light. If this is not the case, then the intensity of light, transmitted by the same tube filled with the solvent only, has to play the role of I_o in Eq. (4.43).
- In the case of a typical light-absorbing solute (i.e., a *dye*), there is not a threshold frequency ν_{th} in the same sense as it was defined for semiconductors. This observation will be clarified later in this section.

As already indicated in Chap. 2, the intensity of the absorbed light is

$$I_{abs}(l) = I_o(1 - e^{-\varepsilon(\nu)lc}) \tag{4.44}$$

The corresponding absorbed energy will be partially dissipated in heat and partially given back radiatively as fluorescence light according to

$$I_{fl}(\nu') = QI_{abs} = QI_o(1 - e^{-\varepsilon(\nu)lc}) \tag{4.45}$$

where Q is the *fluorescence quantum yield* ($0 \leq Q \leq 1$).

As already discussed for silicon (see Sec. 2.1.3), the frequency ν' of the fluorescence emission is smaller than the frequency ν of the exciting light:

$$\nu' < \nu \tag{4.46}$$

Note that the preferential direction of the exciting light is completely lost by the fluorescence light, which is emitted isotropically in all directions. If the product εcl is sufficiently small, then Eq. (4.45) can be approximated by

$$I_{fl}(\nu') \simeq QI_o \varepsilon cl \tag{4.47}$$

which gives a direct linear relationship between solute concentration and fluorescence intensity. A qualitative description of the fluorescence phenomenon at the molecular level is as follows: a photon of frequency ν is absorbed by a molecule M, with the consequent transition of an electron of the molecule from its ground state to an excited one. The excited molecule M* then decays to its ground state either by fluorescence emission or by heat production. The process is described by the kinetic relation

$$\frac{d}{dt}[M^*] = -(k_F + k_H)[M^*] \tag{4.48}$$

where k_F and k_H are the rate constants of the radiation (fluorescence) and radiationless heat processes and [M*] represents the concentration of excited molecules.

In other words, a population of excited molecules M* decays to its ground state according to

$$M^* = M_0^* e^{-t/\tau} \tag{4.49}$$

where
$$\tau = \frac{1}{k_F + k_H} \tag{4.50}$$

It can be easily shown that

$$Q = \frac{k_F}{k_F + k_H} \tag{4.51}$$

The lifetime τ is typically in the order of 10^{-9} to 10^{-7} s, depending both on the molecular structure of the absorbing species and on the physicochemical properties of the solution.

Having introduced a molecular view of the absorption/emission process, we can now consider in some detail the absorption/emission spectra of molecules in solution.

We saw in Chap. 1 that hydrogen atoms absorb photons of discrete frequencies, and consequently their absorption spectrum is made of lines. Semiconductors (e.g., silicon), because of their energy band structure, absorb photons of all the frequencies (up to a certain value) greater than a threshold ν_{th} and therefore their absorption spectrum is flat (up to a certain value), starting from ν_{th} (see Chap. 2). Molecules in solution have absorption spectra made of one or more bell-shaped curves.[3] Let us consider a simple molecule, i.e., a diatomic one which can be approximated by a harmonic oscillator. In the first place we can say that the energy of the molecule will depend on the *electronic state*, i.e., the set of orbitals that the electrons in the molecule occupy.

For a given state, the energy will depend on the distance between the nuclei of the two atoms. For each electronic state, the molecule will have a set of allowed levels of *vibrational energy*, which could be approximately calculated by considering the quantum equivalent of the classical oscillator model.[4]

The energy separation between vibrational levels is of course much smaller than the overall energy separation between electronic levels. Finally, we should also consider a quantized *rotational energy*, with states which correspond to sets of even more closely spaced lines clustered above each vibrational level. A schematic representation of molecular energy levels is given in Fig. 4.3, where a comparison with the room temperature value of kT is given. By making use of the Boltzmann distribution, readers can easily con-

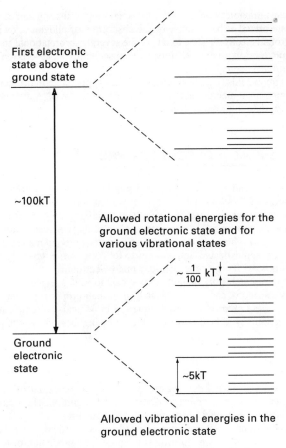

FIGURE 4.3 Schematic representation of typical molecular energy levels. A value of kT corresponding to room temperature is assumed.

vince themselves that at room temperature only the lowest vibrational level of the ground electronic level should be occupied. On the contrary, however, all the rotational levels of this ground level, separate by $1/100\ kT$ only, are expected to be fully occupied.

By transforming differences in energy levels into photon frequencies, it can be appreciated why infrared light is being used to generate *vibrational spectra* of molecules. Infrared photon absorption at room temperature causes transitions from the lowest vibrational level of the ground electronic state to higher vibrational levels of the same ground state. A nonlinear molecule with n atoms will have $(3n-6)$ fundamental modes of vibration. Among the biological molecules (to be considered in the next section), even for a simple substance such as an amino acid, the number of modes is large and consequently a quite complex vibrational spectrum is obtained. Transitions in the visible region of the spectrum are relatively low-energy electronic transitions. They are typical of biological macromolecules containing metal ions. Generally speaking, proteins and nucleic acids, which will be considered in detail in the next sections, present bell-shaped absorption bands in the near ultraviolet (200 to 400 nm). According to the scheme of Fig. 4.3, these bands can be interpreted as being made by the absorption of photons allowing any transition from the ground electronic state to one of the several vibrorotational states of the first excited electronic level, with a probability, and consequently an absorption intensity, dictated by quantum mechanical rules.[3]

Only a few amino acids (e.g., *tryptophan*) present appreciable fluorescence emissions. In most cases, biological macromolecules such as proteins and DNA are made fluorescent by binding appropriate fluorescent dyes to them.

4.4 BIOLOGICAL MOLECULES—PROTEINS

Up to now we have considered water, ions, and small metabolites. In a typical mammalian cell, with a volume of about 4×10^{-9} cm^3, their weight contribution is 70 percent, 1 percent, and 3 percent of the total, respectively. The remaining 26 percent of the weight is roughly shared among proteins (18 percent), nucleic acids (1 percent), lipids (5 percent), and sugars (2 percent). These are the molecules peculiar to living matter. Their organization in a cell is very sophisticated, and it seems to follow a *bottom-up* design, just opposite to the *top-down* process typical of present-day microelectronics.

The amount of research work on these biological molecules, especially on proteins and nucleic acids, is enormous, and a detailed analysis of their properties is out of the scope of this book. Under these premises, a few features of biological molecules, relevant to our *bioelectronic approach,* will be described in the following, beginning with proteins.

4.4.1 Proteins

Proteins are very sophisticated devices that play many fundamental roles in living systems, including catalysis, mechanical resistance, and recognition. The functional properties of proteins depend upon their three-dimensional structures.

The three-dimensional (3-D) structures arise because particular sequences of *amino acids* fold to generate complex architectures in space. The prediction of the 3-D protein structure, starting from the knowledge of its amino acid sequence, is the main issue of *protein engineering*.

As already stated, proteins are made of amino acids. There are 20 amino acids (plus a few rare ones). All of the 20 amino acids have in common a central carbon atom (C_α) to which a hydrogen atom (H), an amino group (NH$_2$), and a carboxyl group (COOH) are attached. A side chain is linked to the C_α atom through its fourth valency (Fig. 4.4). Any of the 20 amino acids differs from the others because of a different side chain.

Amino acids are joined end to end during protein synthesis by the formation of covalent bonds named *peptide bonds.* The bond formation is enzymatically catalyzed. The carboxyl group of one amino acid is *condensed* with the amino group of the next by elimination of a molecule of water, yielding a peptide bond NH—CH—CO. This process is repeated as the chain elongates. The formation of a succession of peptide bonds (from the first-unbonded-amino terminus to the last-unbonded-carboxyl terminus) generates the backbone of a protein, from which the various side chains project (see Fig. 4.4). This sequence of amino acids is the protein *primary structure.* The number of amino acids in a protein can range from a few tens to thousands.

The 20 amino acids are polyprotic acids and they can be divided into three classes, according to the properties of the side chain: class I, with an apolar (i.e., hydrophobic) side chain; class II with a polar side chain; and class III with an acid or basic side chain. The amino acid *glycine* has only a hydrogen atom as a side chain and can be considered to constitute a fourth class. With the exception of glycine, the four groups attached to the C_α atom are chemically different. Therefore all amino acids (except glycine) are chiral forms, which can exist in left (L-) and right (D-) forms. Proteins of eucaryotic cells (i.e., cells provided with a nucleus) are all made of L-amino acids.

The structures of an apolar amino acid (*alanine*) and of an acidic amino acid (*aspartic acid*) are compared in Fig. 4.5. Note that the amino and carboxyl groups of the main chain are shown in their ionized form, i.e., NH_3^+ and COO^-, respectively.

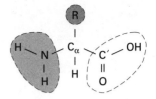

FIGURE 4.4 Schematic representation of an amino acid.

As Fig. 4.5 shows, the side chain of the aspartic acid has an extra carboxyl group. As a consequence, the aspartic acid has an isoelectric point pI smaller than 7 and, at physiological pH (around pH = 7.4), it is negatively charged. Figure 4.6 summarizes the pH-dependent amino acid charge. Other amino acids present an extra amino group (for example, *lysine*), have an isoelectric point greater than 7, and bear a positive charge at physiological pH.

As the reaction schemes of Fig. 4.6 suggest, at a given pH the primary structure of a protein is a complex sequence of negative charges, positive charges, and hydrophobic and polar regions, according to the side chains of the amino acid sequence. Moreover, it should be noted that H and the highly electronegative elements O and N are periodically repeated along the backbone and therefore hydrogen bonding (see Sec. 4.1.3) can be expected. A specific sequence of amino acids in a protein can wind on itself via H bonding, forming an helix (called α *helix* by L. Pauling, who first described it in 1951).

The length of α helices varies considerably in proteins, ranging from 4 or 5 amino acids to over 40. Helix formation/disruption is a phenomenon driven by several physicochemical parameters, such as pH, temperature, ionic strength and medium polarity.

The formation of an α helix is a well-studied phenomenon, and the classical mathematical tools of statistical mechanics are appropriate for describing it. As indicated in Fig. 4.7, a helix turn is formed via hydrogen bonding between O of amino acid i and H of amino acid $i + 4$. This helix turn forces six conformation angles along the chain to assume fixed values. The propagation of the helix implies fixing two angles per turn. This is a clear example of a *cooperative phenomenon*, which is characterized by the fact that the *starting of the process is more energy-demanding than the advancing of it*. In other words, the coil-to-helix transition is a nonlinear phenomenon.

If we plot the fraction of helix turns θ versus any physicochemical parameter s (e.g., pH), by varying where a transition is induced, then we get a sigmoidal curve (Fig. 4.8)

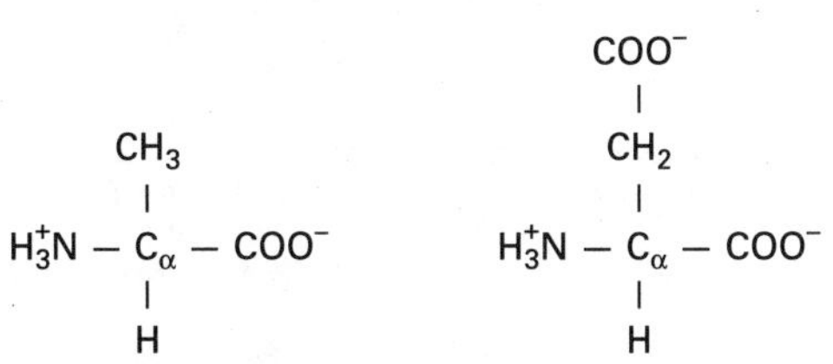

$$\begin{array}{cc} & COO^- \\ & | \\ CH_3 & CH_2 \\ | & | \\ H_3^+N - C_\alpha - COO^- \quad & H_3^+N - C_\alpha - COO^- \\ | & | \\ H & H \end{array}$$

FIGURE 4.5 Structures of *alanine* (left) and *aspartic* acid (right).

$$H_3^+N - C_\alpha - COOH \rightleftharpoons H_3^+N - C_\alpha - COO^- \rightleftharpoons H_3^+N - C_\alpha - COO^- \rightleftharpoons H_2N - C_\alpha - COO^-$$

with side chains CH_2-COOH, CH_2-COOH, CH_2-COO$^-$, CH_2-COO$^-$ respectively, and H on C_α:

AA$^+$ AA^0 AA$^-$ AA^{2-}

Low pH ⟶ High pH

FIGURE 4.6 At pH = pI = 2.98 the net charge of the amino acid (AA) *aspartic acid* is zero (AA^0). The amino acid is positively charged (AA$^+$) at low pH and negatively charged (AA$^-$, AA^{2-}) at high pH. This is a general behavior, which can be applied both to amino acids and proteins.

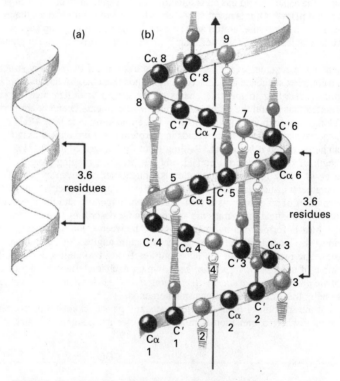

FIGURE 4.7 (a) Idealized diagram of the path of the main chain in a protein α helix. There are 3.6 residues per turn in an α helix, which corresponds to 0.54 nm (0.15 nm per residue); (b) the path with approximate positions for main chain atoms and hydrogen bonds included. (*Adapted from C. Branden and J. Tooze.*[6] *Used by permission.*)

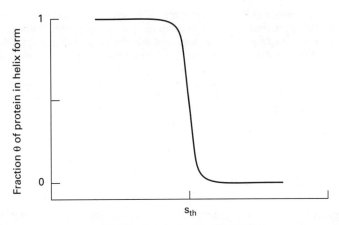

FIGURE 4.8 The coil-to-helix transition is a cooperative phenomenon. The fraction of helicity is "almost" one for s smaller than the threshold value s_{th} and it "jumps" to zero for $s > s_{th}$.

with a threshold value s_{th}.[5] This curve is an approximation of a two-state function, which is the fingerprint of any (electronic or biological) device appropriate for information processing.

Hydrogen bonds can also be formed among several regions of a polypeptide chain. This results in a structure known as β *sheet*. Different regions of a sequence of amino acids can be organized in either of these local regular structures, which represent the secondary structure of a protein. The packing of these structures into one or several compact globular units, called *domains*, determines the tertiary structure of a protein. The sequence of primary, secondary, and tertiary structures is sketched in Fig. 4.9.

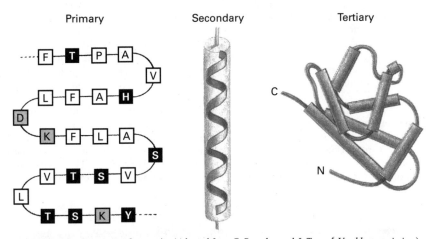

FIGURE 4.9 Structures of a protein. (*Adapted from C. Branden and J. Tooze.*[6] *Used by permission.*)

Protein structures are built up by combinations of secondary structural elements, α helices and β strands, according to a bottom-up design. α helices and β strands typically form the core regions of globular proteins and are connected to the protein surface by loops.[5] A protein may also contain several polypeptide chains arranged in a *quaternary structure*. By formation of tertiary and quaternary structures, amino acids far apart in the primary sequence are brought close together in three dimensions to form a *functional* region (an active site). In conclusion, in a very complex and only partially understood process, the bottom-up procedure organizes structures in such a way as to obtain highly sophisticated functions. A few examples will be described in the discussion of the plasma membrane in Sec. 4.7. Enzyme catalysis will be considered in Chap. 5.

4.5 NUCLEIC ACIDS

There are two kinds of nucleic acids in a living cell: *deoxyribonucleic acid* (DNA) and *ribonucleic acid* (RNA). Both of them bear negative charges at physiological pH. The term *nucleic* suggests their presence in the nucleus of a cell, but they can be found also outside the nucleus of an eucaryotic cell: DNA is present in mitochondria and RNA is present in various forms in the cytoplasm (see Sec. 4.8). A few key features of these biological macromolecules will be summarized in the following.

4.5.1 DNA and RNA Structures

A DNA chain is a long, unbranched polymer composed of four types of subunits, called *deoxyribonucleotides*. Each nucleotide is formed by a *base*, a *five-carbon sugar*, and one *phosphate group*. There are four kinds of bases. They are all nitrogen-containing ring compounds, either purines [the bases *adenine* (A) and *guanine* (G)] or pyrimidines [the bases *cytosine* (C) and *thymine* (T)]. The name *base* arises from the fact that these molecules can combine with H^+ in acidic solutions. Nucleotides are joined together by a covalent bond called *phosphodiester linkage*. The result is a linear chain, which can be identified simply by indicating the bases symbols as in Fig. 4.10.

Specific hydrogen bonding between G (large purine base) and C (smaller pyrimidine base) and between A (large purine base) and T (smaller pyrimidine base) generates base pairing between complementary linear chains. The result is the well-known DNA *double helix*, described in 1953 by Watson and Crick on the basis of x-ray diffraction data obtained by Wilkinson and Franklin. In a DNA molecule two antiparallel strands that are complementary in their nucleotide sequence are paired in a right-handed double helix with about 10 nucleotide pairs per helix turn. This is the so-called *B form* of DNA (see Fig. 4.11)

Two other forms have been described, namely A and Z forms. The B form is considered to be the typical form assumed by DNA inside the nucleus of a living cell.

A typical mammalian cell contains about one meter of DNA (3×10^9 nucleotide pairs). In other words, inside a cell nucleus (diameter in the order of 5 to 10 μm) there is a collection of 6×10^9 symbols, belonging to four categories only (A, C, G, T). Specific subsets of this collection represent the *code* utilized to synthesize different proteins. The process of synthesis involves RNA.

Inside a cell there are three major species of RNA present in several copies, namely: *messenger* RNA (mRNA); *transfer* RNA (tRNA); *ribosomal* RNA (rRNA). Any RNA

$\cdots - A - C - T - A - G - C - \cdots$

FIGURE 4.10 Symbolic representation of a DNA strand.

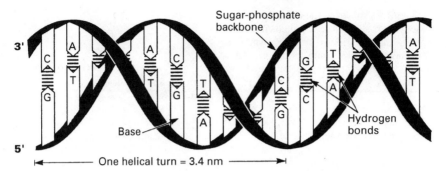

FIGURE 4.11 In a B-DNA molecule two antiparallel strands that are complementary in their base sequence are paired in a right-handed double helix held together by hydrogen bonds.

molecule is formed again by nucleotides, where the sugar is now *ribose* (and not deoxyribose, as in DNA) and one of the pyrimidine bases is *uracil* instead of thymine.

DNA is a single huge macromolecule lying inside the core of the cell as protected as the source code of a very important software program. In contrast, there are several small molecules of RNA (a transfer RNA is in the order of 80 nucleotides long) which behave like executable programs. The three different types of RNA have a different role in the process of translating subsets of DNA (*genes*) into proteins. The understanding of this translation procedure is one of the major scientific achievements of this century. The detailed description of this procedure is obviously out of the scope of this book. A short summary will be given in the next section, for the sake of completeness. The reader interested in a deeper description of protein synthesis is referred to Ref. 7.

4.5.2 Outline of Protein Synthesis

A simplified collection of key points concerning the process follows.

1. The synthesis of proteins involves copying specific regions of DNA (the genes) into polynucleotides of RNA. Molecules of RNA are synthesized by a process known as DNA *transcription*, by which one of the two strands of DNA acts as a template for generating a sequence of RNA, named *messenger* RNA, as symbolically shown in Fig. 4.12.

One can imagine RNA nucleotides floating around the appropriate DNA sequence, where the base-pairing abilities (hydrogen bonding) of incoming nucleotides are tested. Appropriate RNA nucleotides stay near the template long enough to be covalently linked together through the catalytic action of an enzyme known as RNA *polymerase*. When the sequence to be translated is terminated, the mRNA molecule breaks off from the DNA template.

2. Then (in eucaryotes) RNA molecules are spliced to remove *intron* (i.e., noncoding) sequences and only coding sequences (*exons*) are left (see Fig. 4.13).

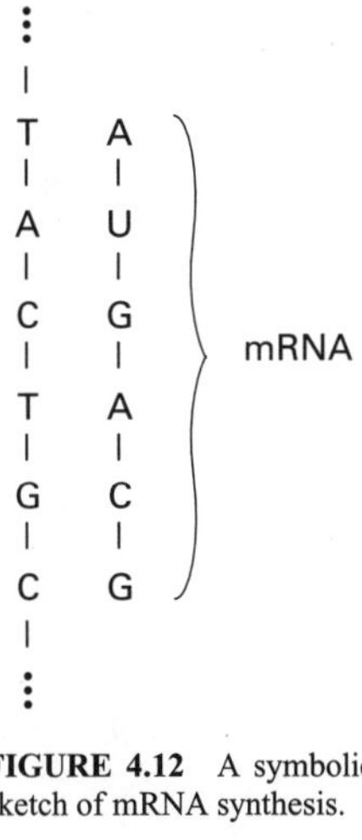

FIGURE 4.12 A symbolic sketch of mRNA synthesis.

3. mRNA moves to the cytoplasm, to interact with molecular devices known as *ribosomes* and to allow protein synthesis. Sequences of mRNA nucleotides are "analyzed" in sets of three and translated into amino acids.

$\cdots - A - \overline{|U - G - A|} - C - G - \cdots$

$\cdots - A - C - G - \cdots$

FIGURE 4.13 Symbolic representation of intron removal.

This three-by-three grouping is the basis of the famous *genetic code* which gives the rule for translating a sequence of three bases into a specific amino acid, e.g., {A—C—G} → threonine. Such sequences are named *codons*. There are $4^3 = 64$ possible codons corresponding to about 20 amino acids. This fact implies that the genetic code is *degenerate* (i.e., the code is redundant). The process of "analysis" is physically performed in connection with the ribosomes and it involves various tRNA molecules (each molecule bearing a different amino acid).

The reader is invited to compare this sophisticated chain of events with the most demanding sequence of processes utilized in the field of silicon technology (i.e., doping, oxidation, patterning, etc.). To stimulate further considerations (and readings[7]) we point out the impression that everything seems to be *loosely coupled* in this biological scheme while, on the contrary, everything is *tightly coupled* in any technological scheme.

4.5.3 A Short Note on Computation with DNA

Hopefully, in the previous sections we have been able to transmit to the reader the impression that cell processes, such as protein synthesis, are indeed very sophisticated ones. The physical procedure of DNA synthesis and the generation and recognition of complementary strands suggest by themselves remarkable combinatory properties: a specific sequence of 10 nucleotides will be complementary to exactly 1 over 4^{10} possible sequences and, in a way, it will "sort it out" by double stranding. Interestingly enough, the possibility of "real" computation with DNA has been seriously considered in the 1990s. In accordance with the bioelectronic approach of this book we very briefly expose the reader to this idea, even if, on the basis of the results available up to 1997, its value (i.e., a breakthrough or simply a clever scientific exercise) is not clear yet.

The idea, suggested by L. Adelman[8] in 1994, is to use the enormous parallelism of solution-phase chemistry to solve a computational problem, which is known as the Hamilton path problem. A path through a graph is said to be Hamiltonian if it visits each vertex exactly once. In a way, this is the starting point for solving more complex problems, such as the well-known traveling salesman problem, which looks for the minimum path. Adelman has approached the Hamiltonian path problem in a biological context where each vertex and edge of the graph can be represented by a short synthetic sequence of nucleotides. The binding together of chosen oligonucleotides representing vertices results in DNA molecules that encode the solution to a Hamiltonian in path problem.

In the experiments performed by Adelman, various copies of different oligonucleotides were generated by taking advantage of the DNA *ligation reaction,* catalyzed by the the enzyme *ligase.* Amplification was obtained by using the polymerase chain reaction. The "computation" required 7 days of laboratory work. Further details can be found in Ref. 8.

4.6 PHOSPHOLIPIDS ORGANIZATION

Several fatty acids are present in a living organism. They include: *tryglicerides, steroids* (e.g., cholesterol), and *phospholipids*. Phospholipids are self-assembling molecules which constitute the major component of cell membranes.

Phospholipid structure is characterized by a polar head (containing a phosphoric acid) and two hydrophobic fatty acid tails. This asymmetrical (hydrophilic-hydrophobic) design makes these molecules perfectly suited to self-assemble themselves in solution into organized structures. This is a clear example of the bottom-up scheme pursued by biological organisms to function.

The polar part of a phospholipid easily makes physical bonds with water dipoles and/or ions present in solution. On the contrary, the hydrophobic tails can make only weak hydrophobic bondings with other fatty acid chains.

On the basis of these simple considerations, three major space organizations of phospholipid assemblies can be qualitatively predicted:

1. An assembly of phospholipids gently dissolved on the surface of a water solution will tend to self organize into a *monolayer* with the heads interacting with the solution and all the tails parallel to each other out in the gas phase on the top of the solution. This can be considered as an example of a molecular insulator, a few nanometers thick (typically 2 to 3 nm), which can be deposited on the surface of a solid material appropriately dipped into the solution.

2. An assembly of phospholipids forced inside a water solution will tend to self-organize into small (4 to 6 nm) drops, with the external surface formed by the phosphorous heads and the core by packed, water-excluding, hydrophobic tails. This self-organized structure is named *micelle*.

3. A self-sealing spheroidal bilayer represents a third way of satisfying the hydrophilic/hydrophobic rule. This is a more complex system which achieves the fundamental result of separating an outer water solution from an inner water solution, with two polar (charged) surfaces facing the two solutions. The result is a system (we could call it a "device") known as *liposome*. Liposomes are of relevance for at least two reasons: First, biocompatible liposomes can be loaded with drugs and then injected into the blood. In this way the drug can be released inside the body through the slowly leaking bilayer of the liposome with a time scale that can be, to some extent, programmed by the experimenter. Second, and more important for the purposes of this book, liposomes are the structural basis for all cellular membranes (see Sec. 4.7). The three described self-organized structures are sketched in Fig. 4.14.

Finally, before moving to the membrane structures, let us introduce *glycolipids*. Like phospholipids, these molecules are composed of a hydrophobic region, containing two long hydrocarbon tails, and a polar region which contains one or more sugar residues. Glycolipids belong to the family of *glycoconiugates* which include also proteins covalently linked to sugars. Glycoconiugates play a fundamental role as *receptors* (i.e., *molecular sensors*) on the outer cell membrane.

4.7 CELL MEMBRANE

Every cell is separated from the outside world by a membrane. A cell membrane represents a further step in the bottom-up design, resulting from the assemblage of lipids, pro-

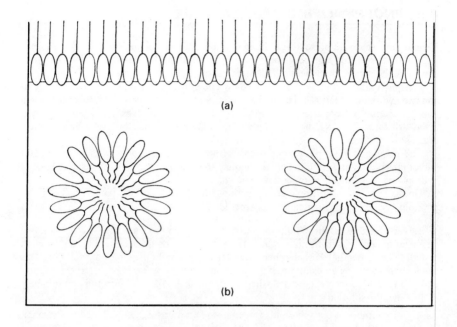
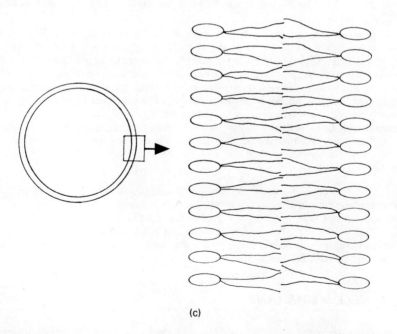

FIGURE 4.14 (a) Lipid monolayer, (b) micelle, and (c) liposome. (*Kindly provided by Maddalena Grattarola.*)

teins, and glycoconiugates. The cell membrane is much more than a simple physical interface. In the language of this book it should be considered a device (better, an array of devices) rather than a kind of material. It is shortly introduced and it will be considered again when dealing with membrane transport (Chap. 7), and with neurons (Chap. 11).

4.7.1 The Cell Membrane Structure

The membrane of a typical cell of a multicellular organism is known as *plasma membrane*. It encloses the cell and maintains the essential differences between the *cytosol* (i.e., the inside of the cell) and the extracellular environment. All the plasma membranes have a common general structure: each is a very thin film of lipid and protein molecules, held together mainly by noncovalent interactions. Cell membranes are dynamic, fluid structures and most of their molecules are able to move about in the plane of the membrane. In some cases this movement can be described as a random one, known as *Brownian motion* (see Chap. 5). The lipid molecules (mostly phospholipids) are arranged as a continuous double layer about 5 nm thick. This is the same structure already described for liposomes. The lipid bilayer furnishes the basic structure of the membrane and plays the role of a relatively impermeable barrier to the passage of most water-soluble molecules. From this point of view, the lipid bilayer can be considered as an *insulator*. Moreover, in consideration of the fact that some of the lipid heads bear a charge and that the membrane is surrounded by two conducting electrolyte solutions, the lipid layer, plus the conducting inner and outer media, can be schematized as a *capacitor*.

Protein molecules "dissolved" in this two-dimensional solvent mediate most of the functions of the membrane, for example, transporting ions across it (*ion channels*), catalyzing membrane-associate reactions (*enzymes*), or acting as sensors of external signals, allowing the cell to change its behavior in response to environmental cues (*receptors*).

Ion channels, enzymes, and receptors act as a multifunctional array of "fluid devices" which continuously reorganize themselves as a function of external physicochemical signals and internal ones, the latter coming from the genes.

Several kinds of phospholipids form, together with cholesterol, the lipid bilayer of any plasma membrane. As a rule, negatively charged lipids are located in the inner monolayer, and therefore there is a significant difference in charge between the two halves of the bilayer.

Another asymmetry in lipid distribution among the two halves is related to glycolipids (see Sec. 4.6). These sugar-containing lipid molecules are found exclusively in the extracellular half of the lipid bilayer, where they are thought to self-associate into microaggregates by forming hydrogen bonds with one another. Microaggregates of glycolipids could have some role in interactions of the cell with its surroundings. The most complex of the glycolipids, the *gangliosides* contain oligosaccharides with one or more sialic acid residues, which give gangliosides a net negative charge.[7] This could imply *electrical effects*: their presence will alter the electrical field across the membrane and the concentrations of ions such as Ca^{2+} at the cell surface. Glycolipids can act in general as membrane receptors.

Figure 4.15 schematically shows the asymmetrical distribution of phospolipids and glycolipids.

4.7.2 Membrane Proteins

Roughly speaking, about 50 percent of the mass of a "typical" plasma membrane is made of proteins. This corresponds to 1 protein molecule for every 50 lipid molecules, which

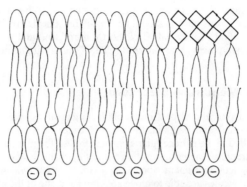

FIGURE 4.15 The asymmetrical distribution of phospholipids and glycolipids in the lipid bilayer. Glycolipids are drawn with double-square polar head groups. Cholesterol (not shown) is thought to be distributed about equally in both monolayers.

are much smaller than the proteins. Like membrane lipids, membrane proteins often have sugar chains attached to them in the extracellular half of the membrane. As a consequence, the surface that the cell presents to the exterior consists largely of carbohydrates, forming a fluid coat known as *glycocalyx*.

Many membrane proteins extend through the lipid bilayer. These proteins present hydrophobic regions (typically organized in α helices) inside the lipid bilayer, interacting with the hydrophobic tails of the lipid molecules, and hydrophilic regions exposed to the electrolyte solution on one or the other side of the membrane. Other membrane proteins are located entirely in the *cytoplasm* (see Sec. 4.8), and are associated with the bilayer by means of covalenty attached fatty acid chains. Similarly, other proteins are entirely exposed to the extracellular side, being attached to the bilayer only by a covalent linkage via sugar groups. All of these proteins, tightly bound to the membrane, are called *integral proteins*.

Some proteins, called *peripheral membrane proteins,* are simply bound to one or the other face of the membrane by noncovalent interactions with other membrane proteins. Most of these proteins can be detached from the membrane by simple exposure to solutions of very high ionic strength and extreme pH, which break protein-protein noncovalent binding but leave the lipid bilayer intact. Figure 4.16 summarizes all the considered cases.

Transmembrane proteins can transport molecules across the membrane, or function on both sides of it. Some cell-surface receptors, for example, are transmembrane proteins that bind signaling molecules in the extracellular space and generate different intracellular signals on the opposite side of the plasma membrane. Several input signals, such as *light* and *neurotransmitters* are recognized and amplified at the plasma membrane level with a common scheme involving similar receptors. These *receptors* are transmembrane proteins with seven helices spanning the lipid bilayer. Light-responding *rhodopsin,* and *acetylcholine*-responding *muscarinic receptors* are two examples of this transmembrane protein family. The signals received by these receptors are *amplified* by a common family of membrane proteins known as *G proteins*. G proteins activate an enzyme at the membrane level, and the final result is the production of a second messenger inside the cell. The described scheme is depicted in Fig. 4.17. Note that receptor, G protein, and enzyme can move independently inside the lipid bilayer.

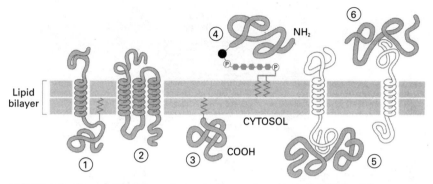

FIGURE 4.16 Six ways in which membrane proteins associate with the lipid bilayer. Most transmembrane proteins are thought to extend across the bilayer as a single α helix (1) or as multiple α helices (2). Some of these "single-pass" and "multipass" proteins have a covalently attached fatty acid chain inserted in the cytoplasmic monolayer (1). Other membrane proteins are attached to the bilayer solely by a covalently attached lipid in the cytoplasmic monolayer (3) or, less often, via an oligosaccharide, to a phospholipid, in the extracellular monolayer (4). Finally, many proteins are attached to the membrane only by noncovalent interactions with other membrane proteins (5 and 6). (*From B. Alberts et al.*[7] *Used by permission.*)

Among the family of transmembrane proteins that transport molecules across the membrane, *ion channels* are protein-based *membrane sensors* which can be activated by chemical or physical inputs. Voltage-gated channels open in response to a change in the intensity of the electric field present inside the plasma membrane. They represent one of the key devices in the operation of neurons, and, in the framework of a "totally mechanistic" viewpoint, they can be considered (together with *synapses*) the molecular basis of thought. A hypothetical view of a voltage-gated channel, taken from the excellent book by Hille,[9] is shown in Fig. 4.18. The cartoon is a kind of mesoscopic view of the channel, where the equivalent volumes of the protein channel are given rather than the detailed sec-

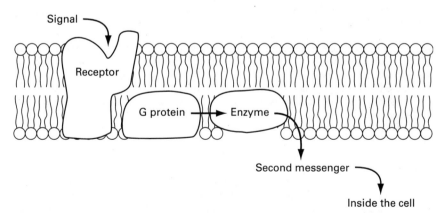

FIGURE 4.17 Sketch of a signaling pathway using G proteins. (*Adapted from B. Hille.*[9] *Used by permission.*)

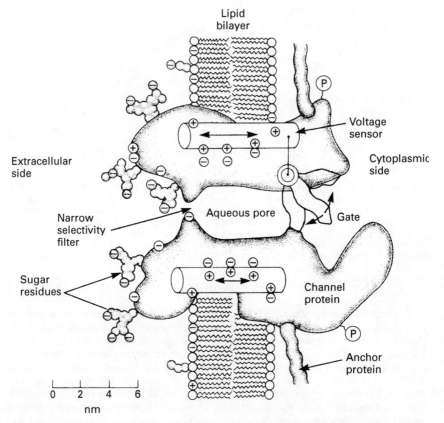

FIGURE 4.18 Working hypothesis for a channel. The channel is drawn as a transmembrane macromolecule with a hole through the center. The external surface of the molecule is glycosylated. (*Adapted from B. Hille.*[9] *Used by permission.*)

ondary structure. The reader should imagine α helix chains repetitively crossing the plasma membrane and connected by long hydrophilic loops projecting into the extracellular and intracellular media.

The protein forming the channel is known to be large, consisting of 1800 to 4000 amino acids, with some hundreds of sugar residues covalently linked to amino acids on the outer surface. The open channel forms a water-filled pore fully extending across the membrane. The pore may narrow in a specific region, where ionic selectivity is established.

Gating requires a conformational change of the pore that moves a "gate" into and out of an occluding position, and it is controlled by a "sensor" (see Fig. 4.18). In the case of a voltage-gated channel, the sensor is a voltage-sensitive one.

We close this section by inviting the engineering-oriented reader to meditate on this system that includes in a volume of a few thousand cubic nanometers a molecular filter (ion selectivity), a molecular sensor, and a molecular actuator (gate).

4.8 AN OVERVIEW OF THE EUCARYOTIC CELL

The *cell* is the functional unit of any living organism. With the exception of bacteria, the cells constituting any organism are called *eucaryotic cells*. They typically comprise an inner region, called *nucleus,* and an exterior region, surrounding the nucleus, called *cytoplasm*. Bacteria also are made of cells (typically a bacterium is just one cell), but these cells are devoid of nucleus. They are named *procaryotic cells*. In principle, any living cell (both eucaryotic and procaryotic) is able to self-replicate. Self-replication may be considered as partial in structures simpler than procaryotic cells, the *viruses,* which are particles consisting of nucleic acids (RNA or DNA) enclosed in a protein coat and able to replicate within a host cell.

Both higher animals and vegetal organisms are made of eucaryotic cells. A "typical" eucaryotic animal cell is structured as follows (see Fig. 4.19):

The *plasma membrane* (see Sec. 4.7) separates the inside of the cell from the outside. The fundamental component of the cytoplasm (the *cytosol*) is a nonideal aqueous solution containing ions (mostly K^+, Na^+, and Cl^-) and organic molecules. The cytoplasm of eucaryotic cells is crossed by a dynamic system of protein filaments known as the *cytoskeleton*. They include *actin filaments* (8 nm diameter), *intermediate filaments* (10 nm diameter), and *microtubules* (25 nm diameter). The cytoskeleton gives the cell shape and the capacity for movement. Flattened membranous sacs and tubes and small membrane-bound vesicles extend throughout the cytoplasm. As a whole, this complex includes the *Golgi apparatus* and the *endoplasmic reticulum*. The Golgi apparatus is involved in the *secretion* of macromolecules. The endoplasmic reticulum (ER) is in physical continuity with the nuclear envelope and it is further divided into smooth and rough ER. Smooth ER function is related to lipid synthesis. The "roughness" of the rough ER is caused by the presence of

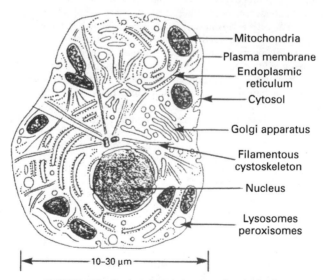

FIGURE 4.19 Section of a generic eucaryotic animal cell.

small (about 50 nm diameter) particles, known as *ribosomes,* which are involved in protein synthesis (see Sec. 4.5.2). The cytoplasm of a metabolically active eucaryotic cell is full of membrane-bound *organelles,* with dimensions in the micrometer range. The most common are *mitochondria, lysosomes,* and *peroxisomes.* Mitochondria have intriguing similarities with the procaryotic cell, including the presence of a small amount of "naked" DNA.[7] Mitochondria can be considered the power plants of all eucaryotic cells and they make energy available by combining oxygen with organic material to produce adenosintriphosphate (ATP) molecules.[7] *Lysosomes* are membrane-bound vesicles containing hydrolytic enzymes dealing with intracellular digestions. *Peroxisomes* are membrane-bound vesicles containing oxidative enzymes. Higher plant cells contain also *chloroplasts,* specialized structures dealing with the highly sophisticated cycle of chlorophyll photosynthesis.

The nucleus can be considered the largest organelle in the eucaryotic cell. It is separated from the cytoplasm by a double membrane containing *pores.* Inside the nucleus, DNA is connected in a complex way with basic proteins (*histones*) to form a complex 3-D structure, known as *chromatin.* A small subregion of the nucleus, known as the *nucleolus,* is RNA-rich and involved in the synthesis of ribosomes.

Cells reproduce by duplicating their content and then dividing in two. This process, of course, happens during the development of multicellular eucaryotic organisms (*embryogenesis*) and it also takes place in the adult individual, to replace cells that are lost for several reasons. The capability of self-reproduction is not equally distributed in the cells forming the various *tissues* of an adult individual. The way a cell duplicates its content and then divides is known as the *cell cycle.* The cell cycle has a number of features common to all kind of reproducing (i.e., *cycling*) cells: it is made of *four successive phases,* indicated as G_1, S, G_2, and M phases. The cell cycle begins with a "preparatory period," the G_1 phase, after which the cell enters the S phase. The symbol S stands for DNA synthesis. This is a process of nuclear DNA replication, which is a necessary prerequisite in order to finally produce two daughter cells with the same genetic content of the mother cell. S phase is followed by a gap, the G_2 phase, which somehow allows the cell to prepare to divide. The division process occupies the subsequent *mitotic* (M) phase, during which nuclear chromatin condenses into visible *chromosomes* and the cell eventually physically divides into two new cells (*cytokinesis*). The subsequent cycle starts again with a presynthesis gap, G_1 phase, which allows the cell to monitor its environment and its own size, up to a decisive step that commits it to DNA replication and completion of a new division cycle. It should be noted that at any instant, most of the billions (10^{13} in the human body) cells in an adult complex organism, such as a mammal, are not proliferating, but in a resting state, performing their specialized function while retired from the cell cycle.

A great deal of information concerning the cell cycle of mammalian cells has been obtained by studying cells isolated from the intact animal and growing *in culture.* Cultured cells offer the researcher the opportunity to study living biological systems under controlled environmental conditions produced in the laboratory. Further analysis of this topic is beyond the purpose of the book. The interested reader can find in Ref. 7 a detailed discussion of this topic and of the related topics of *cell lines, cell transformation,* and *cell senescence.*

PROBLEMS

4.1 In vacuum a dipole with $l = 0.2$ nm and charge $= 0.5q$ is at a distance $R = 1$ nm from a monovalent cation. Both particles are fixed in space.
  a. Calculate the interaction energy for $\theta = 0°$ and $\theta = 90°$.
  b. Compare this energy with kT for $T = 300$ K and $T = 1000$ K

4.2 Reconsider Prob. 4.1 after immersing ion and dipole in a medium with relative permittivity $\varepsilon_r = 80$.

4.3 What is the (a) H^+ ion concentration, (b) pH, (c) OH^- ion concentration of a 1-mM solution of HCl?

4.4 What is the pH of a 10^{-9}-M solution of HCl? (*Warning:* it cannot be pH = 9!)

4.5 Rhodopsin is a seven-helix membrane protein that responds to light. One single absorbed photon of, say, green light finally results in transiently (1 s) preventing the entering of 10^7 Na^+. This results in a change in a rod membrane potential of about 1 mV. Estimate the amplification factor of this chain of events.

REFERENCES

1. J. Israelachvili, *Intermolecular and surface forces,* 2d ed., London: Academic Press, 1992.
2. I. H. Segel, *Biochemical calculations,* 2d ed., New York: John Wiley & Sons, 1975.
3. C. R. Cantor and P. R. Schimmel, *Biophysical chemistry,* Part II, San Francisco: W. H. Freeman, 1980.
4. G. M. Barrow, *Physical chemistry,* 2d ed., New York: McGraw-Hill, 1966.
5. C. R. Cantor and P. R. Schimmel, *Biophysical chemistry,* Part III, San Francisco: W. H. Freeman, 1980.
6. C. Branden and J. Tooze, *Introduction to protein structure,* New York and London: Garland Publishing, 1991.
7. B. Alberts, D. Bray, J. Lewis, M. Raff, K. Roberts, and J. D. Watson, *Molecular biology of the cell,* 3d ed., New York and London: Garland, 1994.
8. L. M. Adleman, "Molecular computation of solutions to combinatorial problems," *Science,* 266: 1021–1024, 1994.
9. B. Hille, *Ionic channels of excitable membranes,* 2d ed., Sunderland, Mass.: Sinauer Associates, 1992.

CHAPTER 5
MOTION IN SOLUTION AND CHEMICAL REACTIONS

We saw in the first three chapters that the number of electrons and holes inside a small semiconductor volume can change because of motion induced by an electric field (*drift*) and a concentration gradient (*diffusion*) and because of appearance/disappearance of particles due to unbalances in the generation-recombination rate (*mass action law*).

We have just seen in Chap. 4 that a small volume of a biological medium (e.g., a cell cytoplasm) typically contains a variety of molecules, including water dipoles, ions, charged macromolecules and neutral macromolecules. Drift, diffusion, and mass action law are again at work and the rules of classical physics will again be utilized to give an approximate picture of what is going on, without making use of quantum physics calculations. The reader is warned that molecules with quite different sizes, charges, concentrations, and other chemical properties participate in the phenomena to be analyzed, making their quantitative description quite complex. Just to give a numerical example of this variety, the diffusion coefficient of a "huge" protein moving in the plane of a membrane can be 10 orders of magnitude smaller than the diffusion coefficient of an ion in solution, which is 3 orders of magnitude smaller than the diffusion coefficient of an electron inside a semiconductor. We will discuss the fact that a protein could present a pH-dependent charge and it could be surrounded by an atmosphere of counter-ions, which makes a precise quantitative description of its drift (named now *electrophoresis*) very difficult.

Finally, we will see that chemical reactions, even still based on the mass action law, can become very complex, especially in the case of autocatalytic biochemical events, and their corresponding differential equations cannot be linearized in the same straightforward way as we did in Chap. 2, when dealing with the time evolution of the minority carriers after a flash of light.

5.1 DIFFUSION IN SOLUTION

Let us consider aqueous diffusion in one dimension for a neutral species. In 1855 Fick described aqueous diffusion flux F_c as equal the product of concentration gradient and a diffusion coefficient D for the diffusing species with a concentration c:

$$F_c = -D \frac{\partial c}{\partial x} \qquad (5.1)$$

The change in time of the amount of molecules inside a very small volume will be then equal to the spatial difference between incoming flux and outcoming flux (see Chap. 3). If we still consider one-dimensional phenomena, then

$$\frac{\partial c}{\partial t} = -\frac{\partial F_c}{\partial x} = D\frac{\partial^2 c}{\partial x^2} \qquad (5.2)$$

F_c is usually a molar flux, with dimensions mole/cm^2-s and c is usually a concentration measured as mole/liter (M).

If we transform molar quantities into molecular ones, then Eq. (5.1) becomes the same equation already deduced in Chap. 3.

Solving Eq. (5.2) yields the solute concentration everywhere in the sample as a function of time, i.e., $c(x, t)$, given appropriate boundary conditions. In the study of the diffusion of macromolecules (when effects due to charge separation are negligible), there are two typical boundary conditions, which consider a sharp boundary in the solute concentration or a thin band of solute, respectively. In the first case, the boundary conditions are

$$c(x, 0) = c_0 \quad \text{if } x < 0 \qquad (5.3a)$$

and

$$c(x, 0) = 0 \quad \text{if } x > 0 \qquad (5.3b)$$

Then the solution of Eq. (5.2) is[1]

$$c(x, t) = \left(\frac{c_0}{2}\right)\left[1 - \left(\frac{2}{\sqrt{\pi}}\right)\int_0^{x/\sqrt{4Dt}} e^{-v^2} dv\right] \qquad \text{for } t > 0 \qquad (5.4)$$

where

$$v^2 = \frac{x^2}{4Dt} \qquad (5.5)$$

No analytical form exists for the integral present in Eq. (5.4), which is called the *probability integral*, and it is tabulated in many handbooks. The corresponding profile of the concentration and of its gradient $\partial c/\partial x$ are sketched in Fig. 5.1.

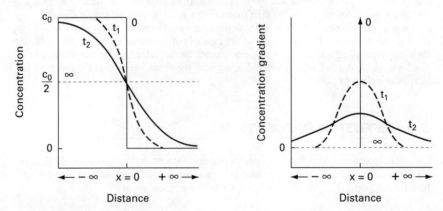

FIGURE 5.1 Solute concentration (left) and concentration gradient (right) at different times ($0, t_1, t_2, \infty$) in a free diffusion experiment starting from a steplike initial concentration.

The case of a thin band of solute implies the boundary condition

$$c(x, 0) = A_0 \delta(0) \tag{5.6a}$$

where A_0 is the total amount of solute (per unit area) present in the thin band and $\delta(0)$ is the *Dirac delta function*, whose value is infinite when its argument is zero and is zero everywhere else.

Conservation of matter implies

$$A_0 = \int_{-\infty}^{+\infty} c(x, t)\, dx \tag{5.6b}$$

This diffusion case is appropriately solved by using *Fourier transforms*. By applying the Fourier operator $\int_{-\infty}^{+\infty} e^{ikx} dx$ to both sides of Eq. (5.2), one obtains

$$\frac{d}{dt} \int_{-\infty}^{\infty} e^{ikx} c(x, t)\, dx = D \int_{-\infty}^{\infty} e^{ikx} \frac{d^2}{dx^2} c(x, t)\, dx \tag{5.7}$$

The final solution is[1]

$$c(x, t) = \frac{A_0 e^{-x^2/4Dt}}{\sqrt{4\pi Dt}} \tag{5.8}$$

As Eq. (5.8) shows, the shape of the initial thin zone becomes *gaussian* and its width broadens according to $\sqrt{4Dt}$ as diffusion proceeds. The concentration distribution given by Eqs. (5.4) and (5.8) are of relevance also for describing dopant diffusion in semiconductors (i.e., solid-state diffusion).

Before reinterpreting the diffusion process at a single-particle level, let us consider another kind of liquid motion, i.e., *convection*. Convection results from movement of the fluid either by forced means (e.g., *stirring*) or from *density gradients* or *temperature gradients*. The mathematical description of mass transport by convection needs the addition of a further term to Eq. (5.2), to take into account the motion of the solution as a whole. This additional term describes the concentration change associated with the replacement, due to flow, of solution in a particular region of space within a stationary coordinate system, by an amount of solution at a different concentration. As a consequence, Eq. (5.2) becomes

$$\frac{\partial c(x, t)}{\partial t} = D \frac{\partial^2 c(x, t)}{\partial x^2} - v_x \frac{\partial c(x, t)}{\partial x} \tag{5.9}$$

where v_x is the velocity of the solution. In many experimental situations convection plays the role of a disturbance to diffusion.

The just-considered equations deal with the behavior of concentrations of particles. Let us now slightly change the point of view and focus instead on *single particles*. Let us begin with a quite idealized one-dimensional problem, known as *one-dimensional random walk*.

Suppose that a particle moves on a straight line in a world where both time and space are discrete entities. At time $t = 0$, the particle is at the origin of its one-dimensional world. Then, every discrete unit of time, it moves a quantity l, with equal probability either to the left ($-l$) or to the right ($+l$). The reader will notice that in this way we introduce something similar to the mean free time and mean free path already considered in Chap. 3. Assume that after $(N-1)$ steps the particle has progressed a distance x_{N-1}. If it takes one more step, the distance from the origin will be either

$$x_N = x_{N-1} + l \tag{5.10}$$

or

$$x_N = x_{N-1} - l \tag{5.11}$$

Averaging Eqs. (5.10) and (5.11) one, of course, obtains zero displacement. Squaring them, one obtains

$$x_N^2 = x_{N-1}^2 + l^2 + 2x_{N-1}l \qquad (5.12)$$

and
$$x_N^2 = x_{N-1}^2 + l^2 - 2x_{N-1}l \qquad (5.13)$$

The average of these two situations is

$$x_N^2 = x_{N-1}^2 + l^2 \qquad (5.14)$$

This is the result for x_N^2 when the distance traveled after $(N-1)$ steps is exactly x_{N-1}. In general, however, we can only expect, for the value of the square of the distance at the $(N-1)$th step, an averaged value $\langle x_{N-1}^2 \rangle$, and thus we write

$$\langle x_N^2 \rangle = \langle x_{N-1}^2 \rangle + l^2 \qquad (5.15)$$

At the beginning of the random walk, i.e., after zero steps, the situation is given by

$$\langle x_0^2 \rangle = 0 \qquad (5.16)$$

After one step, it is

$$\langle x_1^2 \rangle = l^2 \qquad (5.17)$$

After two steps, we have

$$\langle x_2^2 \rangle = \langle x_1^2 \rangle + l^2 \qquad (5.18)$$

and, using Eq. (5.17), we get

$$\langle x_2^2 \rangle = 2l^2 \qquad (5.19)$$

Therefore, in general,

$$\langle x_N^2 \rangle = Nl^2 \qquad (5.20)$$

Equation (5.20) has been obtained for a one-dimensional random walk, but it can be shown to be valid also for three-dimensional random flights.

The mean square distance $\langle x^2 \rangle$ depends upon the time of travel as it follows. The number of steps, N, increases with time and is proportional to it; i.e.,

$$N = bt \qquad (5.21)$$

where b is a constant of proportionality. Hence,

$$\langle x^2 \rangle = btl^2 \qquad (5.22)$$

which may be written

$$\langle x^2 \rangle = \alpha t \qquad (5.23)$$

Let us link this result to a process of diffusion in order to relate the (microscopic) parameter α to the (macroscopic) coefficient of diffusion D.[2,3]

Let us now consider a situation in an aqueous solution where the concentration of the particles of interest is constant in the y-z plane but varies in the x direction. To analyze the diffusion of particles, assume unit area of a reference plane normal to the x direction. This reference plane will be called the *transit plane T*. There is a random walk of particles across T both from left to right and from right to left. On either side of the transit plane, we can consider two planes, L and R, which are parallel to T and placed at a distance $\sqrt{\langle x^2 \rangle}$ from it. In other words, the region under consideration has been split into left and right compartments in which the concentrations of particles are different and designated by c_L and c_R, respectively.

In a time t (seconds), a random-walking particle covers a mean square distance of $\langle x^2 \rangle$, or a mean distance of $\sqrt{\langle x^2 \rangle}$. Thus, by choosing the plane L to be at a distance $\sqrt{\langle x^2 \rangle}$ from T, we ensure that (on the average) all the particles in the left compartment will cross the transit plane in a time t provided they are moving in the left-to-right direction. The number of moles of particles in the left compartment is equal to the volume with unit area $\sqrt{\langle x^2 \rangle}$ of this compartment times the concentration c_L of particles. It follows that the number of moles of particles which make the L to T crossing in t is $\sqrt{\langle x^2 \rangle} c_L$ times the fraction of particles making left-to-right movements. Since the particles are random-walking, right-to-left movements are as likely as left-to-right movements; i.e., only half the particles in the left compartment are moving away toward the right compartment. Thus, in t the number of moles of particles making the L to T crossing is $\frac{1}{2}(\sqrt{\langle x^2 \rangle})c_L$, and, therefore, the number of moles of particles making the L to T crossing in 1s is $\frac{1}{2}(\sqrt{\langle x^2 \rangle}c_L)/t$. Similarly, the number of moles of particles making the R to T crossing in 1s is $\frac{1}{2}(\sqrt{\langle x^2 \rangle}c_R)/t$.

In conclusion, the diffusion flux of particles across the transit plane, i.e., the net number of moles of particles crossing unit area of the transit plane per second from left to right, is given by

$$F_c = \frac{1}{2} \frac{\sqrt{\langle x^2 \rangle}}{t}(c_L - c_R) \tag{5.24}$$

Now, the concentration gradient dc/dx in the left-to-right direction can be written

$$\frac{dc}{dx} = -\frac{c_L - c_R}{\sqrt{\langle x^2 \rangle}} \tag{5.25}$$

or

$$c_L - c_R = -\sqrt{\langle x^2 \rangle} \frac{dc}{dx} \tag{5.26}$$

Introducing Eq. (5.26) into Eq. (5.24) results in

$$F_c = -\frac{1}{2} \frac{\langle x^2 \rangle}{t} \frac{dc}{dx} \tag{5.27}$$

and, by equating the coefficients of this equation with that of Fick's first law [Eq. (5.1)], we obtain

$$\frac{\langle x^2 \rangle}{2t} = D \tag{5.28}$$

or

$$\langle x^2 \rangle = 2Dt \tag{5.29}$$

This is the *Einstein-Smoluchowski equation;* it provides a bridge between the microscopic view of random-walking particles and the coefficient D of the macroscopic *Fick law*.

We will see in the next section that Eq. (5.29) can be considered an appropriate approximation of a more general expression.

5.2 BROWNIAN MOTION

Consider the motion of a macroscopic particle inside a fluid described by

$$m\frac{dv}{dt} = -fv \tag{5.30}$$

where fv is the *frictional force* acting on the particle and f the friction coefficient. The velocity of the particle will approach zero, according to

$$v = v_0 e^{-(f/m)t} \tag{5.31}$$

where v_0 is the velocity at the beginning of the observation.

In the simplest view, frictional forces arise in a fluid because of attractions among the particles of the fluid. In order to push a solid object through the fluid, it is necessary to also move solvent particles with respect to others. From a slightly different perspective, we can imagine a very large number of particles colliding with the macroscopic object while it moves. There are so many colliding particles per unit of time that they can be appropriately described by a single macroscopic parameter f. Now, what happens if the object becomes smaller and smaller? The collisions per unit time should become fewer and fewer. Are we still allowed to think about a "simple" macroscopic friction force? When the dimension reaches the micrometer range, the answer is no. This fact was experimentally observed (but not explained) by the English botanist R. Brown in 1827, who reported the incessant and irregular motion of intracellular granules within pollen cells, which could not be explained by the simple but all too regular mechanism of convection.

Such an observation opened the road to the description of a class of particles, of great relevance for biology, large enough to be viewed one by one with an optical microscope but small enough to be still moved around by thermally driven fluctuations. It was Einstein's ingenuity in 1905 which suggested a treatment of phenomena of this type by combining determinism and randomness. As later proposed by Langevin, the phenomenon can be studied by considering the equation, named the *Langevin equation*:

$$m\frac{dv}{dt} = -fv + A'(t) \tag{5.32}$$

where $A'(t)$ is a *fluctuating force not to be confused with an external deterministic force*. By dividing both sides of Eq. (5.32) by the mass m, we rewrite Eq. (5.32) as

$$\frac{dv}{dt} = -\beta v + A(t) \tag{5.33}$$

The fluctuating force $A(t)$ (normalized to the particle mass m) represents the resultant of the molecular collisions, the usual frictional term having been "extracted." Physically, the "friction" derives from the fact that the momentum transferred from the molecules of the fluid to the particle is, on the average, greater in the direction opposing the motion than in the direction of the motion. This is why there is a progressive slowing-down of a macroscopic body in a fluid, with the consequent dissipation of energy.

In the case of Brownian motion, we have essentially the same phenomenon (molecule-particle collisions), but the appearance is completely different. This is because the Brownian particle is so small and light that it "feels" changes in direction, from the average value of the resultant of the forces, so that it undergoes an erratic motion.

It is usual to define the level of description given by the Langevin equation as a *meso-

scopic one, i.e., a level which is not either microscopic (nanometers downward) or macroscopic (millimeters upward).

The reader should note that the fluctuating force $A(t)$ is not considered as a given function of time, but it is defined only in statistical terms. This introduces a corresponding change in the meaning of the solution of Eq. (5.33), that is, to find the probability distribution of the variable in question, and no longer its law of motion.

Let us now define the statistical properties of the fluctuating force. From experimental observations and from considerations of symmetry, it must have an average value of zero, i.e.,

$$\langle A(t) \rangle = 0 \tag{5.34}$$

We can consider the averages as ensemble averages, calculated over a population of Brownian particles which are not subject to mutual interaction. Observations also demonstrate that the direction and intensity of a change in the motion of a Brownian particle appear to be independent of the characteristics of the preceding change. Let us suppose that the fluctuating force has an infinitely short memory, so that

$$\langle A(t)A(t+s) \rangle = \sigma^2 \delta(s) \tag{5.35}$$

where $\delta(s)$ is the well-known *Dirac delta distribution* already introduced in the previous section.

Equation (5.35) implies

$$\sigma^2 = \langle A^2(t) \rangle \tag{5.36}$$

where by considering σ as a constant we have assumed that the mean square of the fluctuating force does not vary with time.

Physically, *infinitely short* means just longer than the average time between one collision and the next, which could be typically in the order of 10^{-11} s. (The reader can compare this description to that of the motion of an electron inside a silicon lattice given in Chap. 3).

The hypothesis of a δ-correlated fluctuating force, also expressed by the term *white noise,* introduces a "pathological" element into the mathematical description given by the Langevin equation, where $A(t)$ should be considered a white noise with a *gaussian distribution.*

We can solve Eq. (5.33) by a method originally proposed by Ornstein and Uhlenbeck in 1930.[4,5] Formally integrating Eq. (5.33), we obtain

$$v(t) = v_0 e^{-\beta t} + e^{-\beta t} \int_0^t e^{\beta s} A(s)\, ds \tag{5.37}$$

From this expression we can directly obtain the first moment of the distribution

$$\langle v(t) \rangle = v_0 e^{-\beta t} \tag{5.38}$$

Squaring both sides of Eq. (5.37) to calculate the second moment, we obtain

$$v^2(t) = v_0^2 e^{-2\beta t} + e^{-2\beta t} \int_0^t ds \int_0^t e^{\beta(s+q)} A(s)A(q)\, dq + 2v_0 e^{-2\beta t} \int_0^t e^{\beta t} A(s)\, ds \tag{5.39}$$

Averaging, and taking into account Eq. (5.34), we have

$$\langle v^2(t) \rangle = v_0^2 e^{-2\beta t} + \frac{\sigma^2}{2\beta}(1 - e^{-2\beta t}) \tag{5.40}$$

By knowing the statistical properties of the fluctuating force, it would be possible to determine moments of a higher order of the distribution of v. This is not, however, necessary: from Eq. (5.37), $v(t)$ can be seen to be a linear combination of gaussian terms added to a deterministic part. Thus v also has a gaussian distribution, and a knowledge of the first two moments is sufficient for its complete determination. The transition probability density is

$$p(v, t|v_0) = \left[\frac{\beta}{\pi\sigma^2(1-e^{-2\beta t})}\right]^{1/2} \exp\left[-\frac{\beta(v-v_0 e^{-\beta t})^2}{\sigma^2(1-e^{-2\beta t})}\right] \qquad (5.41)$$

The transition density given by Eq. (5.41) has a gaussian form with the average value

$$\langle v(t) \rangle = v_0 e^{-\beta t} \qquad (5.42)$$

and with variance

$$\langle (v(t) - \langle v(t) \rangle)^2 \rangle = \frac{\sigma^2}{2\beta}(1 - e^{-2\beta t}) \qquad (5.43)$$

Physically, Eq. (5.42) implies that a population of Brownian particles, all having an initial velocity v_0, tend to evolve in such a way that the average velocity exponentially approaches zero, while the variance increases and tends asymptotically toward the value $\sigma^2/2\beta$.

The distribution is modified by random collisions between the particles and the fluid molecules; we can see that the asymptotic value of the variance is largely determined by the ratio between the fluctuating force intensity σ^2 and the intensity of the dissipative term.

By imposing that the distribution given by Eq. (5.41) asymptotically tends toward the *Maxwell distribution*,[4] we can identify σ^2/β as $2kT/m$.

It is also possible to obtain the probability distribution for the variable x. In fact, by integrating Eq. (5.37), it follows that

$$x(t) = x_0 + \int_0^t v(s)\,ds = x_0 + \frac{v_0}{\beta}(1 - e^{-\beta t}) + \int_0^t e^{-\beta s}\,ds \int_0^s e^{\beta q} A(q)\,dq \qquad (5.44)$$

from which

$$\langle x(t) \rangle = x_0 + \frac{v_0}{\beta}(1 - e^{-\beta t}) \qquad (5.45)$$

In order to evaluate the variance $\langle x^2 \rangle - \langle x \rangle^2$, we can rearrange Eq. (5.44), integrating by parts and taking into account Eq. (5.45) for the average value. We thus obtain

$$x(t) = \langle x(t) \rangle - \frac{e^{-\beta t}}{\beta}\int_0^t e^{\beta s} A(s)\,ds + \frac{1}{\beta}\int_0^t A(s)\,ds \qquad (5.46)$$

Squaring and averaging, we obtain

$$\langle x^2(t) \rangle - \langle x(t) \rangle^2 = \frac{\sigma^2}{2\beta^3}(-3 + 2\beta t - e^{-2\beta t} + 4e^{-\beta t}) \qquad (5.47)$$

From Eqs. (5.46) and (5.47), a gaussian distribution for the displacements x can be deduced.

If we further assume

$$\langle x(t)\rangle = 0 \tag{5.48}$$

and

$$t \gg \frac{1}{\beta} \tag{5.49}$$

then Eq. (5.47) can be approximated by

$$\langle x^2(t)\rangle \simeq \frac{\sigma^2}{\beta^2} t \tag{5.50}$$

Equation (5.50) is formally coincident with Eq. (5.29) if we assume that $2D = \sigma^2/\beta^2$. It can be also shown[5] that

$$D = \frac{kT}{m\beta} = \frac{kT}{f} \tag{5.51}$$

The Einstein-Smoluchowsky relationship

$$(\langle x^2\rangle)^{1/2} = (2Dt)^{1/2} \tag{5.52}$$

can be utilized in molecular cellular biology to roughly estimate the diffusion coefficient of mesoscopic "particles" such as subcellular structures or even small living cells (e.g., red blood cells).

Conversely, if the diffusion coefficient of a molecular species is known, then approximate answers to questions such as "How long will it take for this molecule to cross the cell cytoplasm?" are quickly suggested by Eq. (5.52).

In closing this section, it is worth mentioning that another approach can be utilized, which directly deals with probability densities instead of single-particle motion. This approach leads to the so-called *Fokker-Planck equation*,[6] which has, in simple cases, the same formal structure of the diffusion equation

$$\frac{\partial p}{\partial t} = \gamma \frac{\partial^2 p}{\partial x^2} \tag{5.53}$$

where p is a probability density and *not* a concentration. Further discussion of this equation, including its links with the integro-differential Boltzmann equation considered in Chap. 3, is out of the aim of the book.

5.3 ELECTROPHORESIS

In a hypothetical experiment, suppose an external force F is applied to a molecule with mass m, in an aqueous solution. We can write

$$m\frac{dv}{dt} = -fv + F \tag{5.54}$$

where fv is the frictional force already considered in the previous section. The solution of the linear differential Eq. (5.54) is

$$v(t) = \frac{F}{f} + e^{-(f/m)t}\left[v_0 - \frac{F}{f}\right] \tag{5.55}$$

Equation (5.55) indicates that the velocity decays exponentially from the initial value v_0 to a constant linear value $v(\infty)$, which is proportional to the applied force.

The rate of approach to a constant velocity depends on the ratio f/m. For a typical macromolecule in a water solution, this ratio is in the order of 10^{-12}s^{-1}. Suppose that the molecule has a charge Q and that we apply a uniform and steady electric field. Then

$$F = QE \qquad (5.56)$$

$$v(t) = \frac{QE}{f} + e^{-(f/m)t}\left[v_0 - \frac{QE}{f}\right] \qquad (5.57)$$

and, for times greater than 10^{-12} s,

$$v = \frac{QE}{f} \qquad (5.58)$$

This is exactly the same linear relationship found in Chap. 3 [see Eqs. (3.3b) and (3.4b)], where the mobility of electrons and holes was introduced, i.e.,

$$v = \mu E \qquad (5.59)$$

The procedure to deduce Eqs. (3.3b) and (3.4b) has been quite different, because it was chosen not to introduce a *frictional coefficient f* for the silicon lattice.

For a spherical macromolecule of radius a and with a charge equal to zq, it can be shown that[1]

$$v = \frac{zq}{6\pi\eta a} E \qquad (5.60)$$

where η is the *viscosity* of the solution. On the basis of Eqs. (5.55) through (5.60), we could conclude that, if the electical field originates from parallel plates and if the molecular shape can be approximated to a sphere, then the molecule will travel in a straight line with a speed which could be completely predicted by Eq. (5.60). Unfortunately, *this is not the case*. First of all, diffusion phenomena should also be taken into account. This can be done either by adding a fluctuating force $A'(t)$ to the equation of motion, along the lines indicated in the previous section, that is,

$$m\frac{dv}{dt} = -fv + qE + A'(t) \qquad (5.61)$$

or by considering fluxes of matter and balancing drift and diffusion of concentrations of molecules.

This latter approach, which has been already used in Chap. 3, gives

$$F_c = -D\frac{dc}{dx} + uzqcE \qquad (5.62)$$

and

$$\frac{\partial c}{\partial t} = D\frac{\partial^2 c}{\partial x^2} - uzq\frac{\partial}{\partial x}(cE) \qquad (5.63)$$

where u is the *apparent mobility*. To transform Eq. (5.62) into Eqs. (3.46) and (3.47), the reader should

1. Set $\mu = zqu$.
2. Multiply both sides by the charge, thereby transforming fluxes into current densities.
3. Transform molar concentrations (mole/liter) into particle concentrations (particles/cm^3).

The reader is also invited to verify that, by imposing $F_c = 0$, an equilibrium Boltzmann distribution is obtained.

In practice, when dealing with electrophoretic experiments (e.g., separation of proteins by an applied electric field), things can be organized in such a way that diffusion is not of great relevance. A more serious difficulty arises over what to call the net charge on a macromolecule.

In any aqueous solution, there are counter-ions (i.e., ions carrying a charge opposite in sign to the macromolecule charge). Some of them may be associated rather tightly with the macromolecule, others more loosely. Let us consider what will happen if an electric field is applied to a negatively charged macromolecule with only enough cations in the solution to neutralize this charged (i.e., no added electrolyte). The macromolecule will tend to migrate in one direction, the counter-ions in the opposite direction. The electrostatic energy involved in any net charge separation is very large, and therefore little net overall motion will actually take place. If there is any net mobility at all, it will be an average of the mobilities of macromolecule and counter-ions. On the other hand, in an electrolyte solution with a high ionic strength, the electrolyte forms an ion atmosphere around the macromolecule; there are two ways to view its effect. The ion atmosphere will effectively reduce the net charge on the macromolecule, because oppositely charged ions will tend to be attracted closely to the macromolecule. Alternatively, we can say that the ion atmosphere sets up an electric field that is felt by the macromolecule and that must be considered along with any applied field in computing its mobility.

It is usual to consider this atmosphere as a continuous distribution and to ignore individual charges. The problem can then be tackled by solving the Poisson-Boltzmann equation. This will not be done here, because similar and more tractable problems will be considered in Chap. 7, where we will deal with electrified interfaces as an introduction to *biosensors*.

Despite the aforementioned complications, *electrophoresis* is a powerful and practical tool in the analysis and separation of proteins and nucleic acids, so long as one does not demand quantitative structural data from it. The most common use of electrophoresis is the separation or qualitative analysis of mixtures, typically proteins, based on their different isoelectric points (see Chap. 4) and sizes. As discussed in Chap. 4, the charge on a protein is a function of pH, and, given a pH, proteins in a mixture could move either toward the cathode if the chosen pH is below the isoelectric point, or toward the anode if it is above, or they will not drift at all if the pH is equal to the isoelectric point.

Historically, electrophoresis methods started around 1933 with *moving-boundary* (or *free*) electrophoresis, where a broad zone of macromolecules in solution was subjected to an electric field.

Because of various artifacts, convection, and experimental difficulties, free electrophoresis has been supplanted by *zonal electrophoresis* where a solid support permeated with buffer is used. Some supports interact only weakly or nonspecifically with the macromolecular solutes; these include paper, thin-layer cellulose, and cellulose acetate. Other materials retard the motion of certain molecules relative to one another; these include polyacrylamide and agarose gels. The use of a solid support makes a quantitative analysis of mobility very difficult, the molecules being forced to trace tortuous paths through the support medium. On the other hand, the separation capability does increase with the complexity of the support structure. To fix ideas, the separation of blood plasma proteins increases from about 5 classes of proteins with cellulose acetate to about 20 with polyacrylamide gel.

The separation capability is further improved with the ingenious method known as *isoelectric focusing,* which makes use of *ampholytes.*[1] Ampholytes are molecules with positive and negative charges—for example, polymers containing numerous amino and carboxyl groups. A mixture of ampholytes with a wide range of isoelectric points is allowed to distribute in a column under the influence of an electric field. This procedure establishes a pH gradient in which each particular ampholyte comes to rest at a position near its isoelectric point. The pH gradient is created on a support medium to block convection. A small amount of protein is added to the system. Each protein migrates until it reaches the pH of its isoelectric point. It will remain there as a sharp band. To analyze the results of the experiment, we can scan the pH gradient for protein absorbance, can stain it, or can cut it up into slices and analyze each of these for pH and protein content, enzymatic activity, or radioactivity. About 40 plasma proteins can be resolved with the method of isoelectric focusing.

So far, electrophoretic techniques have been considered as a separation method based on the pH-dependent charge on proteins (and nucleic acids). As discussed in Chap. 4, there is not any correlation (or, at the best, a very weak one) between *charge* and *number* of amino acids, i.e., molecular weights. Information about molecular weights can be obtained by electrophoresis of proteins in *sodium dodecylsulfate* (SDS). Here, the basic idea is to try to convert all proteins into similar structures that differ only in molecular weight. SDS is an effective protein denaturant and it binds to all proteins qualitatively in the same way, at about 1.4 g per gram of amino acid. There is one negative charge on each SDS molecule. The resulting charge density due to all the SDS in the SDS-protein complex more or less overwhelms variations in the charges of different protein molecules. Hydrodynamically, SDS-protein complexes appear to be prolate ellipsoids or rods with a constant diameter of about 1.8 nm and a length that is a linear function of the molecular weight.

Polyacrylamide gels generally are used as the support medium in SDS electrophoresis. These gels can be prepared at various ratios of solid gel to fluid solution. It is a general observation that the higher the gel concentration, the lower the apparent electrophoretic mobility. Surprisingly enough, the mobilities of a set of similar proteins, extrapolated to zero gel concentration, are the same. The data can be fit to an equation of the form

$$\ln u(c) = -k_x c + \ln u(0) \tag{5.64}$$

where $u(c)$ is the apparent mobility at a gel concentration c, k_x is a constant that depends on the extent of cross-linking of the gel and on the shapes and molecular weights of the particular proteins, and $u(0)$ is a constant for a set of similar kinds of molecules.

The mobility at zero gel concentration should reflect the true electrophoretic mobility of the SDS-protein complexes, not obscured by any interactions with the gel. The fact that $u(0)$ is a constant can be understood as follows: If the protein-SDS complexes have a constant weight percentage of SDS, and if protein charge can be neglected, the net charge zq will be proportional to the molecular weight, and thus proportional to the length l because the complexes are rods. The frictional coefficient also is roughly linear in the molecular weight.[1] The hydrated volume of a rod will be directly proportional to its length. It can be shown[1] that $u(0)$ will be independent of the molecular weight. So it appears that, in SDS electrophoresis, the electric field is simply causing a constant drift to all the molecules, and any separation based on molecular weight is originating from

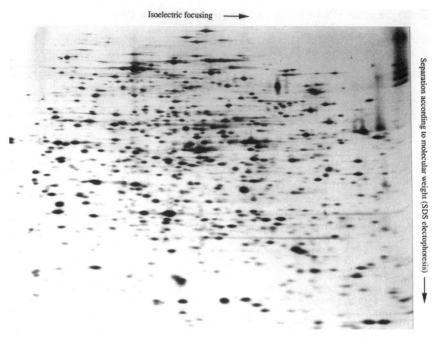

FIGURE 5.2 Two-dimensional electrophoresis. IF and SDS mean isoelectric focusing and sodium dodecylsulfate, respectively. (*Courtesy of P. O'Farrell, J. Biol. Chem. 250: 4007, 1975.*)

the specifics of interaction of different-sized SDS-protein rods with the supporting gel matrix.

The separation power of SDS electrophoresis is based completely on molecular weight, and that of isoelectric focusing is based completely on charge. Then, if the two techniques are combined into a single separation scheme, its resolution will be quite impressive. This can be done by *two-dimensional electrophoresis,* in which one dimension is SDS and the other is isoelectric focusing. The resolution power of this technique being roughly the product of the two independent monodimensional methods, it can easily reach 1000 proteins (see Fig. 5.2), thus coming near to the total number of kinds of proteins present in a cell. Automated digital image analysis techniques help in classifying the thousands of spots generated by 2-D electrophoresis.

5.4 CHEMICAL REACTIONS

The rate at which any chemical species reacts to form another one is proportional to its concentration. In a very simple way this statement defines the *law of mass action.* It allows us to readily translate reaction schemes into differential equations. We already utilized it to write down the recombination term in the generation-recombination processes governing electrons and holes. We will see in the following how to apply it to simple reaction schemes, to enzyme catalysis, and finally to complex biochemical reactions, resulting in sets of nonlinear differential equations.

5.4.1 Kinetic Rate Constants and Equilibrium Constants

For the sake of simplicity, let us use in this section italic capital letters $A, B, C, \ldots$ for indicating both chemical species and their concentration. Then the reaction scheme

$$A \xrightarrow{k_1} B \tag{5.65}$$

means that the chemical species A transforms into the chemical species B with a kinetic rate constant k_1. By the *law of mass action*, the rate of disappearance of A will be proportional to A itself. The rate constant k_1 is the proportionality factor. Therefore, the concentration of A obeys the following differential equation

$$\frac{dA}{dt} = -k_1 A \tag{5.66}$$

By inspection, we immediately see that the dimension of k_1 is (time)$^{-1}$. The chemical species B is generated while A disappears. As indicated by Eq. (5.65), the rate of appearance of B is proportional to A; i.e.,

$$\frac{dB}{dt} = k_1 A \tag{5.67}$$

Taken together, Eqs. (5.66) and (5.67) indicate that the total amount of molecules ($A + B$) is constant; i.e.,

$$\frac{dA}{dt} + \frac{dB}{dt} = \frac{d}{dt}(A + B) = 0 \tag{5.68}$$

or
$$A + B = A(t=0) + B(t=0) = A_0 + B_0 \tag{5.69}$$

where the subscript 0 means the value of the involved quantity at time equal to zero.
By solving Eq. (5.66), we obtain

$$A = A_0 e^{-k_1 t} \tag{5.70}$$

We can transform the two variables A, B into a single one by observing that at any time the amount of A reacted must be equal to the amount of B formed:

$$A(t) = A_0 - x(t) \tag{5.71}$$

$$B(t) = B_0 + x(t) \tag{5.72}$$

$$x(t=0) = x_0 = 0 \tag{5.73}$$

where x plays the role of a reaction variable and it satisfies

$$\frac{dx}{dt} = -k_1 x + k_1 A_0 \tag{5.74}$$

which gives

$$x = x_0 e^{-k_1 t} + e^{-k_1 t} \int_0^t e^{k_1 s} k_1 A_0 \, ds \tag{5.75}$$

or
$$x = A_0 (1 - e^{-k_1 t}) \tag{5.76}$$

which of course is the same result as Eq. (5.70).

Equation (5.65) characterizes a *first-order* irreversible reaction. *Second-order* irreversible reactions obey the scheme

$$A + B \xrightarrow{k_2} C \tag{5.77}$$

which means

$$\frac{dA}{dt} = -k_2 AB \tag{5.78}$$

The dependence of dA/dt on the product of reactants can be made intuitive if we interpret the reaction of A with B as *collisions* of particles of type A with particles of type B. The reader should notice that now the dimensions of k_2 are (concentration-time)$^{-1}$.

By introducing the reaction variable x, we find

$$A = A_0 - x \tag{5.79}$$

$$B = B_0 - x \tag{5.80}$$

and after separation of variables, we obtain

$$\frac{dx}{(A_0 - x)(B_0 - x)} = k_2 \, dt \tag{5.81}$$

This can be integrated using the method of partial fractions to give the result

$$\frac{1}{A_0 - B_0} \ln\left(\frac{B_0 A}{A_0 B}\right) = k_2 t \tag{5.82}$$

under the condition $A_0 \neq B_0$. If we put $B = A$ in Eq. (5.77), we obtain

$$A + A \xrightarrow{k_2} P \tag{5.83}$$

This reaction is the simplest example of polymerization: two *monomers* react to form a *dimer*.

After separation of variables, we obtain

$$\frac{dx}{(A_0 - x)^2} = k_2 \, dt \tag{5.84}$$

which gives

$$\frac{1}{A_0 - x} - \frac{1}{A_0} = k_2 t \tag{5.85}$$

How long does it take for the reactants to react? It is customary to define the half-time $t_{1/2}$ of a reaction as the time required to halve the initial concentration of reactants (i.e., A_0). This definition gives

$$\frac{2}{A_0} - \frac{1}{A_0} = k_2 t_{1/2} \tag{5.86}$$

and

$$t_{1/2} = \frac{1}{k_2 A_0} \tag{5.87}$$

All reactions approach equilibrium.

Up to now we have considered irreversible kinetic processes. Let us consider now a *first-order reversible reaction*.

$$A \xrightleftharpoons[k_{-1}]{k_1} B \qquad (5.88)$$

The law of mass action gives

$$\frac{dA}{dt} = -k_1 A + k_{-1} B \qquad (5.89a)$$

$$\frac{dB}{dt} = k_1 A - k_{-1} B \qquad (5.89b)$$

At equilibrium

$$\frac{dA}{dt} = \frac{dB}{dt} = 0 \qquad (5.90)$$

and

$$A = A_{eq} \qquad (5.91a)$$

$$B = B_{eq} \qquad (5.91b)$$

where the subscript eq indicates the equilibrium value. At any time we can also write

$$A = A_{eq} + \Delta A \qquad (5.92)$$

$$B = B_{eq} + \Delta B \qquad (5.93)$$

where ΔA and ΔB are both zero at equilibrium.

At any time it is also true that

$$A + B = \text{constant} = A_{eq} + B_{eq} \qquad (5.94)$$

Therefore

$$\Delta A = -\Delta B \qquad (5.95)$$

By using Eqs. (5.92) and (5.95), we can write

$$\frac{dA}{dt} = -k_1 A + k_{-1} B \qquad (5.96)$$

in the following form:

$$\frac{d(\Delta A)}{dt} = -k_1(A_{eq} + \Delta A) + k_{-1}(B_{eq} - \Delta A) \qquad (5.97)$$

At equilibrium, Eq. (5.97) becomes

$$0 = -k_1 A_{eq} + k_{-1} B_{eq} \qquad (5.98)$$

Therefore we can write Eq. (5.97) as

$$\frac{d(\Delta A)}{dt} = -(k_1 + k_{-1})\Delta A \qquad (5.99)$$

Finally we have

$$(A - A_{eq}) = (A_0 - A_{eq}) e^{-(k_1 + k_{-1})t} \qquad (5.100)$$

At equilibrium, from Eq. (5.98), we can write

$$k_1 A_{eq} = k_{-1} B_{eq} \qquad (5.101)$$

or
$$\frac{B_{eq}}{A_{eq}} = \frac{k_1}{k_{-1}} = K \qquad (5.102)$$

where K is the *thermodynamic equilibrium constant*.

The equilibrium values of the chemical reactants should satisfy the equilibrium Boltzmann distribution. Therefore

$$B_{eq} = A_{eq} e^{-\Delta G/kT} \qquad (5.103)$$

where ΔG is the free energy[7] associated with the reaction, and

$$K = e^{-\Delta G/kT} \qquad (5.104)$$

The fractional equilibrium value of species A will be given by the sigmoidal expression

$$\frac{A_{eq}}{A_{eq} + B_{eq}} = \frac{1}{1 + e^{-\Delta G/kT}} \qquad (5.105)$$

Let us close this section with a general observation. In describing kinetic reaction schemes we employed the apparatus of differential equations. This may seem to be an obvious choice, but it is not. Chemical reactions deal with *discrete* entities. Discrete systems are described by *stochastic equations* based on *probability theory*. In the case of large, "well-behaved" populations of reactants, the deterministic result obtained by solving differential equations is perfectly appropriate, but this cannot hold when dealing with a limited number of particles, and this situation can happen inside a living cell.

Just to introduce the stochastic approach, let us consider again the simple reaction

$$A \xrightarrow{k} B \qquad (5.106)$$

Let the random variable $X(t)$ be the number of particles in the system at time t. A stochastic model is then defined by the following assumptions, which represent a special case of a birth-death process.[8]

1. The probability of transition $(x + 1) \to x$ in the interval $(t, t + \Delta t)$ is $kx\,\Delta t$.
2. The probability of transition $(x + 1) \to (x - j), j > 0$, in the interval $(t, t + \Delta t)$ is zero.
3. The reverse reaction occurs with probability zero.

Then, a detailed balance gives

$$p_x(t + \Delta t) = k(x + 1)\Delta t\, p_{x+1}(t) + (1 - kx\, \Delta t) p_x(t) \qquad (5.107)$$

where
$$p_x(t) = \text{prob}\{X(t) = x\} \qquad (5.108)$$

and the second term in the right side of Eq. (5.107) takes into account the case of no transition during the considered interval.

By taking the limit $\Delta t \to 0$, we get the differential-difference equation

$$\frac{dp_x}{dt} = k(x + 1) p_{x+1}(t) - kx p_x(t) \qquad (5.109)$$

Equation (5.109) represents the starting point for studying chemical kinetics by means of a stochastic approach. It can be shown[8] that the expectation value of $x(t)$ coincides with the deterministic solution (Eq. 5.70):

$$E\{x(t)\} = x_0 e^{-kt} \tag{5.110}$$

5.4.2 Enzyme Kinetics

Enzymes are a class of proteins. Every living cell is able to make enzymes of many different species. Each enzyme has a unique shape and binds a particular set of other molecules (called *substrates*) in such a way to greatly speed up (up to 10^{14} times!) a particular chemical reaction without having any effect on the equilibrium constant K of the reaction, thus acting as a *catalyst*. The literature on enzymes is enormous. Just to give a few examples, both *condensation* and *hydrolysis* reactions are catalyzed by enzymes. Condensation (or *dehydration*) is the general mechanism to form biological polymers (e.g., peptide bonding in proteins), according to the reaction scheme

$$A\text{—}H + B\text{—}OH \longrightarrow A\text{—}B + H_2O \tag{5.111}$$

This biosynthetic reaction requires the input of chemical energy, which is achieved by coupling it to an energetically favorable hydrolysis [i.e., the reverse of reaction Eq. (5.111)].[9]

Oxidation of glucose to gluconolactone is another relevant example of a reaction catalyzed by the enzyme glucose oxidase. This example will be further considered in Chap. 10, which deals with biosensors.

Like all other catalysts, enzyme molecules themselves are not changed after participating in a reaction and therefore can function over and over again. According to the theory of Michaelis-Menten (1913), the enzyme (concentration E) first reacts reversibly with the substrate (concentration S) to form a complex (concentration C). The reader should note that here we use C instead of $[ES]$ (as in Chaps. 10 and 12) for the sake of simplicity. Then, the complex breaks apart into an altered substrate or *product* (concentration P) and the original enzyme. The last reaction is usually assumed to be irreversible, in which case the overall kinetic scheme is

$$E + S \underset{k_{-1}}{\overset{k_{+1}}{\rightleftarrows}} C \xrightarrow{k_{+2}} E + P \tag{5.112}$$

The law of mass action for the concentrations $E(t)$, $S(t)$, $C(t)$, and $P(t)$ takes the forms

$$\frac{dE}{dt} = -k_{+1}ES + k_{-1}C + k_{+2}C \tag{5.113a}$$

$$\frac{dS}{dt} = -k_{+1}ES + k_{-1}C \tag{5.113b}$$

$$\frac{dC}{dt} = +k_{+1}ES - k_{-1}C - k_{+2}C \tag{5.113c}$$

$$\frac{dP}{dt} = k_{+2}C \tag{5.113d}$$

Standard initial conditions, which apply to the usual investigations of enzymatically controlled reactions, are

$$E(t=0) = E_0 \quad S(t=0) = S_0 \quad C(t=0) = 0 \quad P(t=0) = 0 \tag{5.114}$$

Adding together Eqs. (5.113a) and (5.113c) yields

$$\frac{d}{dt}(E + C) = 0 \qquad (5.115)$$

And, by using the initial conditions (5.114), we find

$$E(t) + C(t) = E_0 \qquad (5.116)$$

In laboratory experiments, it is usually the case that, at the beginning of the experiment, many substrate molecules are present for each enzyme molecule.[10] Under these circumstances we expect that after an initial short transient period there will be a balance between the formation of complex by the union of enzyme and substrate and the breaking apart of complex (either to enzyme and substrate, or to enzyme and product). Because there are so many substrate molecules, this balance will be reached before there is appreciable transformation of substrate into product. Therefore, the calculation of product formation can be carried out under the assumption that $dC/dt = 0$, or, from Eq. (5.113c),

$$k_{+1}ES = (k_{-1} + k_{+2})C \qquad (5.117)$$

This equation is said to result from a *quasi-* or *pseudo-steady-state* hypothesis.

Upon substitution of Eqs. (5.116) and (5.117) into Eq. (5.112b), we obtain the following equation for S:

$$\frac{dS}{dt} = -\frac{k_{+1}k_{+2}E_0 S}{k_{+1}S + k_{-1} + k_{+2}} \qquad (5.118)$$

Equation (5.118) can be solved, subject to the initial condition $S(0) = S_0$, to obtain the concentration S at any time t.

The common value of $|dS/dt|$ and dP/dt is called the *velocity of the reaction* and is denoted by v.

By taking into account the fact that the velocity is usually measured before S decreases appreciably from its initial value S_0, we can write

$$v = \frac{k_{+2}E_0 S_0}{k_m + S_0} \qquad (5.119)$$

where

$$k_m = \frac{k_{-1} + k_{+2}}{k_{+1}} \qquad (5.120)$$

is the *Michaelis constant*.

The maximum value of v, v_{max}, is approached for very large value of S_0. Therefore,

$$v_{max} = \lim_{S_0 \to \infty} v = k_{+2} E_0 \qquad (5.121)$$

In terms of v_{max}, Eq. (5.119) can be written as

$$v = \frac{v_{max} S_0}{k_m + S_0} \qquad (5.122)$$

By inspection, we can see that the Michaelis constant k_m corresponds to the initial substrate concentration at which the reaction velocity reaches its half-maximal value.

Let us conclude this section by discussing the limits of application of the quasi-steady-state approximation [Eq. (5.117)], which is the basis of the subsequent results [Eqs.

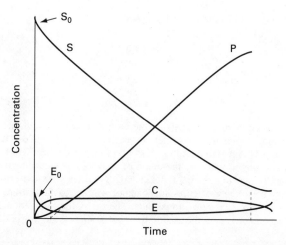

FIGURE 5.3 Progress curve for a catalyzed reaction where the initial reactant (substrate) concentration, S_0, is significantly greater than the initial enzyme concentration, E_0. As the ratio of S_0/E_0 increases, the steady-state region accounts for an increasing fraction of the total reaction time. Dashed vertical lines identify the time interval where pseudo-stationarity (i.e., $dC/dt = 0$) is verified. (*Adapted from L. A. Segel.*[10] *Used by permission.*)

(5.118) to (5.122)]. The approximation means that the time scale for substrate changes is long compared with the time scale for complex changes (see Fig. 5.3).

It can be shown[11] that the time scale for C is

$$k^{-1} = (k_{+1}S_0 + k_{-1} + k_{+2})^{-1} \tag{5.123}$$

and the time scale for S is

$$\frac{k_m + S_0}{k_{+2}E_0} \tag{5.124}$$

Therefore, the pseudo-steady-state is satisfied when

$$\frac{k_{+2}E_0}{k_{+1}} \ll (k_m + S_0)^2 \tag{5.125}$$

This implies a product formation rate k_2 sufficiently small and/or $S_0 \gg E_0$.

5.4.3 Autocatalysis

Suppose that, in the presence of an external supply, a chemical species X is able to increase itself. Then a possible kinetic scheme could be

$$A + X \xrightarrow{k} 2X \tag{5.126}$$

and the rate of formation of X will be

$$\frac{dX}{dt} = kAX \tag{5.127}$$

where it is assumed that A is externally supplied.

The resulting self-stimulated increase of X is given the name *autocatalysis*. From the point of view of thermodynamics, it can occur if the system is sufficiently far from equilibrium. This is a complex and fascinating topic, which allows us to study the formation of *organized structures* from near-uniform initial states. We will limit ourselves to give just an introductory example, which will be not developed in detail. The following (hypothetical) scheme has been proposed by the Brussels School of I. Prigogine in the '70s, and for this reason it is known as the "Brussellator."

$$A \xrightarrow{k_1} X \qquad (5.128a)$$

$$B + X \xrightarrow{k_2} Y + D \qquad (5.128b)$$

$$2X + Y \xrightarrow{k_3} 3X \qquad (5.128c)$$

$$X \xrightarrow{k_4} E \qquad (5.128d)$$

The quantities A, B, D, E, X, Y are assumed to be positive; moreover, A, B, D, E are maintained externally at constant values. As before, the italicized letter represents both the species and its concentration, for simplicity. Adding together Eqs. (5.128), we find

$$A + B \xrightarrow{k} D + E \qquad (5.129)$$

Therefore, X and Y have the role of intermediates, the only function of which is that of mediating the conversion of the reagents A and B into the products D and E.

The reader will note that all the reverse reactions have been ignored. These conditions of irreversibility may be realized by driving the system far from equilibrium.[5,12,13] Equation (5.128c) is a trimolecular autocatalytic reaction, whose biochemical credibility will not be discussed here. If all the reaction constants are put equal to one, then the kinetic equations for the intermediate X, Y are

$$\frac{dX}{dt} = A + X^2Y - BX - X \qquad (5.130a)$$

$$\frac{dY}{dt} = BX - X^2Y \qquad (5.130b)$$

Let us find the stationary states of Eqs. (5.130) by imposing

$$\frac{dX}{dt} = \frac{dY}{dt} = 0 \qquad (5.131)$$

One single steady state is found, having coordinates in the phase plane X-Y:

$$X_0 = A \qquad Y_0 = \frac{B}{A} \qquad (5.132)$$

It can be shown[5,12,13] that, for appropriate combinations of the positive parameters A and B, the steady state is unstable and the system allows a regime of self-sustained oscillations. In other words, a *time-organized structure* does appear (a "limit cycle"). A more rigorous mathematical description of these somewhat obscure statements is beyond the purpose of the book. The reader is referred to books on nonequilibrium thermodynamics.[13]

PROBLEMS

5.1 Suppose that at time $t = 0$ we were able to force a population of membrane receptors on just half of the surface of a spherical cell with diameter 10 μm. We find experimentally that after 20 minutes the receptors are evenly distributed all over the surface. On the basis of this data, make an estimate of the receptor coefficient of diffusion.

5.2 Extend to three dimensions the one-dimensional Einstein-Smoluchowski relation $\langle x^2 \rangle = 2Dt$.

5.3 Electrophoretic measurements of macromolecules are done in the presence of an excess of an inert electrolyte of low molecular weight. Why?

5.4 A gel electrophoresis is performed at pH = 7 with a mixture of proteins A, B, C which have the following isoelectric point, respectively: $pI_A = 5$; $pI_B = 7$; $pI_C = 8$. Indicate the direction (cathode/anode) of motion of each type of protein.

5.5 Find the expression for the time evolution $A(t)$ of the monomer A which undergoes the polymerization process $A + A + A \xrightarrow{k} P$.

5.6 On the basis of the following data:

$$S = 6 \times 10^{-5} \text{ moles/L}; \quad v = 4 \times 10^{-9} \text{ moles/L-s}; \quad v_{max} = 10 \times 10^{-9} \text{ moles/L-s}$$

find the Michaelis constant K_m of the enzyme-catalyzed reaction.

5.7 In 1920 Lotka proposed the following autocatalytic scheme

$$A + X \longrightarrow 2X$$
$$X + Y \longrightarrow 2Y$$
$$Y \longrightarrow P$$

(all the kinetic constants are put equal to 1). Write down the differential equations describing the intermediates X and Y and determine the corresponding stationary points.

REFERENCES

1. C. R. Cantor and P. R. Schimmel, *Biophysical chemistry,* Part II, San Francisco: W. H. Freeman, 1980.
2. A. Einstein, *Investigations on the theory of the Brownian movement,* New York: Dover Publications, 1956.
3. J. O. M. Bockris and A. K. N. Reddy, *Modern electrochemistry,* 3d ed., Vol. 1, Plenum/Rosetta edition, New York: Plenum, 1977.
4. G. E. Uhlenbeck and L. S. Ornstein, "On the theory of the Brownian motion," in *Noise and stochastic processes,* N. Wax, editor, New York: Dover Publications, 1954.
5. R. Serra, G. Zanarini, M. Andretta, and M. Campiani, *Introduction to the physics of complex systems,* Oxford: Pergamon Press, 1986.
6. H. Risken, *The Fokker-Planck equation,* Berlin: Springer-Verlag, 1988.
7. I. Tinoco, Jr., K. Sauer, and J. C. Wang, *Physical chemistry,* Englewood Cliffs, N.J.: Prentice-Hall, 1978.
8. D. A. McQuarrie, *Stochastic approach to chemical kinetics,* London: Methuen, Ltd., 1967.
9. B. Alberts, D. Bray, J. Lewis, M. Raf, K. Roberts, and J. D. Watson, *Molecular biology of the cell,* 3d ed., New York and London: Garland, 1994.

10. I. H. Segel, *Biochemical calculations,* 2d ed., New York: John Wiley & Sons, 1976.
11. L. A. Segel, *Modeling dynamic phenomena in molecular and cellular biology,* New York: Cambridge University Press, 1987.
12. M. V. Volkenshtein, *Biophysics,* Moscow: MIR Publishers, 1983.
13. G. Nicolis and I. Prigogine, *Self-organization in nonequilibrium systems,* New York: John Wiley & Sons, 1977.

PART · 3

JUNCTIONS AND MEMBRANES

CHAPTER 6
SEMICONDUCTOR JUNCTIONS

In Chaps. 1, 2, and 3, we showed that electrons and holes can be considered as classical charged particles within a semiconductor crystal; we also evaluated the carrier concentrations, and we described the three main processes by which carriers move: drift, diffusion, and generation-recombination. With these ideas established, we are now ready to begin our study of a basic structure of the electronic devices: the *semiconductor junction*, that is, the interaction of a semiconductor with another adjacent semiconductor or a metal. The semiconductor-oxide junction will be considered in Chap. 8, in dealing with the MOS structure.

6.1 pn *JUNCTION*

A *pn* junction consists of a semiconductor having a *p*-type region and an *n*-type region separated by a region of transition from one type of doping to the other. Figure 6.1*a* shows the as-fabricated three-dimensional (real) abrupt *pn* junction, while Fig. 6.1*b* shows its one-dimensional approximation. To discuss the properties of the *pn* junction behavior without introducing unessential and complicated details, we make some assumptions.

We consider a *step* or *abrupt pn* junction: this is a junction in which the transition from the *p*-type to the *n*-type semiconductor region occurs abruptly, i.e., over a region of negligible thickness. We also assume that the *metallurgical junction* (at $x = 0$ in Fig. 6.1*b*) or boundary between the *n*-type and *p*-type region is a plane, that all carrier distributions and currents are *uniform* on planes parallel to the boundary plane, and that all currents are directed perpendicular to the boundary plane. This *one-dimensional* model, reasonable for most *pn* junctions, simplifies the equations of state to one spatial variable and facilitates closed-form solutions of the differential equations. Such an assumption is justified if the main variables change rapidly in only one direction.

Deviations and limitations of this *ideal pn* junction model are not taken into account, unless strictly necessary to the aim of the book. The reader can find the description of these effects by referring to Refs. 1 to 6. Here, only the basic physical operation of the ideal *pn* junction is discussed. Now we will examine the *pn* junction under equilibrium conditions.

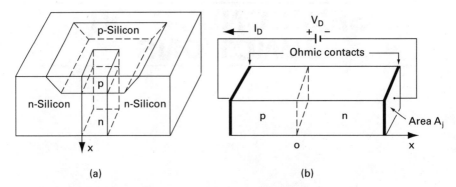

FIGURE 6.1 *pn* junction. (*a*) Real structure and (*b*) one-dimensional approximation. (*From G. W. Neudeck.*[2] *Used by permission.*)

6.2 pn JUNCTION IN EQUILIBRIUM

We assume that no illumination, no radiation, no mechanical stress, and no magnetic fields are present, and that all points of the semiconductors forming the *pn* junction are at the same temperature (understood to be room temperature, unless specifically indicated otherwise). Until further notice, we also assume that the semiconductor materials under discussion are self-contained, with no externally applied voltage or current, and that the electric field is zero in their environment (the assumption of zero electric field will be replaced later on in the discussion). Finally, we will assume that all the above have been satisfied for a long time, so that conditions within the semiconductors have settled. Under these assumptions, the *pn* junction is then said to be in *thermal equilibrium*. Under thermal equilibrium conditions, the hole current and the electron current must *each* vanish at every point in the semiconductor.

Let us imagine that a barrier exists at the boundary plane at $x = 0$ in the structure of Fig. 6.1*b*, so that no carrier (electrons or holes) flows between the *p*-type region and *n*-type region (see Fig. 6.2*a*). The two regions, then, behave like two homogeneous and separated semiconductor materials, whose electron and hole concentrations are uniform and are determined only by acceptor or donor concentrations and by temperature. In addition, we assume that *charge neutrality* exists everywhere in the semiconductor.

Thus, as described in Sec. 2.6, we write for $x > 0$ (*n*-type region)

$$n_{n0} \simeq N_D \qquad (6.1a)$$

$$p_{n0} \simeq \frac{n_i^2(T)}{N_D} \qquad (6.1b)$$

and for $x < 0$ (*p*-type region)

$$n_{p0} \simeq \frac{n_i^2(T)}{N_A} \qquad (6.2a)$$

$$p_{p0} \simeq N_A \qquad (6.2b)$$

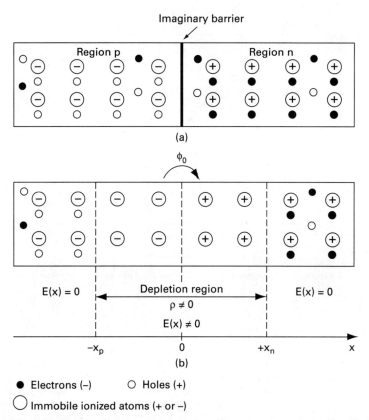

- ● Electrons (−)   ○ Holes (+)
- ◯ Immobile ionized atoms (+ or −)

FIGURE 6.2 What happens when a *pn* junction is formed, assuming (*a*) the presence of an imaginary barrier at $x = 0$ and (*b*) the removal of the imaginary barrier (real situation). (*Kindly provided by Maria Mattioli.*)

where N_D, N_A, and n_i are the donor, acceptor, and intrinsic carrier concentrations, respectively. Clearly, concentration gradients of both holes and electrons must exist at the junction, as indicated in Fig. 6.3.

If the *barrier* (in reality it does not exist) at $x = 0$ is removed, the *pn* junction is formed, and diffusion takes place in its vicinity. Electrons diffuse from the region where they are in high concentration (*n* side) to the region where their concentration is low (*p* side). Each such electron leaves behind a donor atom with a net positive charge, since the donor atoms are initially neutral: such ionized donor atoms are indicated by the ⊕ signs in Fig. 6.2*b*, the circle suggesting a bound charge (i.e., one that cannot move). Similarly, holes diffuse from the *p* to the *n-type region*. They leave behind acceptor atoms with a net negative charge, since the acceptor atoms are initially neutral: such ionized acceptor atoms are indicated by the ⊖ signs in Fig. 6.2*b*.[7]

In the region near the junction, then, ionized atoms are left *uncovered*. From *Gauss' law*, a net charge concentration implies the existence of an electric field and therefore a potential difference. Since the charge is positive on the right-hand side of the junction and negative on the left-hand side of the junction (see Fig. 6.2*b*), the electric field will be di-

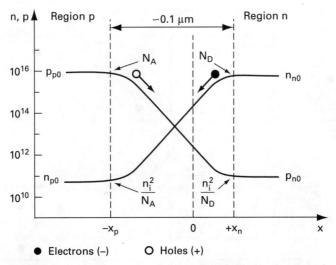

FIGURE 6.3 Approximate distributions of carriers near the boundary plane of a *pn* junction.

rected along the negative x axis; that is, the electric field is negative. Thus, the electric field created by these atoms has such a direction as to inhibit the diffusion of free carriers.

In other words, any flow of charge across the junction is a self-limiting process because the electric field at the junction, which is a direct consequence of the charge transport, increases to exactly the value required to counterbalance the diffusive tendencies of holes and electrons. The presence of this electric field causes a potential barrier ϕ_0 between the two types of semiconductor; this potential is frequently referred to as the *built-in potential*, or the *contact potential*. This latter definition holds because the built-in potential can be thought of as a contact potential between two dissimilar materials.

Therefore, a *space-charge region* where, by diffusion, the neutrality condition is violated would be established at the junction. The space-charge region is more often referred to as *depletion region* in which the space charge is made up of doping ions. Let the depths of this depletion region at the two sides of the junction be $+x_n$ and $-x_p$ as shown in Fig. 6.2b.

6.2.1 Built-in Potential

As mentioned before, when equilibrium is reached, the magnitude of the electric field is such that the tendency of electrons (holes) to *diffuse* from the *n*-type (*p*-type) region into the *p*-type (*n*-type) region is balanced by the tendency of the electrons (holes) to *drift* in the opposite direction under the influence of the built-in electric field.

This equilibrium condition results in zero electron and hole currents at any point of the structure. Thus, we can write

$$J_p = J_{p,\text{drift}} + J_{p,\text{diff}} = 0 \tag{6.3a}$$

$$J_n = J_{n,\text{drift}} + J_{n,\text{diff}} = 0 \tag{6.3b}$$

Taking into account Eqs. (3.46) and (3.47) yields

$$q\left[-D_p \frac{dp(x)}{dx} + \mu_p p(x) E(x)\right] = 0 \qquad (6.4a)$$

$$q\left[D_n \frac{dn(x)}{dx} + \mu_n n(x) E(x)\right] = 0 \qquad (6.4b)$$

Solving for $E(x)$, from Eqs. (6.4) we obtain

$$E(x) = \frac{D_p}{\mu_p} \frac{1}{p(x)} \frac{dp(x)}{dx} = -\frac{D_n}{\mu_n} \frac{1}{n(x)} \frac{dn(x)}{dx} \qquad (6.5)$$

From the electromagnetic field theory, we know that the electric field is the negative of the gradient of the potential, thus the built-in potential ϕ_0 is found, taking into account Eq. (6.5), to be

$$\phi_0 = -\int_{-x_p}^{x_n} E(x)\,dx = -\frac{D_p}{\mu_p} \int_{-x_p}^{x_n} \frac{1}{p(x)} \frac{dp(x)}{dx}\,dx$$

$$= \frac{D_p}{\mu_p} [\ln p(-x_p) - \ln p(x_n)] \qquad (6.6a)$$

$$\phi_0 = -\int_{-x_p}^{x_n} E(x)\,dx = \frac{D_n}{\mu_n} \int_{-x_p}^{x_n} \frac{1}{n(x)} \frac{dn(x)}{dx}\,dx$$

$$= \frac{D_n}{\mu_n} [\ln n(x_n) - \ln n(-x_p)] \qquad (6.6b)$$

As the carrier concentrations outside the depletion region are known [see Eqs. (6.1) and (6.2)], the total *built-in potential* ϕ_0 can be written as

$$\phi_0 = \frac{D_p}{\mu_p} \ln \frac{p_{p0}}{p_{n0}} = \frac{kT}{q} \ln \frac{p_{p0}}{p_{n0}} \approx \frac{kT}{q} \ln \frac{N_A N_D}{n_i^2} \qquad (6.7a)$$

$$\phi_0 = \frac{D_n}{\mu_n} \ln \frac{n_{n0}}{n_{p0}} = \frac{kT}{q} \ln \frac{n_{n0}}{n_{p0}} \approx \frac{kT}{q} \ln \frac{N_A N_D}{n_i^2} \qquad (6.7b)$$

where Einstein relations have been used. The built-in voltage ϕ_0 can be considered as the voltage developed in response to a gradient in the concentration of charged species (i.e., electrons and holes), and it exhibits the logarithmic dependence on concentration shown in Eqs. (6.7). This dependence recalls the *Nernst* potential (see Sec. 7.3.2) where $RT/\mathscr{F}$ (notation used in the electrochemistry and biology literature) takes the place of kT/q used in Eqs. (6.7) according to the notation used in electronics. Because of the logarithmic dependence, wide variations of the product $N_A N_D$ are needed to obtain an appreciable variation of ϕ_0. Built-in potential values are in the range 0.2 to 1V, for typical *pn* junctions.

Inspection of Eqs. (6.7) shows that ϕ_0 can be resolved into two components:

$$\phi_n = \frac{kT}{q} \ln \frac{N_D}{n_i} \qquad (6.8a)$$

where ϕ_n is the potential at the neutral edge of the depletion region in the n-type semiconductor, and

$$\phi_p = -\frac{kT}{q} \ln \frac{N_A}{n_i} \tag{6.8b}$$

where $\phi_p < 0$ is the potential at the neutral edge of the depletion region in the p-type semiconductor.

Thus, the total potential change ϕ_0 from the neutral n-type region to the neutral p-type region is

$$\phi_0 = \phi_n - \phi_p = \frac{kT}{q} \ln \frac{N_D}{n_i} + \frac{kT}{q} \ln \frac{N_A}{n_i} = \frac{kT}{q} \ln \frac{N_A N_D}{n_i^2} \tag{6.9}$$

just as obtained in Eqs. (6.7).

The built-in potential ϕ_0 is an electrostatic potential that develops across the depletion region, and as Eq. (6.9) indicates, depends on the dopant concentration in each region of the junction and on the temperature. It cannot be measured with any voltmeter because ϕ_0 is exactly canceled out by the difference in contact potentials for probes applied to p-type and n-type semiconductors.

The major portion of the potential change occurs in the region with the lower dopant concentration, and the depletion region is wider in the same region (as shown later).

In addition, we note from Eq. (6.9) that the potential at the junction plane ($x = 0$) is not exactly zero unless the junction is symmetrical (that is, $N_A = N_D$).

From Eqs. (6.7) we can also determine the carrier concentrations at the edges of the depletion region (i.e., at $-x_p$ and at x_n). Then

$$n_{p0}(-x_p) = n_{n0}(x_n) e^{-q\phi_0/kT} \tag{6.10a}$$

$$p_{n0}(x_n) = p_{p0}(-x_p) e^{-q\phi_0/kT} \tag{6.10b}$$

These equations are known as *Boltzmann relations* for the *pn* junction, and they can be thought as the exponentiation of Eqs. (6.7).

At this point, the reader may ask if the energy band model described in Chap. 2 can be applied to the *pn* junction, and if it yields the same information relative to ϕ_0, as previously described. To apply the energy band model, recall that at thermal equilibrium the Fermi energy level E_F must be constant throughout the semiconductor, independent of position.

To accommodate this, the conduction band level E_c, the valence band level E_v, and the intrinsic level E_i (located very close to the middle of the forbidden energy gap) may have to "bend" accordingly. This is illustrated in Fig. 6.4, which shows the energy band diagram for a *pn* junction in equilibrium.

The difference between any two among E_c, E_v, and E_i is kept constant with distance. For any point along the horizontal axis in Fig. 6.4, the distance between E_F and E_i, and their relative positions, must be such as to give the correct values of n and p given by the *Maxwell-Boltzmann* statistics, as explained in Chap. 2. According to these statistics, the electron and hole concentrations at equilibrium can be expressed as follows [see Eqs. (2.65) and (2.66)]:

$$n = n_i e^{(E_F - E_i)/kT} \tag{6.11a}$$

$$p = n_i e^{(E_i - E_F)/kT} \tag{6.11b}$$

where n_i is the intrinsic carrier concentration.

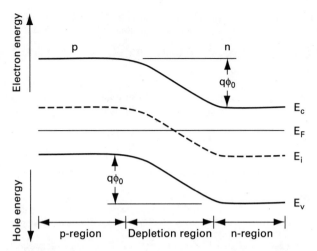

FIGURE 6.4 *pn* junction energy band diagram at thermal equilibrium.

We recall that Eqs. (6.11) hold for nondegenerate semiconductors, i.e., for semiconductors where E_F is not too close to either end of the forbidden energy gap, or, equivalently, when the relation

$$(E_v + 3kT) < E_F < (E_c - 3kT) \tag{6.12}$$

is satisfied.

Let now $\Delta E_c = \Delta E_i$ be the electron potential energy difference between points 1 and 2 of the *pn* junction at equilibrium. In particular consider point 1 in the *n*-type region and point 2 in the *p*-type region. Then, the electrostatic potential difference $\Delta \phi$ between them (potential energy difference per unit charge) will be

$$\Delta \phi = \frac{\Delta E_i}{-q} \tag{6.13}$$

Thus, the electrostatic potential varies in a direction opposite from E_i, E_c, or E_v. From Eq. (6.11a), the electron concentration at the two points considered above will be

$$n_{n0} = n_i e^{(E_F - E_{i1})/kT} \tag{6.14a}$$

$$n_{p0} = n_i e^{(E_F - E_{i2})/kT} \tag{6.14b}$$

Dividing Eq. (6.14a) by Eq. (6.14b) and using Eq. (6.13) gives

$$\frac{n_{n0}}{n_{p0}} = e^{-q\Delta\phi/kT} \tag{6.15}$$

which is the same as Eq. (6.7b). Similarly, one is led to Eq. (6.7a) starting from Eq. (6.11b). The reader can verify this statement as an exercise.

Since in energy band treatments the Fermi potential ϕ_F is defined as

$$\phi_F = \frac{E_F - E_i}{-q} \tag{6.16}$$

then the two components [Eqs. (6.8)] of the built-in potential ϕ_0 are formally indicated by ϕ_{Fn} and ϕ_{Fp}, representing the Fermi potentials of the n-type region and p-type region, respectively.

The above discussion has shown that the energy band theory and the contact potential theory lead to the same results.

Example 6.1 For an Si semiconductor maintained at room temperature and doped with $N_A = 10^{15}/cm^3$ acceptors in the p-type region and $N_D = 10^{15}/cm^3$ donors in the n-type region, Eq. (6.9) gives

$$\phi_0 = \frac{kT}{q} \ln \frac{N_A N_D}{n_i^2} = 25.86 \times 10^{-3} \times \ln \frac{10^{15} \times 10^{15}}{2.1 \times 10^{10}} = 0.576 \text{ V} \quad (E6.1)$$

For the variables not specified, refer to App. A. If the same Si semiconductor is now doped with $N_A = 10^{17}/cm^3$ acceptors and $N_D = 10^{15}/cm^3$ donors, Eq. (6.9) gives $\phi_0 = 0.695$ V. Note that the greater the doping of either region, the greater ϕ_0. This arises from Eq. (6.9) or, equivalently, from the energy band diagram of Fig. 6.4. If the p-type region doping is increased, then E_v must move closer to E_F and $q\phi_0$ must increase. Doping the n-type region to a higher degree, moves E_c closer to E_F, also increasing $q\phi_0$.

Silicon has a band gap $E_g = 1.12$ eV, or in terms of kT units at room temperature, $E_g = 43.08$ kT. To keep the silicon semiconductor from being degenerate, E_F must satisfy Eq. (6.12). If the junction is doped so that E_F is placed at $3kT$ from the band edges on each region of the junction, then we can write from Eq. (6.12)

$$q\phi_0 = 43.08 kT - 6kT = 0.9589 \text{ eV} \quad (E6.2)$$

and therefore $\phi_0 = 0.9589$ V is the maximum value of ϕ_0 at room temperature without having a degenerate silicon semiconductor.

6.2.2 Static Properties of the *pn* Junction

The charge density ρ, the electric field E, and the potential ϕ across the *pn* junction under thermal equilibrium conditions are based on the solution of the *Poisson equation;* thus we rewrite it for the reader's convenience in the one-dimensional form

$$\frac{d^2\phi(x)}{dx^2} = -\frac{dE(x)}{dx} = -\frac{\rho(x)}{\varepsilon_s} = -\frac{q}{\varepsilon_s}(p - n + N_D - N_A) \quad (6.17)$$

where $\varepsilon_s = \varepsilon_0 \varepsilon_r$ is the permittivity of the semiconductor (see App. A). In general E, ϕ, p, n, N_D, and N_A are functions of the position x, except for uniform doping where N_D and N_A are constants.

Poisson equation in the form of Eq. (6.17) is not easy to be solved for most devices, because n and p are in turn functions of E and ϕ, the unknown quantities. Thus, to solve analytically Eq. (6.17), we have to make some simplifications. In particular

1. We assume a *one-dimensional pn* junction with the metallurgical junction at $x = 0$.
2. We assume perfect *ohmic contacts,* far removed from the metallurgical junction.
3. We consider a *uniformly doped pn step* junction, where the doping concentrations change abruptly from N_D to N_A in $x = 0$, as shown in Fig. 6.5a.
4. We assume that the mobile carrier concentrations (n and p) are *small* compared to the donor and acceptor ion concentrations in the depletion region, and that the *pn* junction is charge-neutral elsewhere. In the limit, the region between $-x_p$ and $+x_n$ is completely depleted of mobile carriers, as shown in Fig. 6.5b. This simplification is

SEMICONDUCTOR JUNCTIONS **143**

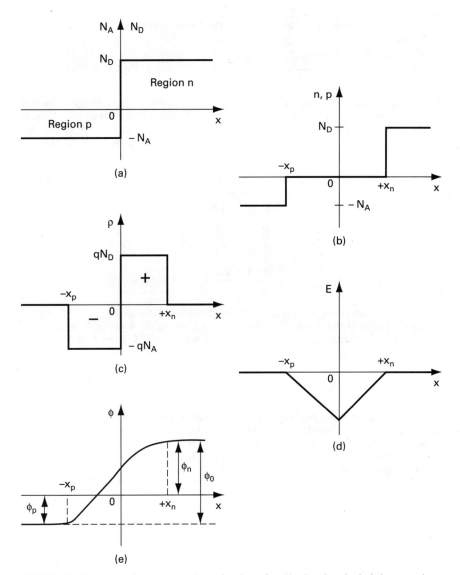

FIGURE 6.5 Properties of a step *pn* junction as functions of position based on the depletion approximation. (*a*) Net doping concentration, (*b*) carrier concentration, (*c*) space charge density, (*d*) electric field, and (*e*) potential.

called the *depletion approximation*. Thus, under the above assumptions, Eq. (6.17) becomes

$$\frac{d^2\phi(x)}{dx^2} = -\frac{\rho(x)}{\varepsilon_s} \simeq -\frac{q}{\varepsilon_s}(N_D - N_A) \qquad (6.18)$$

5. We also assume that the mobile majority-carrier concentrations abruptly become equal to the respective doping concentrations at the edges of the depletion region, as shown in Fig. 6.5b. The charge density is therefore zero everywhere, except in the depletion region, where it takes the value of the ionized doping concentrations, as shown in Fig. 6.5c.

All the approximations are justified because the solution of the Poisson equation does well explain experimental data obtained from real devices.

Under the above assumptions, Eq. (6.18) becomes[1-4]

$$\frac{d^2\phi(x)}{dx^2} = -\frac{dE(x)}{dx} = -\frac{qN_D}{\varepsilon_s} \qquad \text{for } 0 \leq x \leq x_n \qquad (6.19a)$$

$$\frac{d^2\phi(x)}{dx^2} = -\frac{dE(x)}{dx} = \frac{qN_A}{\varepsilon_s} \qquad \text{for } -x_p \leq x \leq 0 \qquad (6.19b)$$

which may be integrated from an arbitrary point inside the *n*-type (*p*-type) depletion region to the edge of the depletion region at x_n ($-x_p$), where the semiconductor becomes neutral and the electric field vanishes. By integrating Eq. (6.19a), we find the electric field in the *n*-type depletion region, i.e.,

$$E(x) = -\frac{qN_D}{\varepsilon_s}(x_n - x) \qquad \text{for } 0 \leq x \leq x_n \qquad (6.20a)$$

The electric field is negative through the depletion region and changes linearly with x, showing its maximum absolute value at $x = 0$ (see Fig. 6.5d). The direction of the electric field toward the left is physically correct, since its force must balance the tendency of the negatively charged electrons to diffuse toward the left, out of the neutral *n*-type region.

By integrating Eq. (6.19b), we find the electric field in the *p*-type depletion region, i.e.,

$$E(x) = -\frac{qN_A}{\varepsilon_s}(x + x_p) \qquad \text{for } -x_p \leq x \leq 0 \qquad (6.20b)$$

The electric field in the *p*-type region is also negative in order to oppose the tendency of the positively charged holes to diffuse toward the right.

As indicated in Fig. 6.5d, the electric field must be continuous at $x = 0$, since there is no layer of charge at that position. Thus from Eqs. (6.20a) and (6.20b), evaluated at $x = 0$, we obtain

$$-\frac{qN_D}{\varepsilon_s}x_n = -\frac{qN_A}{\varepsilon_s}x_p \qquad (6.21a)$$

or

$$N_D x_n = N_A x_p \qquad (6.21b)$$

Equation (6.21b) multiplied by qA_j (where A_j is the *pn*-junction area) states that the total negative charge must equal the total positive charge. The equality is also pointed out in Fig. 6.5c where the "area" $x_p N_A$ must be equal to the "area" $x_n N_D$.

From Eq. (6.21b) it also follows that the width of the depleted region on each side of the junction changes inversely with the magnitude of the doping concentration; the higher the doping concentration, the narrower the depletion region. From Fig. 6.5c, we find $N_D > N_A$, and by Eq. (6.21b), it follows $x_n < x_p$. This is an easy method of relating the doping concentrations to depletion widths.

In a highly asymmetrical junction, where the doping concentration on one side of the junction is much higher than that on the other side, the depletion region enters mainly the lightly doped semiconductor and the width of the depletion region in the heavily doped semiconductor can often be neglected.

The potential function $\phi(x)$ within the depletion region is derived from the definition of potential; that is,

$$\frac{d\phi}{dx} = -E(x) \tag{6.22}$$

Combining Eq. (6.22) with Eq. (6.20a) for the electric field in the n-type depletion region yields

$$\frac{d\phi}{dx} = -E(x) = \frac{qN_D}{\varepsilon_s}(x_n - x) \tag{6.23}$$

Separating variables and integrating gives

$$\int_{\phi(x)}^{\phi_n} d\phi = \phi_n - \phi(x) = \frac{qN_D}{\varepsilon_s} \int_x^{x_n} (x_n - x)\, dx$$

$$= \frac{qN_D}{\varepsilon_s}\left(x_n x - \frac{x^2}{2}\right)\Big|_x^{x_n} \quad \text{for } 0 \leq x \leq x_n \tag{6.24}$$

With some algebraic rearrangements, we obtain in the n-type depletion region

$$\phi(x) = \phi_n - \frac{qN_D}{2\varepsilon_s}(x_n - x)^2 \quad \text{for } 0 \leq x \leq x_n \tag{6.25a}$$

where ϕ_n is given by Eq. (6.8a). Similarly, for the p-type depletion region, we obtain

$$\phi(x) = \phi_p + \frac{qN_A}{2\varepsilon_s}(x + x_p)^2 \quad \text{for } -x_p \leq x \leq 0 \tag{6.25b}$$

where ϕ_p is given by Eq. (6.8b).

Figure 6.5e shows the potential function $\phi(x)$ for the pn step junction under the depletion approximation. By comparing Figs. 6.5e and 6.4, we can observe that the potential function is the horizontal mirror image of the energy band diagram E_c, E_v, or E_i.[2]

Now we are ready to calculate the depletion region width, which represents an important function in the study of the pn junction. As already mentioned for the electric field, since there is no dipole layer at $x = 0$, the potential function must be continuous; that is, $\phi(0^-) = \phi(0^+)$. Taking into account Eqs. (6.25a) and (6.25b), evaluated at $x = 0$, we have

$$\phi_n - \frac{qN_D}{2\varepsilon_s}x_n^2 = \phi_p + \frac{qN_A}{2\varepsilon_s}x_p^2 \tag{6.26}$$

Equations (6.26) and (6.21b) are two equations with two unknown quantities. Solving for x_p from Eq. (6.21b) and substituting the result into Eq. (6.26), we obtain

$$\frac{qN_D^2}{2\varepsilon_s N_A} x_n^2 = -\frac{qN_D}{2\varepsilon_s} x_n^2 + \phi_0 \tag{6.27}$$

where the relation $\phi_0 = (\phi_n - \phi_p)$ has been used.

By solving Eq. (6.27) for x_n, we get

$$x_n = \sqrt{\frac{2\varepsilon_s}{q} \phi_0 \frac{N_A}{N_D(N_A + N_D)}} \tag{6.28a}$$

Similarly, for x_p we get

$$x_p = \sqrt{\frac{2\varepsilon_s}{q} \phi_0 \frac{N_D}{N_A(N_A + N_D)}} \tag{6.28b}$$

The *depletion region* width W can then be determined as

$$W = (x_n + x_p) = \sqrt{\frac{2\varepsilon_s \phi_0}{q(N_A + N_D)}} \left(\sqrt{\frac{N_A}{N_D}} + \sqrt{\frac{N_D}{N_A}} \right) \tag{6.29}$$

With some algebraic rearrangements, Eq. (6.29) can be written in the more convenient form

$$W = (x_n + x_p) = \sqrt{\frac{2\varepsilon_s}{q} \phi_0 \frac{N_A + N_D}{N_A N_D}} \tag{6.30}$$

Equation (6.30) states that the depletion region width depends most strongly on the semiconductor region with the lighter doping, and varies approximately as the inverse square root of the smaller doping concentration.

Often, in practice, we must consider the case of a one-side step junction of the type n^+p (or np^+), i.e., a step junction for which $N_D \gg N_A$ (or $N_A \gg N_D$); these relations imply $x_n \ll x_p$ (or $x_p \ll x_n$), which means that practically all the depletion region extends into the p side (or n side).

6.2.3 Linearly Graded *pn* Junction

The step or abrupt *pn* junction considered in the previous sections is not an adequate representation for a real *pn* junction, where impurities are thermally diffused or ion-implanted. A better approximation is the *linearly graded pn* junction, where the net doping concentration $(N_D - N_A)$ varies linearly (nearly a straight line) from the *p*-type semiconductor to the *n*-type semiconductor. Assuming that *p* impurities are thermally diffused into an *n*-type semiconductor, where the straight line has a slope of $-a$ (see Fig. 6.6a), then the net impurity concentration can be written as

$$N_A(x) - N_D = -ax \tag{6.31}$$

where a is called the *grading constant* and has the units number/cm^4. Under the depletion approximation (see Fig. 6.6b), we can write

$$\rho = qax \quad \text{for } -x_p \leq x \leq x_n \tag{6.32a}$$

$$\rho = 0 \quad \text{elsewhere} \tag{6.32b}$$

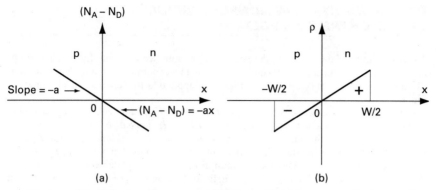

FIGURE 6.6 Linearly graded *pn* junction. (*a*) Net doping concentration and (*b*) space charge density.

The procedures to calculate E, ϕ, and W are similar to those used for the step-junction approximation. The reader should find this solution as a useful exercise, taking into account that the symmetry of the linearly graded charge density implies $x_n = x_p$, or equivalently that the depletion region width must be symmetrical about the metallurgical junction at $x = 0$ which implies $x_n = x_p = W/2$.

Example 6.2 Calculate the total depletion width W and the depletion widths in each region of a silicon *pn* junction at room temperature. The doping concentrations are $N_A = 10^{17}/\text{cm}^3$, $N_D = 10^{15}/\text{cm}^3$. For the other parameters refer to App. A.

Answer First of all, we calculate the built-in voltage ϕ_0 from Eq. (6.9). Thus,

$$\phi_0 = \frac{kT}{q} \ln \frac{N_A N_D}{n_i^2} = 25.86 \times 10^{-3} \times \ln \frac{10^{15} \times 10^{17}}{2.1 \times 10^{20}} = 0.695 \text{ V} \quad (E6.3)$$

From Eq. (6.30), we obtain

$$W = (x_n + x_p) = \sqrt{\frac{2\varepsilon_s}{q} \phi_0 \frac{N_A + N_D}{N_A N_D}}$$

$$= \sqrt{\frac{2 \times 1.053 \times 10^{-12}}{1.602 \times 10^{-19}} \times 0.695 \frac{10^{15} + 10^{17}}{10^{17} \times 10^{15}}}$$

$$= 0.96 \times 10^{-4} \text{ cm} = 0.96 \text{ μm} \quad (E6.4)$$

From charge conservation, given by Eq. (6.21b), we have $N_A x_p = N_D x_n$. Also

$$W = (x_n + x_p) = x_n \left(1 + \frac{N_D}{N_A}\right) = x_n(1 + 0.01) \quad (E6.5)$$

Hence

$$x_n = \frac{W}{(1 + 0.01)} = \frac{0.96}{1.01} = 0.95 \text{ μm} \quad (E6.6)$$

and

$$x_p = (W - x_n) = (0.96 - 0.95) = 0.01 \text{ μm} \quad (E6.7)$$

Since the *p*-type region doping is higher than the *n*-type region doping, then it follows $x_n > x_p$. The junction depletes further into the lightly doped semiconductor. Another way to calculate x_n and x_p is to use Eqs. (6.28).

6.3 pn JUNCTION IN NONEQUILIBRIUM: EFFECT OF THE BIAS VOLTAGE

The discussion and the equations derived in the previous sections hold for the static properties of a *pn* junction at thermal equilibrium. The most common deviation from the equilibrium condition is caused by a bias voltage V_D applied to the junction; this is shown in Fig. 6.7 for $V_D > 0$.

We can distinguish two situations: a *positive voltage* V_D is applied to the *p*-type region of the junction (the junction is said to be *forward-biased*); a *negative voltage* V_D is applied to the *p*-type region of the junction (the junction is said to be *reverse-biased*). To analyze the static properties of the junction when a bias voltage V_D is applied to it, we will again make use of the approximations stated for the equilibrium analysis, together with the assumption that no (or negligible) voltage drop occurs in the bulk *p*- and *n*-type regions. By inspection of Fig. 6.7, since V_D is opposite in polarity to V_j, it must reduce the voltage V_j across the depletion region. This becomes evident when we write the loop equation for the structure of Fig. 6.7, taking into account that $V_j = (V_n + V_p) = \phi_0$ at thermal equilibrium. Thus,

$$V_j = V_n - V_D + V_p = \phi_0 - V_D \tag{6.33}$$

under the assumption that the applied voltage V_D appears entirely across the junction. Furthermore, the solutions of the Poisson equation obtained for equilibrium conditions in the previous sections also apply to the junction under bias. All that needs to be changed is to replace ϕ_0 by $(\phi_0 - V_D)$.

In forward bias conditions, since $(\phi_0 - V_D)$ is less than ϕ_0, the electric field [Eqs. (6.20)], the potential [Eqs. (6.25)], and the depletion region widths [Eqs. (6.28) and (6.30)] values reduce. This can be deduced, for example, from Eq. (6.30), which becomes

$$W = (x_n + x_p) = \sqrt{\frac{2\varepsilon_s}{q}(\phi_0 - V_D)\frac{N_A + N_D}{N_A N_D}} \tag{6.34}$$

where ϕ_0 has been replaced by $(\phi_0 - V_D)$, according to what has previously been stated. An important characteristic of forward bias is that, for these equations to be valid, the requirement

$$V_D < \phi_0 \tag{6.35}$$

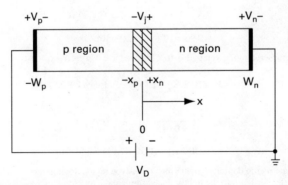

FIGURE 6.7 One-dimensional representation of a *pn* junction under the bias voltage V_D.

must be verified. If Eq. (6.35) is not satisfied, the Kirchhoff voltage law (KVL) is violated. The only way to remove this restriction is to consider the voltage drops in the bulk regions, as it will be described in Sec. 6.9.3.

In reverse bias conditions, Eq. (6.33) states that the applied voltage $V_D < 0$ adds to ϕ_0, and therefore increases V_j. Moreover, since V_D is negative, Eq. (6.35) is *always* verified: in this case there are no restrictions on KVL. The increase of the junction voltage V_j, due to reverse bias, leads to an increase of the static quantities. This effect can be verified on the depletion region width, by inspection of Eq. (6.34). Obviously, if $V_D = 0$ the equilibrium relations are again obtained.

The forward and reverse bias considerations apply both for a step junction and for a linearly graded junction.

The above results can be qualitatively deduced also from the energy band model, as already pointed out in Sec. 6.2.1. This approach, however, is not strictly necessary to the aim of the book, and therefore we do not consider it. The reader interested in this approach can refer, for example, to Ref. 2.

6.4 CURRENT-VOLTAGE CHARACTERISTICS OF THE pn JUNCTION

The use of the continuity equations [Eqs. (3.101a) and (3.101b)], together with the concept of excess-carrier lifetime as derived from the SHR generation-recombination model (see Sec. 3.6), allow us to obtain the expressions for currents in a *pn* junction under forward and reverse bias.

Solutions of the continuity equation in the neutral regions provide the carrier concentrations in terms of position and time. We can obtain expressions for the current by using Eqs. (3.46) and (3.47), which express carrier flow in terms of carrier concentrations.

The *total current* consists, in general, of the sum of four components: hole and electron drift currents and hole and electron diffusion currents.

We now consider a *pn* junction with a constant cross-sectional area A_j, connected to a voltage bias source as shown in Fig. 6.7, and let the carrier concentrations within the junction be influenced *only* by the applied voltage. The applied voltage V_D then appears in part across the neutral regions, and in part across the junction. As the voltage drops across the neutral regions are ohmic (current times resistance), they are very small at low currents, and can be considered negligible up to relatively high values of bias. Thus we neglect these drops from the present analysis of the *ideal pn* junction (we shall remove these assumptions later).

If we assume then that V_D appears entirely across the junction, the total junction voltage will be $(\phi_0 - V_D) = V_j$ [see Eq. (6.33)].

If V_D is positive (*forward bias*), it reduces the barrier to the diffusion flow of majority carriers across the junction. The reduced barrier, in turn, allows a flow of holes from the *p*-type region into the *n*-type region and of electrons from the *n*-type region into the *p*-type region. When these carriers enter the neutral regions, they become minority carries and are neutralized by majority carriers that enter the neutral regions from the ohmic contacts. The minority carriers injected across the depletion region then tend to diffuse into the bulk of the neutral region. If V_D is negative (*reverse bias*), it increases the barrier height to the diffusion of the majority carriers. As equilibrium is disturbed, minority carriers near the depletion region tend to be depleted. The majority carrier concentration is then reduced.[1,2]

From the above statements, it follows that the minority-carrier concentrations determine what currents flow in a *pn* junction. The majority carriers behave only as providers of the injected minority-carrier current or as charge neutralizers in the neutral regions.

Accordingly, we are going to find in the next sections solutions of the continuity equations for the minority-carrier concentrations in each of the neutral regions.

6.4.1 Boundary Values of Minority-Carrier Concentrations

To obtain the solutions of the continuity equations we first must relate the boundary values of the minority-carrier concentrations to the applied bias voltage V_D. To do this, we make two additional assumptions: first, that the applied bias leads to *low-level injection* conditions in the neutral regions and, second, that the applied bias is small enough so that the detailed balance between majority and minority carrier concentrations across the junction regions is not appreciably disturbed (insignificant recombination within the depletion region) so that the minority current density at $-x_p$ is the same as the majority current density at $+x_n$.

The second assumption allows us to use Eqs. (6.10) across the junction, where the potential difference is given by Eq. (6.33). Both of these assumptions are valid when V_D is very small ($|V_D| \ll \phi_0$).

Thus, the concentrations at the depletion region boundaries are

$$p_n(x_n) = p_p(-x_p)e^{-q(\phi_0 - V_D)/kT} \tag{6.36a}$$

$$n_p(-x_p) = n_n(x_n)e^{-q(\phi_0 - V_D)/kT} \tag{6.36b}$$

Under the condition of low-level injection, i.e., in each of the neutral regions the minority-carrier concentration is much less than the majority-carrier concentration, we may write that $p_p(-x_p) \simeq p_{p0}$ and $n_n(x_n) \simeq n_{n0}$. Therefore, Eqs. (6.36) become

$$p_n(x_n) = p_{p0}(-x_p)e^{-q(\phi_0 - V_D)/kT} \tag{6.37a}$$

$$n_p(-x_p) = n_{n0}(x_n)e^{-q(\phi_0 - V_D)/kT} \tag{6.37b}$$

Taking into account Eqs. (6.10), we can eliminate ϕ_0 from the above equations. Then Eqs. (6.37) become

$$p_n(x_n) = p_{n0}(x_n)e^{qV_D/kT} \tag{6.38a}$$

$$n_p(-x_p) = n_{p0}(-x_p)e^{qV_D/kT} \tag{6.38b}$$

Equations (6.38) are known as the *law of the junction,* in that they relate the minority-carrier concentration enhancement or diminution at the edge of the depletion region with the polarity of the applied bias voltage V_D. Figure 6.8 shows the carrier concentrations under forward and reverse bias. We can deduce, for example, that under forward bias, holes moving to the *n*-type region and electrons to the *p*-type region become *excess* minority carriers. This because the concentration is now greater than the thermal equilibrium level.

If we define the excess concentrations by

$$n'_p \equiv n_p - n_{p0} \tag{6.39a}$$

$$p'_n \equiv p_n - p_{n0} \tag{6.39b}$$

then the *excess minority-carrier* concentrations at the boundaries in terms of their thermal equilibrium values are

$$n'_p(-x_p) = n_{p0}(-x_p)(e^{qV_D/kT} - 1) \tag{6.40a}$$

$$p'_n(x_n) = p_{n0}(x_n)(e^{qV_D/kT} - 1) \tag{6.40b}$$

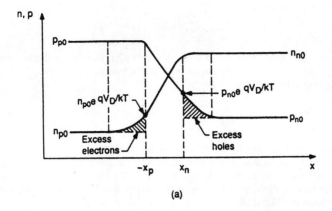

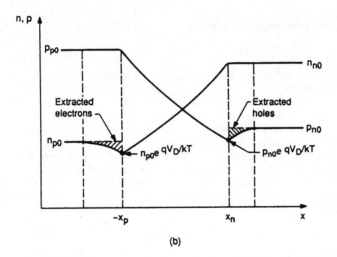

FIGURE 6.8 Carrier concentration in a *pn* junction under (*a*) forward bias and (*b*) reverse bias.

Equations (6.40) are important results that we shall use to define specific solutions for the continuity equations for minority carriers in the neutral regions near the *pn* junction.

6.4.2 Ideal *pn* Junction Analysis

Taking into account what was described in the previous sections, we can now consider solutions of the continuity equations [see Eqs. (3.101)] in the neutral regions.

We consider first the excess holes injected into the *n*-type region, where bulk recombination through generation-recombination centers prevails. Thus, the component

$$\left.\frac{\partial p}{\partial t}\right|_{\substack{\text{thermal}\\R-G}}$$

in Eq. (3.101b) can be then expressed through Eq. (3.92a) because of the assumption of low-level injection. Also, assuming operation is under dark conditions [see Eq. (3.110b)], the continuity equation becomes

$$\frac{\partial p_n}{\partial t} = D_p \frac{\partial^2 p_n}{\partial x^2} - \frac{p_n - p_{n0}}{\tau_p} \tag{6.41}$$

where the subscript n indicates that the holes are in the n-type region.

Assuming a constant donor concentration along x, as frequently happens in practice, and considering the steady-state condition ($\partial p/\partial t = 0$), Eq. (6.41) can be rewritten as a total differential equation in terms of the excess concentration p'_n, which was defined in Eq. (6.39b). Equation (6.41) reduces to

$$0 = D_p \frac{d^2 p'_n}{dx^2} - \frac{p'_n}{\tau_p} \tag{6.42}$$

which has the following general solution

$$p'_n(x) = A e^{-(x-x_n)/\sqrt{D_p \tau_p}} + B e^{(x-x_n)/\sqrt{D_p \tau_p}} \tag{6.43}$$

where A and B are constants to be determined by the boundary conditions.

The characteristic length $\sqrt{D_p \tau_p}$ in Eq. (6.43) is called the *minority-carrier diffusion length for holes* and is usually indicated by L_p. (The diffusion length of an electron in a p-type region is indicated by L_n.) Comparing the diffusion length to the neutral region width, we can consider two limiting configurations, according to two forms of Eq. (6.43): the *long-base diode* and the *short-base diode*.

The word *diode* is commonly employed to designate the *pn* junction as an electronic device for industrial use. The area of application of this device is widespread (e.g., in rectifier circuits, digital logic circuits, and variable capacitors). The reader often will find in the electronic literature that the terms *diode* and *pn junction* are used interchangeably to indicate either the physical structure or the real electronic component.

6.4.3 Long-Base Diode

The first limiting configuration (that of the *long-base diode*) takes place when the lengths W_n and W_p of the n- and p-type regions are much longer than the diffusion lengths L_p and L_n, respectively. In this case, in the n-type region, for example, practically all the injected holes recombine before traveling completely across it, and L_p represents the average distance traveled in the neutral region before an injected hole recombines. Since p'_n must decrease with increasing x, the constant B in Eq. (6.43) must be zero. The constant A in Eq. (6.43) is determined by applying Eq. (6.40b). Thus, Eq. (6.43) becomes

$$p'_n(x) = p_{n0}(e^{qV_D/kT} - 1)e^{-(x-x_n)/L_p} \tag{6.44}$$

as shown in Fig. 6.9.

Now we can find the expression for hole current. The hole current flows only by diffusion, since we have assumed that the electric field is negligible in the neutral regions, therefore, from Eqs. (3.46) and (6.44) we obtain[7]

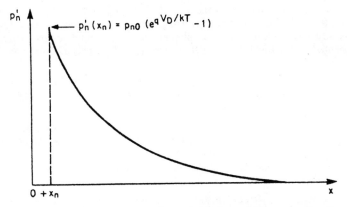

FIGURE 6.9 Variation of hole concentration in the neutral *n*-type region of a long-base diode under forward bias V_D.

$$J_p(x) = -qD_p \frac{dp_n'}{dx} = qD_p \frac{p_{n0}}{L_p}(e^{qV_D/kT} - 1)e^{-(x-x_n)/L_p}$$

$$= qD_p \frac{n_i^2}{N_D L_p}(e^{qV_D/kT} - 1)e^{-(x-x_n)/L_p} \quad (6.45)$$

The hole current exhibits a maximum at $x = x_n$ and decreases away from the junction because the hole gradient decreases as carriers are lost by recombination.

We can obtain the minority-carrier electron current that is injected into the *p*-type region by an analogous analysis used to obtain Eq. (6.45). Under the assumption that the neutral *p*-type region length $W_p \gg L_n = \sqrt{D_n \tau_n}$, then[7]

$$J_n(x) = qD_n \frac{n_i^2}{N_A L_n}(e^{qV_D/kT} - 1)e^{(x+x_p)/L_n} \quad (6.46)$$

Since the origin for x has been chosen at the physical junction (see Fig. 6.7), x will be negative through the *p*-type region. Hence J_n decreases from the junction as J_p does in the *n*-type region. We obtain the expression for the *diode total current J_D* by summing the minority-carrier components at $-x_p$ and $+x_n$, respectively, as expressed in Eqs. (6.45) and (6.46). Thus,

$$J_D = J_p(x_n) + J_n(-x_p) = qn_i^2 \left(\frac{D_p}{N_D L_p} + \frac{D_n}{N_A L_n}\right)(e^{qV_D/kT} - 1)$$

$$= J_S(e^{qV_D/kT} - 1) \quad (6.47)$$

where J_S is called the *reverse saturation current density*.

6.4.4 Short-Base Diode

The second limiting configuration (that of the *short-base diode*) takes place when the lengths W_n and W_p of the *n*- and *p*-type regions are much shorter than the diffusion lengths

L_p and L_n, respectively. In this case, there is a little recombination in the neutral regions, and practically, all the injected minority carriers recombine at the ohmic contacts at either end of the structure. We can obtain the solutions for this configuration by approximating the exponentials in Eq. (6.43) by the first two terms of a Taylor series expansion. This approximation yields

$$p'_n(x) = C + D\frac{(x - x_n)}{L_p} \tag{6.48}$$

where C and D are constants to be determined by the boundary conditions. Because of the ohmic contact assumption at $x = W_n$, we can write $p'_n(W_n) = 0$, while the boundary condition at $x = x_n$ is provided by Eq. (6.40b) as for the long-base diode analysis. The solution for the excess hole concentration in the n-type region is then

$$p'_n(x) = p_{n0}(e^{qV_D/kT} - 1)\left(1 - \frac{x - x_n}{W'_n}\right) \tag{6.49}$$

where $W'_n = (W_n - x_n)$ is the neutral n-type region length (Fig. 6.7). Equation (6.49) indicates that the excess hole concentration decreases linearly with distance across the n-type region (see Fig. 6.10).

The assumption $W_n \ll L_p$ implies that all the holes diffuse across the n-type region before recombining (no recombination takes place in this region). The assumption of no recombination in the n-type region is also obtained by letting the lifetime approach infinity in Eq. (6.42). The differential equation that follows has a linear solution. A linearly varying concentration implies that the hole current is constant through the n-type region. Then[7]

$$J_p(x_n) = -qD_p\frac{dp'_n}{dx}\bigg|_{x=x_n} = qD_p\frac{p_{n0}}{W'_n}(e^{qV_D/kT} - 1)$$

$$= qD_p\frac{n_i^2}{N_D W'_n}(e^{qV_D/kT} - 1) \tag{6.50}$$

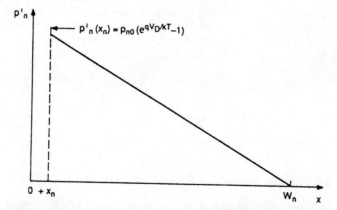

FIGURE 6.10 Variation of holes in the neutral n-type region of a short-base diode under forward bias V_D.

The minority-carrier electron current that is injected into the p-type region can be found by an analogous analysis used to obtain Eq. (6.50). If $W_p \ll L_n$, then[7]

$$J_n(-x_p) = qD_n \frac{n_i^2}{N_A W'_p}(e^{qV_D/kT} - 1) \qquad (6.51)$$

As in the long-base diode analysis, the *total current* in the short-base diode is due to electrons injected into the p-type region and holes injected into the n-type region. Then,

$$J_D = J_p(x_n) + J_n(-x_p) = qn_i^2\left(\frac{D_p}{N_D W'_n} + \frac{D_n}{N_A W'_p}\right)(e^{qV_D/kT} - 1)$$

$$= J_S(e^{qV_D/kT} - 1) \qquad (6.52)$$

where J_S is called the *reverse saturation current density*.

Comparing Eqs. (6.47) and (6.52), we notice that they are similar. In the long-base diode, the characteristic length is the minority-carrier diffusion length; in the short-base diode, it is the length of the neutral region. In practice, a diode can be approximated by a combination of these two limiting configurations; that is, it can be short-base in the p-type region and long-base in the n-type region or vice versa.[1]

Equations (6.47) and (6.52) can be written in the unique form, known as the *Shockley ideal diode equation*,

$$I_D = I_S(e^{qV_D/kT} - 1) \qquad (6.53)$$

where I_S is the *reverse saturation current*, whose expression depends on the diode configuration (long- or short-base), and I_D is the current flowing through the diode. The I_D-V_D characteristics predicted by Eq. (6.53) are shown in Fig. 6.11a on a linear scale. Figure 6.11b shows the circuit symbol of the diode.

Example 6.3
1. Compute the room temperature I_D-V_D characteristics for a p^+n silicon short-base diode with the following characteristics: $W'_n = 10^{-3}$ cm, $W'_p = 10^{-4}$ cm, $N_A = 10^{19}$ cm^{-3}, $N_D = 2 \times 10^{16}$ cm^{-3}, $A_j = 10^{-4}$ cm^2.
2. Assuming that a voltage $V_D = 0.65$ V is applied to the diode, find the current and evaluate the voltage drops in the bulk n- and p-type regions. Refer to App. A for the values of the parameters not specified.

answer
1. First of all we have to calculate the reverse saturation current I_S which, from the hypothesis ($N_A \gg N_D$), simplifies to

$$I_S \simeq qA_j n_i^2 \frac{D_p}{W'_n N_D} = 1.602 \times 10^{-19} \times 10^{-4} \times 2.1 \times 10^{20}$$

$$\times \frac{12.41}{10^{-3} \times 2 \times 10^{16}} = 2.1 \times 10^{-15} \text{ A} \qquad (E6.8)$$

Thus, the I_D-V_D characteristics of the diode under analysis can be found from the relation [Eq. (6.53)]

$$I_D = I_S(e^{qV_D/kT} - 1) = 2.1 \times 10^{-15}(e^{V_D/25.86 \times 10^{-3}} - 1) \qquad (E6.9)$$

with V_D used as a parametric value.

2. For the applied forward bias $V_D = 0.65$ V, from the above relation, we obtain $I_D = 0.173$ mA. The resistance of the bulk regions is given by the relation

$$R = \rho \frac{L}{A_j} \qquad (E6.10)$$

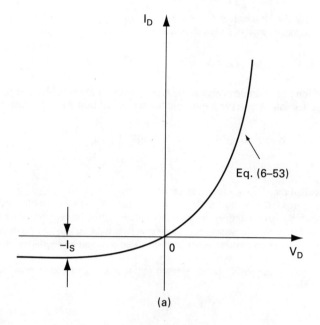

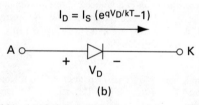

FIGURE 6.11 (a) Ideal diode I_D-V_D characteristic on a linear scale and (b) circuit symbol of the diode.

where L is the length of the resistor (region), A_j is its area, and ρ its resistivity, which can be evaluated from Eq. (3.19). Thus, for the n-type region, being $n \gg p$, Eq. (3.19) reduces to

$$\rho = \frac{1}{q\mu_n n} = \frac{1}{1.602 \times 10^{-19} \times 1350 \times 2 \times 10^{16}} = 0.5 \; \Omega\text{-cm} \quad \text{(E6.11)}$$

Thus, the resistance of the n-type region is

$$R_n = 0.5 \, \frac{10^{-3}}{10^{-4}} = 5 \; \Omega \quad \text{(E6.12)}$$

Analogous arguments hold for the resistance of the p-type region, where $p \gg n$. Then we find $\rho = 1.3 \times 10^{-3}$ Ω-cm and $R_p = 1.3 \times 10^{-3}$ Ω.

The voltage drops in the neutral n- and p-type regions are

$$V_n = I_D R_n = 0.173 \times 10^{-3} \times 5 = 0.86 \text{ mV} \quad \text{(E6.13}a\text{)}$$

$$V_p = I_D R_p = 0.173 \times 10^{-3} \times 1.3 \times 10^{-3} = 0.22 \; \mu\text{V} \quad \text{(E6.13}a\text{)}$$

6.5 CHARGE STORAGE IN THE pn JUNCTION

Charge storage in the depletion region due to the doping concentrations and charge storage due to the minority-carrier charges injected into the neutral regions give rise to two small-signal (see Chap. 9) capacitances usually called the *junction capacitance* (denoted C_j) and the *diffusion capacitance* (denoted C_d), respectively.

6.5.1 Junction Capacitance

The *junction capacitance*, often also called the *depletion* or *transition capacitance*, is associated with the charge dipole formed by the ionized donors and acceptors in the junction depletion region. In particular, the junction capacitance relates the changes in the charge at the edges of the depletion region to changes in the junction voltage. Thus, if the voltage applied to the *pn* junction is changed by a small dV_D, then the depletion-region charge will change by an amount dQ'_J, where Q'_J is the charge per unit area. Hence the depletion region or junction capacitance per unit area can be defined as

$$C'_J = \frac{dQ'_J}{dV_D} \tag{6.54}$$

Since

$$Q'_J = qN_D x_n = qN_A x_p \tag{6.55}$$

in the depletion approximation, then when all the impurities in the depletion region are assumed to be ionized, Eq. (6.54) becomes

$$C'_J = qN_D \frac{dx_n}{dV_D} = qN_A \frac{dx_p}{dV_D} \tag{6.56}$$

From Eqs. (6.28) and (6.30), it follows that

$$\frac{dx_n}{dV_D} = \frac{1}{N_D}\sqrt{\frac{\varepsilon_s}{2q(\phi_0 - V_D)(1/N_A + 1/N_D)}} \tag{6.57}$$

and

$$C'_j = \sqrt{\frac{q\varepsilon_s}{2(\phi_0 - V_D)(1/N_A + 1/N_D)}} = \frac{C'_j(0)}{\sqrt{1 - V_D/\phi_0}} \tag{6.58}$$

where $C'_j(0)$ is the capacitance at equilibrium, i.e., at $V_D = 0$. Figure 6.12 shows a plot of the junction capacitance as a function of the bias voltage based on Eq. (6.58).

Using Eq. (6.30) into Eq. (6.58) yields

$$C'_j = \frac{\varepsilon_s}{W} \tag{6.59}$$

which is the general relationship for a parallel-plate capacitor. It is valid for an arbitrarily doped junction.

6.5.2 Diffusion Capacitance

The *diffusion capacitance*, often also called the *storage capacitance*, is associated with the excess minority-carrier charge injected into the neutral regions under forward bias.

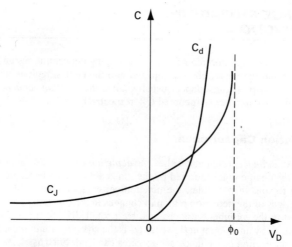

FIGURE 6.12 Junction capacitance C_j and diffusion capacitance C_d as a function of bias voltage V_D.

The total injected minority-carrier charge per unit area stored in the neutral n-type region can be evaluated by integrating the excess hole concentration from the edge of the depletion region across the neutral region:

$$Q'_p = q \int_{x_n}^{W_n} p'_n(x)\, dx \qquad (6.60)$$

An equation similar to Eq. (6.60) can be written for the excess electron charge per unit area Q'_n in the neutral p-type region.

Let us first consider the *ideal long-base diode*. Introducing Eq. (6.44) into Eq. (6.60) and integrating from the edge of the depletion region across the neutral region, we obtain the charge Q'_p due to holes stored in the n-type region as

$$Q'_p = qL_p p_{n0}(e^{qV_D/kT} - 1) \qquad (6.61)$$

By using the expression for the hole current [Eq. (6.45)] at $x = x_n$ in Eq. (6.61), we obtain

$$Q'_p = \frac{L_p^2}{D_p} J_p(x_n) = \tau_p J_p(x_n) \qquad (6.62)$$

Equation (6.62) indicates that the amount of stored charge can be calculated as the product of the current density and the minority-carrier lifetime.

Let us now consider the *ideal short-base diode*. Introducing Eq. (6.49) into Eq. (6.60) and integrating from the edge of the depletion region across the neutral region, we find the charge Q'_p due to holes stored in the n-type region to be

$$Q'_p = \frac{q(W_n - x_n)}{2} p_{n0}(e^{qV_D/kT} - 1) \qquad (6.63)$$

where W_n is the length of the n-type region. Using again the expression for the hole current [Eq. (6.50)], we obtain

$$Q'_p = \frac{(W_n - x_n)^2}{2D_p} J_p = \tau_{dp} J_p(x_n) \tag{6.64}$$

As for the charge storage in the depletion region, the variation of stored minority-carrier charge in the neutral regions under forward bias gives rise to a small-signal capacitance, usually called the *diffusion capacitance* (denoted C_d). The contribution of the stored holes in the n-type region to the diffusion capacitance per unit area C'_d is

$$C'_{dp} = \frac{dQ'_p}{dV_D} = \frac{q^2}{kT} \frac{D_p}{L_p} \tau_p p_{n0} e^{qV_D/kT} \tag{6.65}$$

for the *long-base diode*, and

$$C'_{dp} = \frac{dQ'_p}{dV_D} = \frac{q^2}{kT} \frac{D_p}{W_n - x_n} \tau_{dp} p_{n0} e^{qV_D/kT} \tag{6.66}$$

for the *short-base diode*.

Similar relations can be obtained for C'_{dn}, which takes into account the electron storage in the p-type region. These two components can then be added to define the total diffusion capacitance per unit area C'_d in a straightforward way. The reader can verify this statement as an exercise.

Under reverse bias, storage in the neutral regions is negligible, and C_j dominates. Under forward bias, even if C'_j increases (because the depletion-region width decreases), the exponential dependence of C'_d makes it dominant. This is shown in Fig. 6.12.

Example 6.4 Find (1) ϕ_0 and $C_j(0)$ for an n^+p junction diode with $N_D = 10^{20}/\text{cm}^3$, $N_A = 10^{16}/\text{cm}^3$, and 20×20 $\mu\text{m}^2 = 400 \times 10^{-8}$ cm^2 in area; (2) C_j for $V_D = -5$V. For the parameters not specified refer to App. A.

Answer

1. From Eq. (6.9), we calculate ϕ_0. Thus,

$$\phi_0 = \frac{kT}{q} \ln \frac{N_A N_D}{n_i^2} = 25.86 \times 10^{-3} \times \ln \frac{10^{20} \times 10^{16}}{2.1 \times 10^{20}} = 0.93 \text{ V} \tag{E6.14}$$

From Eq. (6.58), being $V_D = 0$, we obtain

$$C_j(0) = A_j \sqrt{\frac{q\varepsilon_s}{2\phi_0} \frac{N_A N_D}{N_A + N_D}} = 4 \times 10^{-6} \sqrt{\frac{1.602 \times 10^{-19} \times 1.053 \times 10^{-12}}{2 \times 0.93} \times \frac{10^{20} \times 10^{16}}{10^{20} + 10^{16}}} = 0.12 \text{ pF} \tag{E6.15}$$

2. Using the second formulation of Eq. (6.58),

$$C_j(V) = \frac{C_j(0)}{\sqrt{1 - \frac{V_D}{\phi_0}}} \tag{E6.16}$$

we obtain

$$C_j(-5 \text{ V}) = \frac{0.12}{\sqrt{1 + \frac{5}{0.93}}} = 47.5 \text{ fF} \tag{E6.17}$$

6.6 TRANSIENT BEHAVIOR OF THE pn JUNCTION

The discussions of the electrical and physical properties of the *pn* junction in the previous sections have been concerned with the static or steady-state characteristics. We now consider the switching transients in the *pn* junction (the action of the diode turning on and off).

The ideal diode equation [Eq. (6.53)] relates the diode current to the diode voltage; however, the diode current can also be related to the excess minority-carrier charge (charge control analysis). Let us consider, for example, a long-base diode with one region of the junction much more heavily doped than the other. We assume that $N_A \gg N_D$; then $p_{n0} \gg n_{n0}$ and $Q'_p \gg Q'_n$. Thus we consider the minority-carrier holes in the neutral *n*-type region.

Under these conditions, Eq. (6.47) when its exponential term is eliminated by substitution from Eq. (6.61), yields

$$J_D = Q'_p \frac{D_p}{L_p^2} = \frac{Q'_p}{\tau_p} \tag{6.67}$$

Equation (6.67) relates, on a static basis, the diode current density to the excess minority-carrier charge and the excess minority-carrier lifetime τ_p. In the steady state, the current density J_D provides holes to the neutral *n*-type region at the same rate as they are being lost by recombination.

To obtain a dynamic, or time-dependent relation, we must take into account the time rate of change of the excess minority-carrier charge. Then an equation for the time-dependent current density can be formulated as

$$J_D(t) = \frac{Q'_p}{\tau_p} + \frac{dQ'_p}{dt} \tag{6.68}$$

Equation (6.68), usually referred to as the *charge-control equation* for the *pn*-junction diode, states that the current density $J_D(t)$ provides holes to the neutral *n*-type region at the rate at which holes are being lost by recombination plus the rate at which the stored charge increases. Equation (6.68) can be used to calculate the switching times for a *pn*-junction diode.

6.7 CONSIDERATIONS ON THE IDEAL pn JUNCTION

Equation (6.53) has been obtained by assuming the following approximations:

1. *Uniform doping* and electrical neutrality prevail in the *p*-type and *n*-type regions as shown in Fig. 6.5.
2. The built-in potential and applied voltages are sustained entirely across the depletion region (*depletion approximation*).
3. Throughout the depletion region, the Boltzmann relations [Eqs. (6.10)] are valid (*Boltzmann approximation*).
4. The injected minority-carrier concentrations are small compared with the majority-carrier concentrations in the *n*- and *p*-type regions (*low-level injection approximation*).

5. *No generation-recombination* processes exist in the depletion region, i.e., electron and hole currents are considered constant through this region.

Though it can be shown that these assumptions are reasonable for many practical situations of real diodes, there are, however, differences over a significant range of bias for which the ideal diode equation [Eq. (6.53)] becomes inaccurate, whether for forward bias or for reverse bias.

In the following sections we review the main mechanisms that cause the performance of a *pn* junction (diode) to differ from the predictions of the idealized analysis. (A broad analysis of these mechanisms can be found in Refs. 1 to 6). In particular we will consider the following mechanisms:

1. Internal breakdown associated with high reverse bias voltage.
2. Carrier generation-recombination in the depletion region.
3. High-level injection conditions.
4. Voltage drops in the *p*-type and *n*-type regions.

6.8 REVERSE BIAS: DEVIATIONS FROM THE IDEAL DIODE BEHAVIOR

At large reverse voltages, the I_D-V_D characteristics for real diodes departs from the theoretically predicted relationship [Eq. (6.53)] or shape (Fig. 6.11). This deviation is due to carrier multiplication effects taking place in the depletion region when a critical value of electric field (breakdown field) is exceeded, as well as carrier generation in the depletion region.

In particular, at high fields, semiconductors can suddenly become highly conductive because of the internal generation of extra free carriers. When this takes place in a reverse-biased *pn* junction, there is a sudden increase in current. Therefore, the phenomenon is referred to as *breakdown*.

6.8.1 Junction Breakdown

Two mechanisms of diode breakdown can be recognized: the *avalanche breakdown* and *Zener breakdown*.

Avalanche Breakdown. A thermally generated carrier (e.g., an electron) is accelerated by the electric field in the depletion region and acquires kinetic energy. If this kinetic energy is at least equal to the energy bandgap E_g, when the electron collides with a lattice atom, the kinetic energy it loses can be sufficient to break a covalent bond. The result is that, in addition to the original electron, a new electron-hole pair has now been generated. These carriers can be accelerated by the electric field and, on acquiring sufficient kinetic energy ($\geq E_g$), they can generate another electron-hole pair when colliding with lattice atoms. Thus, each new carrier can, in turn, produce additional carriers through collision and action of breaking bonds. This mechanism of *avalanche multiplication* is usually referred to as *avalanche breakdown*. The resulting high reverse current is shown in Fig. 6.13 (dashed curve).

The values of the avalanche breakdown field E_{BR} for silicon and gallium arsenide are

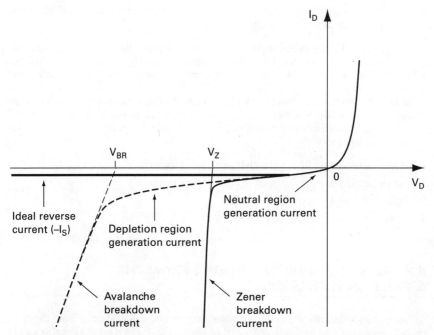

FIGURE 6.13 Diode characteristic showing the effects of avalanche breakdown (dashed line) and Zener breakdown (solid line). Deviation from ideal behavior due to carrier generation in the depletion region, and the ideal reverse current ($-I_S$) are also shown.

about 3×10^5 and 4×10^5 V/cm, respectively. The dependence of the breakdown voltage V_{BR} on the device properties can be expressed as

$$|V_{BR}| = E_{BR}^2 \frac{\varepsilon_s}{2q}\left(\frac{1}{N_A} + \frac{1}{N_D}\right) \tag{6.69}$$

Equation (6.69) shows that we have to reduce the doping concentrations to increase the breakdown voltage. The low doping concentrations imply that the depletion region width is large.

Zener Breakdown. The I_D-V_D characteristic of a diode when Zener breakdown takes place is shown in Fig. 6.13 (solid line). The abrupt increase of the current is due to the *tunneling* effect. Tunneling can occur when charge carriers with a particular energy are separated from empty allowed states at the same energy level by a short physical distance (i.e., less than 5 nm). This condition can occur in a reverse-biased *pn* junction with high doping concentrations on both regions of the junction. The high doping concentrations imply that the depletion region width is small [see Eq. (6.30)]. Though under reverse bias the depletion region width increases, the real separation between the conduction and valence bands decreases. When this gap reaches about 5 nm, the energy levels occupied by electrons in the valence band of the *p*-type region will be aligned with the empty levels in the conduction band of the *n*-type region. Under these conditions, the electrons can *tunnel* from the *p*- to the *n*-type region of the junction.[6] The voltage V_Z at which this phenomenon occurs, is called the *Zener breakdown voltage*. It is generally much lower than that at which avalanche breakdown takes place. We note that if the doping concentrations are such as to make the zero-bias depletion region width large, it will not be possible to increase the reverse bias to V_Z and so to get the necessary physical properties (separation of the energy levels less than 5 nm) for tunneling, because avalanche breakdown will occur first.[6]

With respect to avalanche breakdown, Zener breakdown does not involve collisions of carriers with lattice atoms as does avalanche breakdown.

6.8.2 Carrier Generation in the Depletion Region

The reverse current shown in Fig. 6.13 does not saturate at $-I_S$ as would be indicated by the ideal diode equation [Eq. (6.53)], obtained by assuming that no net generation-recombination rate is present in the depletion region. Under reverse-bias operation, generation in the depletion region becomes important because here the carrier concentrations are less than their thermal equilibrium values ($pn < n_i^2$). Each thermally generated carrier contributes to the current in the external circuit. Therefore, a current in excess of $-I_S$ flows through the diode. If G is the generation rate of electron-hole pairs, the additional reverse current can be calculated as[1–3]

$$I_{R\text{-}G} = -qA_j \int_{-x_p}^{x_n} G\, dx = -qA_j \int_{-x_p}^{x_n} \frac{n_i}{\tau_p + \tau_n}\, dx = -qA_j \frac{n_i}{\tau_p + \tau_n} W \qquad (6.70)$$

Equation (6.70) indicates that $I_{R\text{-}G}$ is a function of V_D, since W depends on V_D [Eq. (6.34)]. The expression for G can be obtained from Eq. (3.89) under the above-mentioned conditions, taking into account that the resulting negative value of R implies a net generation of carriers.

The total diode current is then

$$I_D = I_S(e^{qV_D/kT} - 1) + I_{R\text{-}G} \qquad (6.71)$$

The reader may verify that the generation current is significantly larger than the ideal component I_S by comparing the two reverse components ($I_{R\text{-}G}$ and I_S) of a silicon p^+n diode, for example, at room temperature and using typical values.

6.9 FORWARD BIAS: DEVIATIONS FROM THE IDEAL DIODE BEHAVIOR

Figure 6.14 shows the plot of the ideal and real diode characteristics under forward bias on a semilogarithmic scale. An empirical formula that well simulates, in each region, the curves of Fig. 6.14 can be written by modifying the ideal diode equation [Eq. (6.53)] as follows:

$$I_D = I_S(e^{qV_D/nkT} - 1) \qquad (6.72)$$

where we have introduced the parameter n, called the *ideality factor*, which gives a measure of how close to ideal are the conditions under which the real physical device has been manufactured.

As shown in Fig. 6.14, when $n = 1$, the physical data matches the ideal equation. For most present-day silicon devices, we have $1.0 < n < 1.06$ over 5 or 6 decades of current.[2]

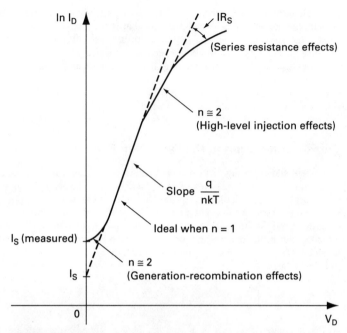

FIGURE 6.14 Forward-bias deviations from the ideal behavior of the diode on a semilogarithmic scale.

Three regions of nonideal behavior are shown in Fig. 6.14. At very small currents (near the extrapolated $V_D = 0$ intercept), the measured current is larger than the ideal equation-predicted I_S. For large currents the slope decreases and eventually no specific slope can be determined. In the following sections we will briefly examine each region.

6.9.1 Carrier Recombination in the Depletion Region

One assumption that defined the ideal diode equation was that the hole and electron current components were constant through the depletion region.

Like the neutral regions, the depletion region contains generation-recombination centers. Since injected carriers, under forward bias, must cross this region, some of them can be lost by recombination. As under reverse bias, where generation of carriers in the depletion region leads to excess current above the reverse saturation current value indicated by the ideal diode equation, also under forward bias we expect a recombination current component which algebraically adds to the ideal current.

The reader should bear in mind that for the ideal diode, the currents injected were determined by the excess carriers at the edges of the depletion region [Eqs. (6.40)], and then by the applied voltage V_D. Thus, at a given value of applied voltage V_D, the carriers, which entirely cross the depletion region, are those injected; in this situation more carriers must enter the depletion region because some are lost due to recombination. The result is a larger total current than indicated by the ideal diode equation at a given value of V_D. Thus, a recombination current component is to be added to the ideal diode diffusion currents.

Deriving the recombination current is possible (see, for example, Ref. 1); however, we only give the result in terms of the modified diode equation, i.e.,

$$I_D = I_S(e^{qV_D/nkT} - 1) + \frac{qA_j n_i}{2\tau_0} W(e^{qV_D/2kT} - 1) \tag{6.73}$$

When the recombination current prevails at low levels of current, the second term of Eq. (6.73) is larger than the first, providing the slope of $q/2kT$ shown in Fig. 6.14. Equation (6.73) is also valid for reverse bias; in this case, the additional component reduces to Eq. (6.70).

6.9.2 High-Level Injection Conditions

We have seen in Sec. 6.4.1 that, under low-level injection conditions, the total minority-carrier concentrations injected into the regions outside the depletion region are much less (and therefore negligible) than the equilibrium majority-carrier concentrations present there. For the pn junction under forward bias, the largest minority-carrier concentration resides at the depletion region edges. At high-level injection, the excess minority-carrier concentrations approach the equilibrium majority concentrations. If the bulk region has to maintain charge neutrality, the majority carrier concentrations must also increase above their equilibrium value. Thus, the recombination term definition, approximated by Eq. (3.92) at low-level injection, is not valid any longer at high-level injection; a redefinition of carrier lifetime is required.[2] The effect of high-level injection can be also modeled by setting the ideality factor to about $n = 2$ in the empirical relation given by Eq. (6.72).

A rough justification of the slope of $q/2kT$ can be derived by considering the law of the junction [see Eqs. (6.38)], applied, for example, to the n-type region of the diode. Thus, multiplying both sides of Eq. (6.38a) by n_{n0}, and taking into account that under high-level injection conditions it holds $p_n \simeq n_{n0}$, we can write

$$p_n n_{n0} \simeq p_n^2 = p_{n0} n_{n0} e^{qV_D/kT} = n_i^2 e^{qV_D/kT} \tag{6.74a}$$

and then

$$p_n \simeq n_i e^{qV_D/2kT} \tag{6.74b}$$

The resulting current is then approximately proportional to $\exp(qV_D/2kT)$ as shown in Fig. 6.14.

At high-level injection, the effect associated to the nonzero resistance of the neutral regions should also be taken into account. This effect will be described in the next section.

6.9.3 Voltage Drops in the *p*-Type and *n*-Type Regions

In obtaining the ideal diode equation, we assumed that the electric field in the neutral n- and p-type regions was about zero and that no voltage drop occurred across the ohmic contacts; these are valid approximations at low current levels. However, at high current levels, the neutral region resistance can cause an appreciable voltage drop and the applied voltage V_D is higher than the voltage across the depletion region. Also, the metal-silicon contacts (see Sec. 6.11) can behave like small resistors, adding to the voltage drop across the n- and p-type neutral regions. Usually these two effects are included into a resistor R_S, the *series resistance*. Figure 6.14 shows the effect of R_S on the I_D-V_D characteristics.

The voltage drops can be evaluated by integrating the electric field E over the neutral regions. Then, if V'_D is the voltage applied to the diode and V_D is the voltage drop across the *pn* junction, we can write

$$V'_D = V_D + \int_{\substack{\text{neutral} \\ \text{regions}}} (-E)\, dx \tag{6.75a}$$

Because V_D is related to the total current I_D by Eq. (6.53), which is the result of the idealized model, Eq. (6.75a) can be written as

$$V'_D = \underbrace{\frac{kT}{q} \ln\left(1 + \frac{I_D}{I_S}\right)}_{V_D} + \underbrace{R_S I_D}_{V_{RS}} \tag{6.75b}$$

where R_S is the series resistance given by

$$R_S = \frac{1}{I_D} \int_{\substack{\text{neutral} \\ \text{regions}}} (-E)\, dx \tag{6.76}$$

Equation (6.75b) suggests that, for small forward currents, the total diode voltage V'_D should vary logarithmically with I_D in accordance with the idealized model. For large forward currents, on the contrary, the voltage should increase linearly with I_D because $I_D R_S$ increases faster than $[(kT/q) \ln(1 + I_D/I_S)]$, and thus V'_D dominates, as shown in Fig. 6.14.

Example 6.5 Consider a reverse-biased *pn* diode illuminated by a photon flux per unit area Φ_0. Assume the applied reverse bias value is less than the value for avalanching. Light absorption in the semiconductor produces hole-electron pairs. Pairs generated in the depletion region or within a distance from it less than the diffusion length of it will eventually be separated by the electric field, leading to current flow in the external circuit as carriers drift across the depletion region.

Under the assumptions that the modulation frequency of the incident light is low, the thermal generation current can be neglected, and the surface *p*-type region is much thinner than $1/\alpha$, α (cm^{-1}) being the absorption coefficient, which depends on the incident light wavelength, calculate the total current density through the diode under steady-state conditions. Let the hole-electron generation rate be given by

$$G(x) = \Phi_0 \alpha e^{-\alpha x} \tag{E6.18}$$

Answer The total current density through the reverse-biased depletion region is

$$J_{\text{tot}} = J_{\text{drift}} + J_{\text{diff}} \tag{E6.19}$$

where J_{drift} is the drift current density due to the carriers generated in the depletion region and J_{diff} is the diffusion current density due to the carriers generated in the bulk of the semiconductor and diffusing into the reverse-biased junction. Now we evaluate these two components.

The drift current density J_{drift} is given by

$$J_{\text{drift}} = -q \int_{-x_p}^{x_n} G(x)\, dx = q\Phi_0 (1 - e^{-\alpha W}) \tag{E6.20}$$

where W is the depletion-region width. For $x > W$, the minority-carrier concentration (holes) in the semiconductor is determined by the one-dimensional diffusion equation

$$D_p \frac{\partial^2 p_n}{\partial x^2} - \frac{p_n - p_{n0}}{\tau_p} + G(x) = 0 \tag{E6.21}$$

where D_p is the diffusion coefficient for holes, τ_p the lifetime of excess carriers, and p_{n0} the equilibrium hole concentration. The solution of Eq. (E6.21) under the boundary conditions $p_n(\infty) = p_{n0}$ and $p_n(W) = 0$ is

SEMICONDUCTOR JUNCTIONS

$$p_n = p_{n0} - (p_{n0} + C_1 e^{-\alpha W}) e^{(W-x)/L_n} + C_1 e^{-\alpha x} \tag{E6.22}$$

with
$$L_p = \sqrt{D_p \tau_p} \tag{E6.23}$$

and
$$C_1 \equiv \left(\frac{\Phi_0}{D_p}\right) \frac{\alpha L_p^2}{1 - \alpha^2 L_p^2} \tag{E6.24}$$

The diffusion current density J_{diff} is

$$J_{\text{diff},p} = -qD_p \frac{\partial p_n}{\partial x}\bigg|_{x=W} = q\Phi_0 \frac{\alpha L_p}{1 + \alpha L_p} e^{-\alpha W} + qp_{n0} \frac{D_p}{L_p} \tag{E6.25}$$

and the total current density J_{tot} is obtained as

$$J_{\text{tot}} = q\Phi_0 \left(1 - \frac{e^{-\alpha W}}{1 + \alpha L_p}\right) + qp_{n0} \frac{D_p}{L_p} \tag{E6.26}$$

Under normal operating conditions, the term involving p_{n0} in the above relation is small compared to the other term, so that the total photocurrent is proportional to the photon flux. The above results hold for low modulation frequency of the incident light. In fact, as the modulation frequency of the incident light is increased, the measured photocurrent is reduced and a phase shift between the photon flux and the photocurrent takes place.

6.10 pn JUNCTION (DIODE) MODELS

The equations obtained in the previous sections link the external observables of current and voltage to the internal semiconductor parameters of the diode.

In analyzing circuits containing diodes, it is useful to be able to model the diode in terms of its *equivalent circuit* components. The more accurate the model is, the more adequately it represents the physical device behavior. High accuracy in a model, on the other hand, implies a large number of complex sets of equations (each describing a particular phenomenon or effect in the real device) to be solved. This is easily possible by using the so-called circuit simulation programs. The program which we refer to in this book is SPICE, probably the most widely used circuit simulator. A comprehensive description of SPICE and of the built-in models for the most common electronic devices can be found in Ref. 7.

The *pn*-junction (diode) models presented in this section are then those implemented in SPICE. The models are shown in Fig. 6.15. We will briefly examine each of them.

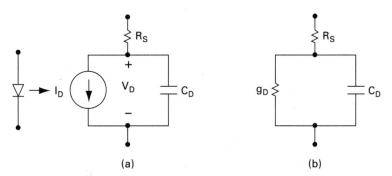

FIGURE 6.15 Diode models. (*a*) Static and large-signal and (*b*) small-signal.

The *static model* of the diode is defined by the nonlinear voltage-dependent current source I_D (representing the *pn* junction) whose value is determined by Eq. (6.72), here rewritten for convenience:

$$I_D = I_S(e^{qV_D/nkT} - 1) \tag{6.77}$$

in series with a resistor R_S (modeling the resistances of the neutral *p*- and *n*-type regions and the high-level injection effects).

The *large-signal model* of the diode (which accounts for charge storage effects), includes capacitors associated with the ionic charge in the depletion region, and, under forward bias, with the storage of the injected minority carriers outside the depletion region. Thus, charge storage in the diode is modeled by the voltage-dependent capacitor per unit area,

$$C_D' = \frac{dQ_D'}{dV_D} = \underbrace{\tau_T \frac{qI_S}{nkT} e^{qV_D/nkT}}_{C_d'} + \underbrace{\frac{C_j'(0)}{\left[1 - \dfrac{V_D}{\phi_0}\right]^m}}_{C_j'} \tag{6.78}$$

Charge storage in the depletion region is taken directly from Eq. (6.58) where *m* is the junction grading coefficient (*m* = 0.5 for a step junction, *m* = 0.33 for a linearly graded junction).

Charge storage due to minority carrier injection across the junction is described by the exponential term in Eq. (6.78), which relates the diode current to the minority carrier charge and lifetime, as given in Eq. (6.65). In Eq. (6.78) the transit time parameter τ_T is equal to τ_n or τ_p, depending on which of these parameters applies.

The models of the diode discussed so far maintain the strictly nonlinear nature of the physics of the device, and therefore are able to represent, both statically and dynamically, the behavior of the diode for both types of bias voltage V_D. In some circuit situations, however, the characteristics of the device must be represented only in a restricted range of currents and voltages.

In particular, for small variations around the operating point (V_D^*, I_D^*) which is usually fixed by a constant source, the nonlinear characteristics of the diode can be *linearized*, so that the incremental current of the diode becomes proportional to the incremental voltage, if the variations are sufficiently small. The linearization of the nonlinear relationship of Eq. (6.72) can be obtained by means of an expansion in Taylor series truncated after the first-order term. Consequently we introduce a small-signal conductance as far as the static behavior is concerned[7]:

$$g_D = \left.\frac{dI_D}{dV_D}\right|_{Op} = \frac{qI_S}{nkT} e^{qV_D^*/nkT} \tag{6.79}$$

and a small-signal capacitance (which plays an important role in determining the frequency response of the diode)[7]:

$$C_D = \left.\frac{dQ_D'}{dV_D}\right|_{Op} = \tau_D g_D + C_j(0)\left(1 - \frac{V_D}{\phi_0}\right)^{-m} \tag{6.80}$$

The *small-signal linearized model* for the diode is shown in Fig. 6.15*b*.

6.11 MS JUNCTION

In the *pn* junction described in the previous sections, an *n*-type and a *p*-type semiconductor were joined together. An equivalent and possibly somewhat simpler junction can be formed if a semiconductor is joined together with a metal. We know that if two materials, characterized by different Fermi energy levels (and therefore not in thermal equilibrium with each other) are placed sufficiently close to each other to interact, electrons will flow from the material with the higher Fermi energy to the material with the lower Fermi energy.

This statement is the basis for the metal-semiconductor (MS) junction theory, which predicts

1. *Blocking* contacts and *rectifying* behavior (diode configuration) for *n*-type semiconductors if the metal work function $q\phi_m$ is greater than the semiconductor work function $q\phi_s$
2. *Ohmic* behavior (ohmic contact configuration) if $q\phi_s$ is greater than $q\phi_m$.

The inverse is true for metal junctions with *p*-type semiconductors. These two operational modes of the MS junction will be described in this section.

6.11.1 Energy-Band Diagram of the MS Junction

A simplified energy band diagram of an ideal metal-semiconductor (*n*-type) junction in thermal equilibrium is shown in Fig. 6.16.

The quantities $q\phi_m$ and $q\phi_s$ are called the metal and the semiconductor *work functions*, respectively, the work function being defined as the energy difference between the vacuum level E_{vac} and the Fermi level E_F. Another important quantity is the *electron affinity* $q\chi_s$ of the semiconductor which corresponds to the energy difference between the vacuum level E_{vac} and the bottom of the conduction band E_c. Typical values for electron affinity range from 2 to 6 eV for metals and from 3 to 5 eV for semiconductors (e.g., for silicon, $q\chi_s = 4.05$ eV). At equilibrium, the Fermi level must be constant throughout the entire metal-semiconductor system. On the other hand, in general, each insulated material has a different Fermi level, and then we expect, once the junction is formed, that a depletion region of width x_n (for the structure of Fig. 6.16) is established at the semiconductor surface.

In the ideal MS junction shown in Fig. 6.16, an energy barrier (*Schottky barrier* for the electrons in the metal)

$$q\phi_{Bn} = q(\phi_m - \chi_s) \tag{6.81}$$

appears between the metal and the *n*-type semiconductor. This barrier height is essentially independent of the doping in the semiconductor. For a *p*-type semiconductor, the Schottky barrier height, for the holes in the metal, is given by

$$q\phi_{Bp} = E_g - q(\phi_m - \chi_s) \tag{6.82}$$

where E_g is the energy gap.

Thus, we can write

$$q(\phi_{Bn} + \phi_{Bp}) = E_g \tag{6.83}$$

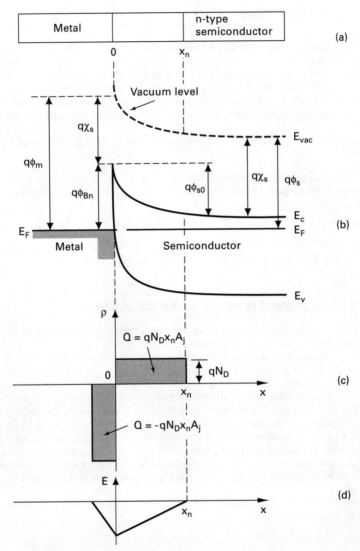

FIGURE 6.16 MS junction. (*a*) Idealized structure, (*b*) idealized equilibrium energy band diagram, (*c*) charge density distribution, and (*d*) electric field. The physical junction is at $x = 0$.

In practice, the Schottky barrier heights are quite different from those predicted by Eqs. (6.81) and (6.82), and some models have been developed to explain the mechanisms determining Schottky barriers.

However, many properties of Schottky barriers can be understood independently of the exact mechanism determining the barrier height; that is, we can simply determine the effective barrier height from experimental data.[4]

Once the Schottky barrier height is known, the variation of the charge density ρ, the electric field E, and the potential ϕ in the semiconductor depletion region (formed because of the difference of the work function of the two materials) can be found by using the depletion approximation. Consider an n-type semiconductor with a donor concentration N_D and a depletion region width x_n. Defining $x = 0$ as the metal-semiconductor interface, we have inside the depletion region ($0 < x < x_n$), in analogy to the pn junction,

$$\rho = qN_D \tag{6.84}$$

$$E(x) = -\frac{qN_D}{\varepsilon_s}(x_n - x) \tag{6.85}$$

$$\phi(x) = -\frac{qN_D}{2\varepsilon_s}(x_n - x)^2 = -\phi_{s0}\left(1 - \frac{x}{x_n}\right)^2 \tag{6.86}$$

where ϕ_{s0} is the surface potential across the depletion region, defined as

$$\phi_{s0} = \phi(x_n) - \phi(0) = \frac{qN_D}{2\varepsilon_s}x_n^2 \tag{6.87}$$

The potential function $\phi(x)$ can be thought as the horizontal mirror image of the energy band diagram of Fig. 6.16b.

From Fig. 6.16b we also see that the surface potential ϕ_{s0} is related to the barrier potential by the relation

$$q\phi_{s0} = q\phi_{Bn} - (E_c - E_F) = q(\phi_m - \phi_s) \tag{6.88}$$

The surface potential ϕ_{s0} is usually called the *built-in voltage* of the MS junction in analogy with the corresponding quantity ϕ_0 in a pn junction. Thus, in equilibrium, the potential barrier seen by the electrons in the conduction band of the semiconductor is equal to ϕ_{s0}, while the potential barrier seen by the electrons in the Fermi level of the metal is equal to ϕ_{Bn}. Analogous considerations hold for holes when a p-type semiconductor is used for the MS junction.

Still, in analogy to the pn junction analysis, the depletion region width, with no voltage applied, is given by

$$x_n = \sqrt{\frac{2\varepsilon_s \phi_{s0}}{qN_D}} \tag{6.89}$$

If a positive voltage V_D is applied to the metal with respect to the semiconductor, the surface potential becomes

$$\phi_{ST} = \phi_{s0} - V_D \tag{6.90}$$

A reverse bias V_D applied to the MS junction adds to ϕ_{s0} and widens the depletion-region width x_n, which extends entirely into the semiconductor. On the other hand, a forward bias V_D applied to the MS junction subtracts from ϕ_{s0} and narrows the depletion-region width x_n, allowing the electrons to flow from the n-type semiconductor into the metal. These mechanisms are identical to those encountered in the analysis of the pn junction.

From the similarity to the pn junction, we can write the junction capacitance per unit area for the MS junction as

$$C'_j = \frac{\varepsilon_s}{x_n} = \sqrt{\frac{q\varepsilon_s N_D}{2\phi_{ST}}} \tag{6.91}$$

In this case, as for the *pn* junction, the plot of the square of the reciprocal of the C_j versus the bias voltage V_D is a straight line if the doping concentration in the semiconductor is uniform. The slope of the straight line can be used to obtain the doping in the semiconductor, and the intercept of the straight line with the abscissa equals ϕ_{s0}.

6.11.2 Current-Voltage Characteristic I_D-V_D

Unlike the *pn* junction, where minority carriers are relevant for conduction, current transport in the MS junction is mainly due to majority carriers. The dependence of current I_D on applied voltage V_D in an MS junction can be obtained by integrating the equations for carrier diffusion and drift across the depletion region near the contact. This approach (*diffusion theory*), which was first taken by Schottky,[8] assumes that the depletion region width is sufficiently large in respect of the mean free paths of the carriers in the semiconductor, and that the electric field is less than that at which the carrier drift velocity saturates.

An alternative physical approach (*thermionic theory*), adopted first by Bethe[9] and based on carrier emission from the metal which cross over the barrier, assumes that the depletion region width is small in respect to the mean free paths of the carriers in the semiconductor. A derivation of the I_D-V_D relation based on this latter approach is possible. However, as both the theories lead to the same I_D-V_D dependence, here we choose to deal with the diffusion approach in order to give continuity and consistence with the *pn* junction analysis developed in the previous sections of this chapter. We present the results in terms of the modified current-voltage equation. Then, if we consider the MS junction of Fig. 6.16 under bias, and we refer to the one-dimensional electron flow through the barrier region, from Eq. (3.47) we can write

$$J_n(x) = q\left[n(x)\,\mu_n E(x) + D_n\frac{dn(x)}{dx}\right] = qD_n\left[-\frac{q}{kT}n(x)\frac{d\phi(x)}{dx} + \frac{dn(x)}{dx}\right] \quad (6.92)$$

where J_n is the electron (*majority carriers*) current density in the depletion region. Equation (6.92) can be rewritten in an integrated form, and evaluated at the edges of the depletion region ($x = 0$ and $x = x_n$). Thus, multiplying both sides of Eq. (6.92) by the integration factor $\exp(-q\phi(x)/kT)$, and using the depletion region edges as integration limits, we obtain

$$J_n\int_0^{x_n} e^{-q\phi(x)/kT}\,dx = qD_n[n(x)e^{-q\phi(x)/kT}]_0^{x_n} \quad (6.93)$$

In writing Eq. (6.93) we have assumed that the current J_n does not depend on position (this is true at steady-state conditions) and thus we have placed it outside of the integral.

The boundary conditions on $\phi(x)$ are

$$q\phi(0) = -q\phi_{Bn} \quad (6.94a)$$

$$q\phi(x_n) = -(E_c - E_F) - qV_D \quad (6.94b)$$

when Eqs. (6.86) and (6.90) are referred to. Boundary conditions for n are also necessary in order to evaluate Eq. (6.93). Thus, from Eqs. (2.34), we write

$$n(0) = N_c e^{-q\phi_{Bn}/kT} \quad (6.95a)$$

$$n(x_n) = N_c e^{-(E_c - E_F)/kT} \quad (6.95b)$$

SEMICONDUCTOR JUNCTIONS

assuming that the electron concentrations at the depletion region edges maintain their equilibrium values.

Thus, integration of Eq. (6.93) can be performed, obtaining for $q\phi_{ST} \gg kT$ (i.e., for reverse bias and low forward bias),

$$J_n = J_{SD}(e^{qV_D/kT} - 1) \tag{6.96}$$

where
$$J_{SD} = \frac{q^2 D_n N_c}{kT} \sqrt{\frac{2qN_D(\phi_{s0} - V_D)}{\varepsilon_s}} e^{-q\phi_{Bn}/kT} \tag{6.97}$$

is the *saturation current density* of the Schottky barrier diode.

Equation (6.96) is formally similar to that obtained from the *pn* junction analysis and it points out the rectifying property of the Schottky barrier, even if the two saturation current expressions are quite different.

Equation (6.97) shows that J_{SD} depends on the applied voltage. However, the square-root dependence of J_{SD} on voltage is weak compared to the term that is exponential in voltage in Eq. (6.96). It is, therefore, possible to approximate the current-voltage characteristic by writing, instead of Eq. (6.96),

$$J_n = J'_{SD}(e^{qV_D/nkT} - 1) \tag{6.98}$$

where J'_{SD} is independent of voltage and n is taken to be a constant having a value that is usually found experimentally to be between 1.02 and 1.15.[1]

The same I_D-V_D dependence expressed by Eq. (6.96) can be obtained by using the thermionic theory. As stated before, we present the results in terms of the modified I_D-V_D equation only.

Thus, from the thermionic approach, we have

$$J_n = J_{ST}(e^{qV_D/kT} - 1) \tag{6.99}$$

where
$$J_{ST} = A^* T^2 e^{-q\phi_{Bn}/kT} \tag{6.100}$$

is the *saturation current density* of the Schottky barrier, and

$$A^* = \alpha \frac{4\pi m_n^* q k^2}{h^3} \tag{6.101}$$

is the *Richardson constant,* measured in A/(cm^2-K^2), α is an empirical factor that accounts for deviations from the simple theory, h is the Planck constant, and k is the Boltzmann constant.

The expressions for the current density obtained from the diffusion and thermionic theory [see Eqs. (6.96) and (6.99)] are very much alike. The saturation current density J_{SD} obtained from the diffusion theory, however, varies more rapidly with voltage (but it is less temperature-dependent) than the saturation current density J_{ST} obtained from the thermionic theory.

The general conclusion of both theories is that the current flowing through the MS junction is caused by the emission of majority carriers. As a consequence, carrier storage effects in the structure, due to storage of minority carriers, are negligible, and therefore the MS junction provides very fast switching. This conclusion is certainly true under low-level injection conditions, where experimental data shows that the minority carrier current component has negligible values compared to the majority-carrier component. The current-voltage characteristic of a Schottky barrier diode is shown in Fig. 6.17*a*.

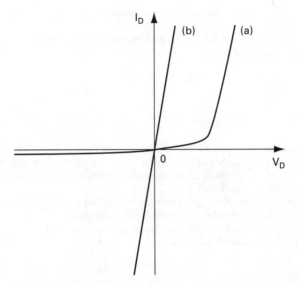

FIGURE 6.17 MS junction I_D-V_D characteristic. (*a*) Schottky barrier behavior and (*b*) ohmic contact behavior.

6.11.3 MS Junction as an Ohmic Contact

An *ohmic contact* has a linear current-voltage characteristics (see Fig. 6.17*b*) and a very small resistance that is negligible compared to the resistance of the active region of the MS diode. Ideally, the ohmic contact to an *n*-type semiconductor should be made by using a metal with a lower work function than that of the semiconductor. A practical way to obtain a low resistance ohmic contact is to increase the doping near the metal-semiconductor interface to a very high value so that the depletion region caused by the Schottky barrier becomes very thin and the current transport through the barrier is enhanced by tunneling. Under these conditions, current flows in either direction. What about is sufficient to the aim of the book; more information can be found, for example, in Ref. 4.

PROBLEMS

6.1 Calculate built-in voltage, depletion region widths, and maximum electric field in a silicon plane-abrupt *pn* junction at 300 K for doping concentrations $N_A = 10^{15}$ atoms/cm³ and $N_D = 10^{16}$ atoms/cm³. Assume a reverse bias of –10 V. For the parameters not specified, refer to App. A.

6.2 Derive the forward voltage V_D of a p^+n junction diode as a function of temperature at a given current density J_D. Could a p^+n junction be used as a thermometer?

6.3 Calculate the theoretical saturation current I_S for an abrupt silicon diode at 300 K with area A_j

SEMICONDUCTOR JUNCTIONS   175

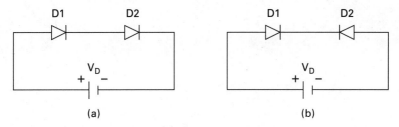

FIGURE 6.18 p^+n diode (D_1) and *Schottky* diode (D_2) connected in (*a*) series and (*b*) back to back.

= 2500 μm², $N_A = 10^{16}$ atoms/cm³, $N_D = 10^{18}$ atoms/cm³, $L_n = 10^{-3}$ cm, $L_p = 10^{-4}$ cm. Refer to App. A for the values of the quantities not given.

6.4   A silicon *pn* junction has a resistivity of 0.2 Ω-cm and 2 Ω-cm for the uniformly doped *p*- and *n*-type regions, respectively. Using for μ_n, μ_p, and n_i the values given in App. A, calculate: (*a*) the built-in voltage of the junction, (*b*) the diode saturation current at room temperature (the minority carrier lifetimes for carriers in the *p*- and *n*-type regions are 20 μs and 40 μs, respectively, the sample cross section is 0.05 cm²), and (*c*) the temperature dependence (and plot) of the saturation current, neglecting, for simplicity, the temperature dependence of mobility (see App. A for the value of E_g).

6.5   A p^+n diode (D_1) and a Schottky diode (D_2) are connected as shown in Fig. 6.18*a* and *b*, and operate at room temperature. Assuming: the Richardson constant, $A^* = 5$ A/cm²-K²; a Schottky barrier height, $\phi_{Bn} = 0.9$ eV; a cross-sectional area of both diodes, $A_j = 10^{-2}$ cm²; a length of the *n*-type region, $W_n = 3$ μm; a doping concentration of the *n*-type region, $N_A = 10^{15}$/cm³; the hole lifetime in the *n*-type region, $\tau = 10^{-6}$ s. Using, for the quantities not specified, the values indicated in App. A, assuming an applied voltage $V_D = 0.9$ V, and assuming ideal diode equations, calculate the electric current and voltage drops across the diodes.

REFERENCES

1. R. S. Muller, T. I. Kamins, *Device electronics for integrated circuits,* New York: John Wiley & Sons, 1977.
2. G. W. Neudeck, *The pn junction diode,* from *Modular Series on Solid State Devices,* Vol. I, Reading, Mass.: Addison-Wesley, 1983.
3. S. M. Sze, *Physics of semiconductor devices,* New York: John Wiley & Sons, 1981.
4. M. Shur, *Physics of semiconductor devices,* Englewood Cliffs, N.J.: Prentice-Hall, 1990.
5. J. K. Hess, *Advanced theory of semiconductor devices,* Englewood Cliffs, N.J.: Prentice-Hall, 1988.
6. D. L. Pulfrey and N. G. Tarr, *Introduction to microelectronic devices,* Englewood Cliffs, N.J.: Prentice-Hall, 1989.
7. G. Massobrio and P. Antognetti, *Semiconductor device modeling with SPICE,* 2d ed., New York: McGraw-Hill, 1993.
8. W. Schottky, *Naturwissenschaften,* 26, 843, 1938.
9. H. A. Bethe, "Theory of boundary layer of crystal rectifiers," Report 43-12, MIT Radiation Laboratory, Cambridge, Mass., 1943.

CHAPTER 7
SOLID-ELECTROLYTE JUNCTIONS AND MEMBRANE TRANSPORT

When two different pieces of matter are brought into contact, very often new physicochemical properties originate at their interface. These properties can be foreseen by the designer, as in the case of the junction between a p-type semiconductor and an n-type semiconductor (as described in the previous chapter), or they can rather be a complex problem to deal with, as in the case of an artificial structure to be interfaced to the human body (i.e., a prosthesis). Broadly speaking, the appropriate operation of all kinds of hybrid bioelectronic devices, e.g., biosensors, critically rests on the precise characterization of the junction between biological and artificial structures.

To make a junction, two solids (as described in Chap. 6), or a solid and a liquid solution (as we will describe in this chapter) are brought into contact. Another interesting situation is produced when two liquid regions are separated by a thin structure designed (by human beings or nature) to selectively distribute specific molecules among the two liquid regions. These "thin structures" are named *membranes* (the structure of a biological membrane has been already introduced in Chap. 4). Examples of membranes relevant to two quite separate research fields include the ion-selective membrane of a neuron and the artificial membranes designed for dialysis purposes. To summarize, solid-liquid junctions and membranes are two very important topics for any student or researcher dealing with bioelectronics and, more generally, with bioengineering.

In the following, the basic properties of solid-liquid interfaces and the transport properties of membranes will be considered in detail.

7.1 ELECTRODE-ELECTROLYTE INTERFACES

Electrical potential differences can develop across the boundary between a solid phase and a liquid phase, in particular an electrolyte. There are several ways in which this potential difference can arise. If one of the phases is an electronic conductor, i.e., an electrode, and the other is an ionic conductor (electrolyte), electron-transfer reactions can occur at the boundary and lead to the development of a potential difference. Alternatively, the electronic conductor can be deliberately charged by a flow of electrons from an external source of electricity. Even without deliberate charging or steady electron-transfer reactions, a potential difference can develop across a solid-electrolyte boundary, typically due

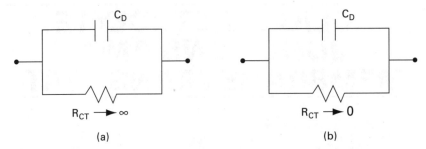

FIGURE 7.1 Equivalent circuit of (*a*) ideally polarizable interface and (*b*) ideally nonpolarizable interface. R_{CT} = charge-transfer resistor and C_D = double-layer capacitor.

to an initial (very small and transient) charge exchange between electrode and electrolyte solution.

In consideration of the vastness of the topic, we will start with a classification of different electrode-electrolyte interfaces. Then, the general scheme of the Poisson-Boltzmann equation will be considered. Finally, the relevance of this topic to the study of biosensors (see Chap. 10) will be briefly discussed.

7.1.1 Nonpolarizable and Polarizable Interfaces

To polarize an interface means to alter the potential difference across it. Thus, an ideally *nonpolarizable* interface is characterized by the fact that the potential difference across it is virtually fixed. On the contrary, in an ideally *polarizable* (or *blocking*) interface, the potential difference changes as a consequence of any variation of the potential difference across the whole system which includes the interface (the system consisting of an electrode immersed in an electrolyte solution, a "reference" electrode, and a power supply). A way to visualize the properties of polarizable/nonpolarizable interfaces is to make use of the simple equivalent circuit representation of Fig. 7.1.

Suppose the capacitor-resistor configuration in Fig. 7.1 is connected to a source of potential difference. Then, if the resistor is very high, the capacitor charges up to the value of the potential difference set by the source; this is the behavior of a polarizable interface (Fig. 7.1*a*). On the other hand, if the resistance in parallel with the capacitor is low, then any attempt to change the potential difference across the capacitor is compensated by charge leaking through the low-resistance path; this is the behavior of a nonpolarizable interface (Fig. 7.1*b*). This description can be generalized to the frequency domain by writing down the frequency-dependent equivalent impedance Z of the circuit, in the presence of a periodic potential:

$$Z(\omega) = \frac{R_{CT}}{j\omega R_{CT} C_D + 1} \quad (7.1)$$

where R_{CT} stands for charge-transfer resistance, C_D for double layer capacitance, and $\omega = 2\pi \times$ frequency. The meaning of the resistance and capacitance will be clarified in the following.

Examples of nonpolarizable interfaces are the so-called *reference electrodes* (e.g., the calomel electrode and the Ag/AgCl electrode). These electrodes guarantee a constant potential drop at their interface and consequently they act as a "reference."

At the other extreme, the classic example of a polarizable interface is the interface between mercury (Hg) and electrolyte solution. Mercury is a liquid at ordinary temperatures. For (solid) metals other than mercury, ideal polarizability (i.e., infinite resistance R_{CT}) is not obtained. Only the softer metals, such as lead, tin, and gallium have large (e.g., 1 V) ranges of applied potential in which the interface can be considered an ideal capacitor. Let us briefly analyze in the following sections the behavior of the various kinds of interfaces.

7.1.2 An Ideally Polarizable Electrode (Hg) in an Electrolyte Solution

In an ideally polarizable interface (connected to an external voltage source), a separation of charges happens at the interface between electrode and solution and a potential difference develops across the interface. The separation of charge in the metal implies redistribution of electrons, the separation of charge in the electrolyte implies redistribution of ions (which can be either hydrated or bare) and of water dipoles.

It should be underlined that no electrons leave the ideally polarized metal to cross the interface and no ions and water dipoles leave the electrolyte to cross the interface. Redistribution of charge on one side is immediately paralleled by redistribution of charge on the other side. This is the behavior of a capacitor and the resulting interface is known as an *electrified interface*. The charging of this interface can be controlled by an appropriate external circuit including a voltage source. Electrons inside the metal are free to move as a kind of gas (see Chap. 1). The charge distribution in the electrolyte is more complex: bare ions can approach the surface of the metal, as water dipoles do. They form a kind of layer, known as the *inner Helmholtz plane* (IHP) (von Helmholtz was one of the first to introduce the concept of electrified interface, around 1879). Hydrated ions form a kind of second layer, known as *outer Helmholtz plane* (OHP). The planes IHP and OHP are sketched in Fig. 7.2.

In summary, as a result of immersing an ideally polarizable metal (e.g., Hg) in an electrolyte, a redistribution of charge occurs both in the metal and in the electrolyte. The charge at the metal side of the interface is *partially* balanced by the IHP and OHP planes. This is a *partial* balance because diffusion processes take place inside the electrolyte; therefore also a *diffuse excess of charge* has to be taken into account. This diffuse layer, where the potential decays exponentially, is known as the *Gouy-Chapman (G-C) layer*. A classical way of describing this diffuse layer is by solving the Poisson-Boltzmann equation. This will be done in Sec. 7.2.

7.1.3 A Solid Metal in an Electrolyte Solution

There is always leakage of charge across the interface with metals other than mercury.

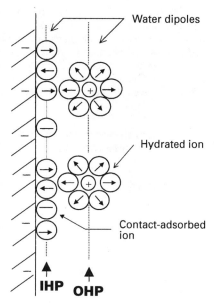

FIGURE 7.2 The inner Helmholtz plane (IHP) and the outer Helmholtz plane (OHP) are shown. The diffuse layer (Gouy-Chapman) is not shown.

Thus, as already mentioned, only mercury and a few of the softer metals can be considered as ideally polarizable over a large (e.g., 1 V) range of potentials. In all the other cases, exchange of electrons does happen across the interface as a result of *reduction* or *oxidation* of dissolved species in solution. A reaction such as

$$O + ne \rightleftarrows R \tag{7.2}$$

with O the reactant and R the product on a Pt electrode, for example, will allow a current to flow (in electrochemical books this current is named *faradaic*). The movement of electrons (*ne*) should be imaged as a kind of "hopping," better described by quantum rather than classic physics.[1] Under equilibrium conditions, electrons cross the electrified interface in both directions. The result is that there is no net current, but two processes, an *electronation* (or reduction) and a *de-electronation* (or oxidation), continue to occur, *at the same rate,* in the presence of an equilibrium potential difference $\Delta\phi_e$ across the interface, which is characteristic of the specific redox process.

In conclusion, at *equilibrium* (i.e., almost immediately after immersing a metal in solution without any external difference in potential applied) there are two currents equal in magnitude and opposite in direction and, as a result, no net current. It can be shown[1,2] that their magnitude i_0 depends exponentially on the potential difference $\Delta\phi_e$. A *net* current can then be produced if the system is taken out of equilibrium, i.e., if an overpotential drop η is added to the equilibrium potential drop resulting in the overall potential difference

$$\Delta\phi = \Delta\phi_e + \eta \tag{7.3}$$

The net current takes the form[1,2]

$$I = I_0[e^{(1-\beta)\eta zq/kT} - e^{-\beta\eta zq/kT}] \tag{7.4}$$

where the *overpotential* η is the "extra part" by which the potential of the electrode departs from that at equilibrium; β is a factor greater than zero but less than unity, known as the *symmetry factor*.[1,2] Of course, if $\eta = 0$, then $I = 0$. Equation (7.4) is known as the *Butler-Volmer* equation, and it represents the starting point for any study concerning non-ideal polarized electrodes in electrolyte solutions. Similarities with the diode equation [Eq. (6.53)] should be evident to the reader. By the way, an electrical equivalent circuit of Eq. (7.4) is given by two diodes connected in reverse-parallel, as shown in Fig. 7.3.

7.1.4 A Semiconductor-Insulator Structure in an Electrolyte Solution

Inside a semiconductor facing an electrolyte, the charge carriers will redistribute according to the rules described in Chap. 6. By adding an insulator (e.g., SiO_2 or Si_3N_4), an elec-

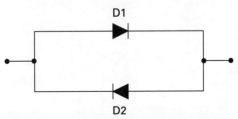

FIGURE 7.3 Two diodes connected in reverse-parallel.

trolyte-insulator-semiconductor (EIS) structure is generated, which is analogous to the metal-oxide-semiconductor (MOS) capacitor, considered in detail in Chap. 8. The two structures are shown for comparison in Fig. 7.4. Note that, in order to polarize the EIS structure, a reference electrode—that is, a virtually nonpolarizable interface—must be inserted to close the circuit.

The presence of an insulator in contact with the electrolyte solution introduces a new kind of charge distribution, which is caused by the formation of *surface groups* that transform the EIS structure into a *pH sensor*. Let us consider this process in some detail.

Let us first consider an SiO_2 insulator exposed to an aqueous solution interacts with H^+ ions in the following way:

$$[SiOH_2^+] \underset{k_1^-}{\overset{k_1^+}{\rightleftharpoons}} [SiOH] + [H^+]_s \tag{7.5a}$$

and

$$[SiOH] \underset{k_2^-}{\overset{k_2^+}{\rightleftharpoons}} [SiO^-] + [H^+]_s \tag{7.5b}$$

The subscript s in $[H^+]_s$ means that the concentration of protons is *near* the surface of the insulator, and $[SiOH_2^+]$, $[SiOH]$, and $[SiO^-]$ are the concentrations of the proton *binding sites* present on the oxide surface. Under equilibrium conditions, the kinetic reactions (7.5a), (7.5b) result in the equilibrium constants

$$K_+ = \frac{[SiOH][H^+]_s}{[SiOH_2^+]} \tag{7.6a}$$

and

$$K_- = \frac{[SiO^-][H^+]_s}{[SiOH]} \tag{7.6b}$$

By multiplying together the two equilibrium constants, we obtain

$$K_+ K_- = \frac{[SiO^-][H^+]_s^2}{[SiOH_2^+]} \tag{7.7}$$

Concentrations of binding sites can be transformed into *fractions of sites* θ_+, θ_0, θ_- giving

$$\frac{[SiO^-]}{[SiOH_2^+]} = \frac{\theta_-}{\theta_+} \tag{7.8}$$

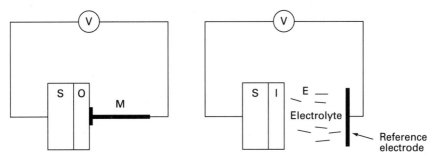

FIGURE 7.4 (*a*) Metal-oxide-semiconductor (MOS) structure and (*b*) electrolyte-insulator-semiconductor (EIS) structure. A reference electrode is present to close the circuit of the EIS. Note that the electrolyte solution plays the role of the conductor.

where θ_-, θ_+, and θ_0 satisfy the constraint

$$\theta_- + \theta_+ + \theta_0 = 1 \tag{7.9}$$

Moreover, under equilibrium conditions, the concentration of protons at the surface of the insulator can be related to the one in the bulk via the Boltzmann equation

$$[H^+]_s = [H^+]_b \, e^{q(\phi_b - \phi_0)/kT} \tag{7.10}$$

where $(\phi_b - \phi_0)$ is the potential drop between electrolyte bulk and insulator surface. By assuming $\phi_b = 0$, Eq. (7.10) reduces to

$$[H^+]_s = [H^+]_b \, e^{-q\phi_0/kT} \tag{7.11}$$

By substitution of Eqs. (7.8) and (7.11) into (7.7), we obtain

$$K_+ K_- = \frac{\theta_-}{\theta_+} [H^+]_b^2 \, e^{-2q\phi_0/kT} \tag{7.12}$$

and, by taking logarithms of both sides and dividing by 2, we obtain

$$\frac{1}{2} \ln (K_+ K_-) = \frac{1}{2} \ln \frac{\theta_-}{\theta_+} + \ln [H^+]_b - \frac{q\phi_0}{kT} \tag{7.13}$$

Note that

$$\ln [H^+]_b \simeq 2.303 \log [H^+]_b = -2.303 \, \text{pH}_b \tag{7.14}$$

Then Eq. (7.13) can be written in pH terms as follows (we drop the subscript b from now on):

$$\frac{1}{2} \ln (K_+ K_-) = \frac{1}{2} \ln \frac{\theta_-}{\theta_+} - 2.303 \, \text{pH} - \frac{q\phi_0}{kT} \tag{7.15}$$

Equation (7.15) holds true for *any* pH value, and, for a given insulator, the ratio θ_-/θ_+ is a given function of pH. In other words, for a given insulator, there is a specific pH value for which

$$\theta_- = \theta_+ \tag{7.16}$$

This specific value is known as the *point of zero charge* of the material and is indicated as pH_{pzc}.

By setting $\phi_0 = 0$ for $\text{pH} = \text{pH}_{pzc}$ (since only differences in potential are ever physically meaningful), we obtain

$$\frac{1}{2} \ln (K_+ K_-) = -2.303 \, \text{pH}_{pzc} \tag{7.17}$$

Thus, we can finally write Eq. (7.17) as

$$2.303(\text{pH} - \text{pH}_{pzc}) = \frac{1}{2} \ln \frac{\theta_-}{\theta_+} - \frac{q\phi_0}{kT} \tag{7.18}$$

or

$$\phi_0 = -2.303 \, \frac{kT}{q} \Delta \text{pH} + \frac{1}{2} \frac{kT}{q} \ln \frac{\theta_-}{\theta_+} \tag{7.19}$$

where

$$\Delta \text{pH} = \text{pH} - \text{pH}_{pzc} \tag{7.20}$$

At room temperature we have

$$2.303 \frac{kT}{q} \simeq 59 \text{ mV} \qquad (7.21)$$

Equation (7.19) relates the potential ϕ_0 and pH and is the basis for utilizing an EIS structure as a pH sensor. The value of ϕ_0 can be deduced by knowing the profile of the potential in the electrolyte, and this will be considered in more detail in the next section. The same procedure followed for SiO_2, can then be utilized for other insulators, such as Si_3N_4. Unfortunately, for Si_3N_4 the reactions are slightly more complicated. On the other hand, Si_3N_4 is much more appropriate than SiO_2 for designing a silicon-based pH meter, so it is worth considering it. At the surface of the Si_3N_4 insulator, *silanol sites, tertiary amine sites, secondary amine sites,* and *primary amine sites* are present. The picture can be approximated by assuming that only silanol sites and basic primary amine sites are present on the surface of the insulator after oxidation.[3,4] Thus the equilibrium constants are

$$K_+ = \frac{[SiOH][H^+]_s}{[SiOH_2^+]} \qquad (7.22a)$$

$$K_- = \frac{[SiO^-][H^+]_s}{[SiOH]} \qquad (7.22b)$$

$$K_{N+} = \frac{[SiNH_2][H^+]_s}{[SiNH_3^+]} \qquad (7.22c)$$

Concentrations of binding sites can again be transformed into fractions of sites as follows:

$$K_+ = \frac{\theta_0}{\theta_+}[H^+]_s \qquad (7.23a)$$

$$K_- = \frac{\theta_-}{\theta_0}[H^+]_s \qquad (7.23b)$$

$$K_{N+} = \frac{\theta_{N0}}{\theta_{N+}}[H^+]_s \qquad (7.23c)$$

with the normalization conditions

$$\theta_+ + \theta_- + \theta_0 = \frac{N_{sil}}{N_s} \qquad (7.24)$$

$$\theta_{N+} + \theta_{N0} = \frac{N_{nit}}{N_s} \qquad (7.25)$$

$$\theta_0 + \theta_+ + \theta_- + \theta_{N+} + \theta_{N0} = 1 \qquad (7.26)$$

where N_{sil} and N_{nit} are the numbers of silanol sites and primary amine sites per unit area, respectively, and N_s is the total number of available binding sites per unit area. The charge density σ_0 of the surface sites on the insulator is given by[4]

$$\sigma_0 = qN_s(\theta_+ + \theta_{N+} - \theta_-) \qquad (7.27)$$

By combining Eqs. (7.22) and (7.27), we get

$$\frac{\sigma_0}{qN_s} = \left(\frac{[H^+]_s^2 - K_+K_-}{[H^+]_s^2 + K_+[H^+]_s + K_+K_-}\right)\frac{N_{sil}}{N_s} + \left(\frac{[H^+]_s}{[H^+]_s + K_{N+}}\right)\frac{N_{nit}}{N_s} \quad (7.28)$$

As for the previously considered SiO_2 insulator, $[H^+]_s$ can then be related to the concentration of protons in the bulk, $[H^+]_b$, via the equilibrium Boltzmann equation:

$$[H^+]_s = [H^+]_b\, e^{-q\phi_0/kT} \quad (7.29)$$

where ϕ_0 is the potential of the electrolyte-insulator interface, referred to the bulk value of the potential. The condition of charge neutrality for the EIS system is

$$\sigma_d + \sigma_0 + \sigma_s = 0 \quad (7.30)$$

where σ_d is the charge density diffuse in the electrolyte, σ_0 is the charge density on the insulator surface, and σ_s is the charge density inside the semiconductor.

As already anticipated for the SiO_2 insulator, the above equations represent the starting point to relate the potential ϕ_0 to the pH of the electrolyte solution. This will be further considered in Sec. 7.2 and then again in Chap. 10, in dealing with ion-sensitive field-effect transistors (ISFETs) and related biosensors.

7.1.5 Colloidal Particles

The *sizes* of the materials in contact with the electrolyte have not quantitatively entered the picture so far. Basically, we can assume that we dealt with single, macroscopic (i.e., at least in the millimeter range) electrodes. The picture becomes quite different if we consider many micrometer-sized particles, each of them generating an electrified interface around itself. Such particles are known as *colloidal particles*. They include inert objects, for example, a suspension of metallic microspheres and living objects such as red blood cells. Micrometer-sized particles belong to the category of "mesoscopic" objects undergoing Brownian motion (see Chap. 5).

Let us consider a population of metallic spheres. As already noticed in Chap. 5, the smaller they are (in the range of micrometers), the more they react to the thermal collisions from the ions and water molecules of the electrolyte; they undergo a random walk through the solution. Large (centimeter-sized) spheres, too, exchange momentum with the particles of the solution, but their masses are huge compared with those of ions or molecules, so that the velocities imparted (to the spheres) from such collisions are essentially zero.

Once the microspheres begin to move in a *Brownian fashion* in the solution, some of them collide with each other. Many aspects of colloidal chemistry are clarified by a consideration of this subject. Each metal sphere feels its environment through its charged interface; each sphere is enveloped in a double layer. All the concepts of the electrified interface developed so far are of relevance to the colliding microspheres. For the sake of simplicity, we will not take into account that we are now dealing with spherical surfaces and not planes. Considering dilute solutions and no contact-adsorbing ions, one can visualize each metal sphere surrounded by a G-C region of diffuse charge. Note, however, that the *G-C layers* of both colliding spheres contain charges of the same sign. Thus, there is Coulomb repulsion as the two spheres come close. The repulsion energy depends on the distance r between the spheres and varies with distance in the same way as the G-C potential. This dependence on distance is approximately given by $\phi_0 e^{-\chi r}$, where χ is the inverse of the *Debye length* and it increases with the ion concentrations (see Sec. 7.2).

Double layers interact with double layers and the metal of one sphere also interacts with the metal of the second sphere. There is what is called the *Van der Waals attraction,* essentially a dispersion interaction, which depends on r^{-6}, and the electron overlap *repulsion,* which varies as r^{-12} (Ref. 1). These interactions between the bulk of the two colloidal metal spheres shall be represented together by a term $(-Ar^{-6} + Br^{-12})$, where A and B depend essentially on the chemical composition of the phase which is dispersed in the solution.

The total interaction between the two metal spheres can then be considered composed of two parts, (1) the surface, or double-layer, interaction determined by the Gouy-Chapman potential $\phi_0 e^{-\chi r}$ and (2) the volume, or bulk, interaction $(-Ar^{-6} + Br^{-12})$. The interaction between double layers becomes repulsive as the particles approach. The bulk interaction leads to an attraction unless the spheres get too close, when there is a sharp repulsion (Fig. 7.5). The total interaction energy U_{total} depends on the interplay of the surface (double-layer) and volume (bulk) effects and may be represented as

$$U_{\text{total}} = \phi_0 e^{-\chi r} + (-Ar^{-6} + Br^{-12}) \tag{7.31}$$

This approximate formula contains information concerning what happens when two colloidal particles collide. Consider one type of energy-distance diagram (Fig. 7.5). It is seen that, for the first type of behavior where the electrostatic repulsion predominates, the net energy U_{total} is always positive; this means that two metal spheres under this condition cannot stick together stably. Note from Fig. 7.5 that, if the spheres did not wrap themselves in double layers, the interaction between the particles themselves, neglecting the double-layer repulsion, would predominate and have a minimum in a negative potential energy region corresponding, therefore, to a favoring of the aggregation of colloidal particles.

Thus, particles of colloidal dimensions survive aggregation into macroscopic phases only because their boundaries develop electrified interfaces. The repulsion between double layers is the key to the stability of colloids.

The structure of an electrified interface and therefore the potential drop across it de-

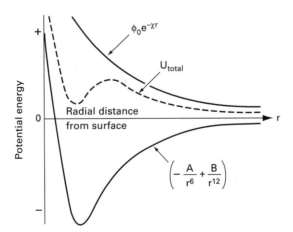

FIGURE 7.5 The energy of interaction between two colloidal particles as a function of their distance apart, in the case of conditions favoring stability of the colloid. (*From O. M. Bockris and A. K. N. Reddy.*[1] *Used by permission.*)

pends on the composition of the electrolyte. The diffuse region can be reduced in thickness and the potential made to fall sharply by concentrating the solution with the addition of some electrolyte. Moreover, contact-adsorbing ions can be added to increase the IHP contribution. All this means that one has, by variation of the solution composition, an indirect control over the double-layer contribution and therefore the total interaction energy for two colloidal particles. In this way, one can control the stability of the colloids.

Aggregation happens by lowering Gouy-Chapman potentials at the r_{min} distance. This is obtained by adding more electrolyte to the solution. As a consequence, χ increases (see Sec. 7.2.1), and, since $\phi = \phi_0 e^{-\chi r}$, ϕ falls more sharply with distance. In other words, the Gouy-Chapman region is compressed, and the total interaction curve becomes negative and shows a minimum at r_{min} (Fig. 7.6). The colloid *has lost its stability*. This is known as *coagulation* or *flocculation*.[1]

Flocculation can be brought about in another way. By contact adsorption of ions, most of the potential drop across the interface can be made to occur between the metal and the IHP. Thus, by the addition of contact-adsorbing ions, the value of ϕ_0 can be reduced without significantly changing the concentration of the bulk electrolyte. The effect of this will be qualitatively similar to that shown in Fig. 7.6 and is shown in Fig. 7.7. The value of U_{total} again comes into the negative potential energy region; i.e., a stable configuration of particles in contact may exist, and a flocculation thus again occurs.

The characteristic behavior of the colloidal state is that double-layer interactions are as significant as bulk interactions. This condition can therefore be realized in all systems where the surface-to-volume ratios are high, i.e., at microscopic dimensions.

A colloidal suspension consisting of discrete, separate particles immersed in a continuous phase is known as a *sol*. A colloidal suspension can also consist of filamentous particles (i.e., macromolecules) dispersed in solution.

Instead of having one phase discontinuous and in the form of separate particles, it is possible to have the phase as a continuous matrix with pores of very fine dimensions through it. This is a porous mass, or membrane, also known as a *gel*. In such membranes, interactions inside the pores become highly dependent on double-layer interactions.

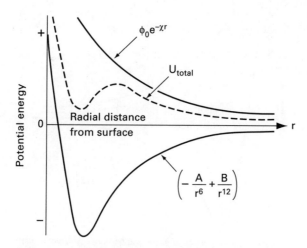

FIGURE 7.6 The energy of interaction between two colloidal particles as a function of their distance apart, for conditions favoring coagulation. (*From O. M. Bockris and A. K. N. Reddy.*[1] *Used by permission.*)

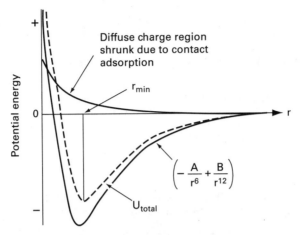

FIGURE 7.7 The effect of the contact adsorption of ions on the condition of the stability of a colloid. (*From O. M. Bockris and A. K. N. Reddy.*[1] *Used by permission.*)

Sols and gels are of great relevance in biological processes. A living cell is separated from the outside by a membrane and inside it can be viewed as a collection of colloidal particles held in suspension by interacting Gouy-Chapman layers. An example of this is given by the electrochemical mechanism of the clotting of blood.

The reader should be aware of the fact that, in comparison to the initial example of metallic microspheres, biological membranes are much more sophisticated objects. With specific reference to interfaces, biological membranes allow us to introduce one more way of charging a surface, i.e., by the adsorption (*binding*) of ions from solution onto a previously uncharged surface. This is the case, for example, of the binding of Ca^{2+} onto the zwitterionic headgroups (i.e., dipole ions) of the surfaces of the lipid bilayer. As a final consideration on living cells, let us conclude that electrified interfaces are essential for them, but charge transfer through them is even more important. This statement will be expanded in the last section of this chapter, when we will consider the flux of matter through membranes.

7.2 SOLUTION OF THE POISSON-BOLTZMANN EQUATION UNDER VARIOUS BOUNDARY CONDITIONS

It should be clear from the previous section that one of the key points for the characterization of an electrode-solution junction is the calculation of the profile of the electrical potential inside the electrolyte. The classical way to obtain this is to solve the *Poisson-Boltzmann equation*. We will do it in the following by first (Sec. 7.2.1) obtaining an expression for $\phi(x)$ from $x = 0$ to $x \to \infty$. The reader should be already aware of the fact that, in accordance with the arguments developed in the previous section, a more appropriate picture should consider solving the Poisson-Boltzmann equation from the *outer Helmholtz plane* to infinity. Moreover, in most practical problems, it is of interest to calculate the potential profile between *two* (or several, as already seen in the case of colloidal particles) electrodes separated by a *finite* distance. This will be considered in Sec. 7.2.2.

7.2.1 Electrical Potential Profile in an Electrolyte in the Presence of an Electrode

Let us consider a lamina in the electrolyte parallel to the electrode and at a distance x from it in the absence of any charge transfer (Fig. 7.8). As for the *pn* junction, the charge density ρ_x can be expressed in two ways: (1) In terms of the Poisson equation, which, for the x dimension in rectangular coordinates reads

$$\rho_x = -\varepsilon_0 \varepsilon_r \frac{d^2 \phi(x)}{dx^2} \tag{7.32}$$

where $\phi(x)$ is the potential difference between the lamina and the bulk of the solution (taken as $\phi(x)_{x \to \infty} = 0$) and (2), in terms of the Boltzmann distribution,

$$\rho_x = \sum_i c_i z_i q = \sum_i c_i^b z_i q e^{-z_i q \phi / kT} \tag{7.33}$$

where c_i and c_i^b are the concentrations of the ith ionic species in the lamina and in the bulk of the solution, respectively, z_i is the valence of the species i, and q is the electronic charge. The factor $z_i q \phi(x)/kT$ represents the ratio of the electrical and thermal energies of an ion at the distance x from the electrode. From the two expressions for the charge density ρ_x in Eqs. (7.32) and (7.33), we obtain the *Poisson-Boltzmann equation* (see also Sec. 8.7)

$$\frac{d^2 \phi}{dx^2} = -\frac{1}{\varepsilon_0 \varepsilon_r} \sum_i c_i^b z_i q e^{-z_i q \phi / kT} \tag{7.34}$$

A simple transformation can now be used, namely

$$\frac{d^2 \phi}{dx^2} = \frac{1}{2} \frac{d}{d\phi} \left(\frac{d\phi}{dx} \right)^2 \tag{7.35}$$

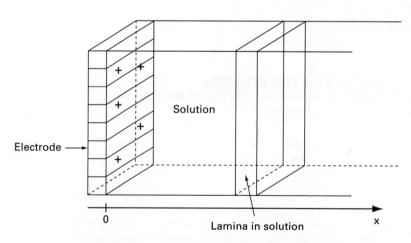

FIGURE 7.8 A lamina in the solution, parallel to a plane electrode.

This identity can be used in the differential Eq. (7.34) to give

$$\frac{d}{d\phi}\left(\frac{d\phi}{dx}\right)^2 = -\frac{2}{\varepsilon_0\varepsilon_r}\sum_i c_i^b z_i q e^{-z_i q\phi/kT} \tag{7.36}$$

or

$$d\left(\frac{d\phi}{dx}\right)^2 = -\frac{2}{\varepsilon_0\varepsilon_r}\sum_i c_i^b z_i q e^{-z_i q\phi/kT} d\phi \tag{7.37}$$

By integrating Eq. (7.37), we obtain

$$\left(\frac{d\phi}{dx}\right)^2 = \frac{2\,kT}{\varepsilon_0\varepsilon_r}\sum_i c_i^b e^{-z_i q\phi/kT} + \text{constant} \tag{7.38}$$

The integration constant can be evaluated by assuming that, deep in the bulk of the solution, i.e., at $x \to \infty$, not only is $\phi(x) = 0$, but the field $d\phi/dx$ is also zero, i.e.,

$$\phi|_{x\to\infty} = \left.\frac{d\phi}{dx}\right|_{x\to\infty} = 0 \tag{7.39}$$

Under these conditions,

$$\text{Constant} = -\frac{2kT}{\varepsilon_0\varepsilon_r}\sum_i c_i^b \tag{7.40}$$

and, therefore

$$\left(\frac{d\phi}{dx}\right)^2 = \frac{2kT}{\varepsilon_0\varepsilon_r}\sum_i c_i^b(e^{-z_i q\phi/kT} - 1) \tag{7.41}$$

Considering the simplest case of one $z{:}z$-valent electrolyte, then

$$|z_+| = |z_-| = z \tag{7.42}$$

and

$$c_+^b = c_-^b = c^b \tag{7.43}$$

Therefore Eq. (7.41) becomes

$$\left(\frac{d\phi}{dx}\right)^2 = \frac{2kT}{\varepsilon_0\varepsilon_r} c^b(e^{zq\phi/kT} - 1 + e^{-zq\phi/kT} - 1) \tag{7.44}$$

or

$$\left(\frac{d\phi}{dx}\right)^2 = \frac{2kT}{\varepsilon_0\varepsilon_r} c^b(e^{zq\phi/2kT} - e^{-zq\phi/2kT})^2 \tag{7.45}$$

Since

$$e^x - e^{-x} = 2\sinh(x) \tag{7.46}$$

hence Eq. (7.45) becomes

$$\left(\frac{d\phi}{dx}\right)^2 = \frac{8kT}{\varepsilon_0\varepsilon_r} c^b \sinh^2\left(\frac{zq\phi}{2kT}\right) \tag{7.47}$$

From Eq. (7.47), we can obtain the field $d\phi/dx$ in the solution by taking square roots on both sides. To decide which root is to be taken, we recall that, at the positively charged electrode, $\phi > 0$, $\sinh(\phi) > 0$ and $d\phi/dx < 0$, while, at the negatively charged electrode,

$\phi < 0$, sinh $(\phi) < 0$, and $d\phi/dx > 0$. Hence, only the negative root of Eq. (7.47) corresponds to the physical situation, i.e.,

$$\frac{d\phi}{dx} = -\left(\frac{8kTc^b}{\varepsilon_0\varepsilon_r}\right)^{1/2} \sinh\left(\frac{zq\phi}{2kT}\right) \tag{7.48}$$

From Eq. (7.32) we can derive the diffuse charge density in the electrolyte. To fix ideas, and in analogy with the procedure utilized for the MOS capacitor (see Chap. 8), let us assume that the electrode was charged by connecting it to an external source of electricity and that a charge density σ_e was added to the electrode. Then, the diffuse charge in the solution can be obtained as follows. We can choose a gaussian box of unit area, extended from $x = 0$ to $x \to \infty$, where ϕ and $d\phi/dx = 0$. Now, the charge density $\sigma_d = (-\sigma_e)$ inside this box will be given by Gauss' law

$$\sigma_d = \varepsilon_0\varepsilon_r \frac{d\phi}{dx}\bigg|_{x=0} \tag{7.49}$$

Thus, by using Eq. (7.48) we obtain

$$\sigma_d = -(8\varepsilon_0\varepsilon_r c^b kT)^{1/2} \sinh\left(\frac{zq\phi_0}{2kT}\right) \tag{7.50}$$

Starting from Eq. (7.48), the potential drop in the diffusion layer can also be estimated. To make the derivation simpler, let us approximate (a complete solution can be found in Ref. 5)

$$\sinh\left(\frac{zq\phi}{2kT}\right) \approx \frac{zq\phi}{2kT} \tag{7.51}$$

Then,

$$\frac{d\phi}{dx} \approx -\left(\frac{8c^b kT}{\varepsilon_0\varepsilon_r}\right)^{1/2} \frac{zq\phi}{2kT} \tag{7.52}$$

or

$$\frac{d\phi}{dx} \approx -\left(\frac{2c^b(zq)^2}{\varepsilon_0\varepsilon_r kT}\right)^{1/2} \phi \tag{7.53}$$

The square root of the quantity in parentheses on the right-hand side of Eq. (7.53) is the inverse of a length and it is customary to indicate it with the Greek letter χ. The inverse of χ is known as the *Debye length*.

In terms of χ, Eq. (7.53) becomes

$$\frac{d\phi}{dx} = -\chi\phi \tag{7.54}$$

and, by integration, we obtain

$$\phi = \phi_0 e^{-\chi x} \tag{7.55}$$

where ϕ_0 is the value of the potential at the electrode surface ($x = 0$).

The reader should note that the Poisson-Boltzmann equation considers ions as *point charges*. As a consequence, they are allowed to concentrate on the surface of the electrode up to unreasonable values. As suggested by Fig. 7.2, a more realistic picture can be obtained by assuming that the outer Helmholtz plane, which crosses the centers of the first row of hydrated ions, identifies a plane of maximum approach for hydrated ions. Thus, the diffuse charge density should be calculated by solving the Poisson-Boltzmann equation from the OHP and not from the electrode surface. The layer delimited by electrode and OHP is referred to as the *Stern layer*.

In closing this section, let us finally address the question of the numerical value of the water relative dielectric permittivity ε_r. Its value is dependent on the electric field strength of the environment. *Bulk* water, where the average electric field strength can be assumed to be zero and no permanent dipole orientation is present, has a corresponding ε_r value around 80. On the other hand, on a charged electrode, water dipoles would orient and attain saturation orientation if the charge density on the electrode is large enough. This oriented water is sometimes termed the *primary hydration sheath* of the electrode and its dielectric permittivity can be assumed to have a numerical value around 6. Here, fully oriented water dipoles and adsorbed ions form the IHP. Away from the electrode, but still near it, water will be partially oriented, most of its dipoles contributing to the hydration sheath of ions. A mean value around 40 can be assumed for its dielectric permittivity. This is the value of ε_r which can be assigned to the OHP.[1]

7.2.2 Two Charged Planar Surfaces in Water

As a last example, in the following we shall consider the ion distribution between two similarly charged planar surfaces in water, where (apart from H^+ and OH^- ions) the only ions in the solution are those that have come off the two surfaces (that is, *no added* electrolyte is present). Such systems occur when, for example, colloidal particles (see Sec. 7.1.5) or bilayers with ionizable groups interact in water. To find the ion distribution, we solve again the Poisson-Boltzmann equation. To do so, we need two boundary conditions. In Sec. 7.2.1, when considering *one* electrode facing a *semi-infinite electrolyte solution*, we chose that both the potential ϕ and its space derivative $d\phi/dx$ approached zero as $x \rightarrow \infty$ [Eq. (7.39)]. In the present case, one boundary condition follows from the symmetry requirement that the electric field must vanish at the midplane between the two surfaces (see Fig. 7.9).

The second boundary condition follows from the requirement of overall *electroneutrality*, i.e., that the total charge of the ions (which came from the electrodes) in the gap between the two electrodes must be opposite to the charge on the surfaces. If σ_l is the surface charge density on each surface, d is the distance between the surfaces and l is a thickness which takes into account the finite size of the ions approaching the surfaces (Stern layer, see the discussion at the end of the previous section and in Ref. 5), then the condition of electroneutrality implies that

$$\sigma_l = -\int_0^{(d/2)-l} \rho \, dx = \varepsilon_0 \varepsilon_r \int_0^{(d/2)-l} \left(\frac{d^2\phi}{dx^2} \right) dx \quad (7.56)$$

or

$$\sigma_l = \varepsilon_r \varepsilon_0 \left(\frac{d\phi}{dx} \right)\bigg|_{(d/2)-l} = -\varepsilon_r \varepsilon_0 E_{s+l} \quad (7.57)$$

where E_{s+l} is the value of the electric field on a plane inside the solution at a distance l from the surface s. Note that this value is independent of the gap width d.

Turning now to the calculation of the ionic distribution, we can write down the Boltzmann distribution of the charge density ρ as

$$\rho = \rho_0 e^{-zq\phi/kT} \quad (7.58)$$

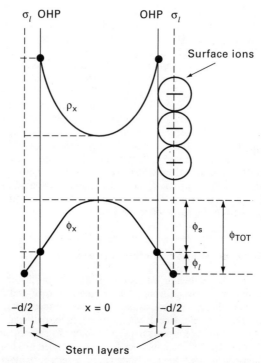

FIGURE 7.9 Stern layers of thickness l at each surface dividing the planes of fixed charge density σ_l from the boundary of the aqueous solution. (*Adapted from Israelachvili.[5] Used by permission.*)

ρ_0 being the charge density at the midplane, where $d\phi/dx = 0$ by symmetry and we may set also $\phi = 0$ (since only *differences* in potential are ever physically meaningful). Differentiating Eq. (7.58) and utilising the Poisson-Boltzmann equation, we obtain

$$\frac{d\rho}{dx} = -\frac{zq\rho_0}{kT} e^{-zq\phi/kT}\left(\frac{d\phi}{dx}\right) = \frac{\varepsilon_0 \varepsilon_r}{2kT} \frac{d}{dx}\left(\frac{d\phi}{dx}\right)^2 \qquad (7.59)$$

hence
$$\rho_x - \rho_0 = \int_{\rho_0}^{\rho(x)} d\rho = \frac{\varepsilon_r \varepsilon_0}{2kT} \int_0^{d\phi/dx} d\left(\frac{d\phi}{dx}\right)^2 = \frac{\varepsilon_r \varepsilon_0}{2kT}\left(\frac{d\phi}{dx}\right)^2 \qquad (7.60)$$

or
$$\rho_x = \rho_0 + \frac{\varepsilon_r \varepsilon_0}{2kT}\left(\frac{d\phi}{dx}\right)^2 \qquad (7.61)$$

which gives ρ at any point x in terms of ρ_0 at the midplane and $(d\phi/dx)^2$ at x.

Starting from the results obtained so far, an expression for the *pressure* existing between two charged surfaces in water can be derived. Moreover, similar equations can be obtained for two charged surfaces in the presence of *added* electrolytes. The interested reader can find treatment of these topics in Ref. 5.

7.3 MEMBRANE TRANSPORT

The characterization of solute movement across a barrier separating two solutions is a topic of great relevance in several scientific fields dealing with biomedical and biotechnological issues. The concept of "barrier" can be used to indicate such different objects as an inert homogeneous material or the highly inhomogeneous membrane of a living neuron. In the following we will begin with continuous transport by introducing the *Nernst-Planck equation* and the so-called constant-field equation. Diffusive potentials and equivalent electric circuits will be then considered. Finally, a short description of discontinuous transport will be given.

7.3.1 The Nernst-Planck Equation

Let us consider the one-dimensional motion of a solute i crossing in the x direction a homogeneous membrane separating two homogeneous solutions in the same solvent and at the same temperature. In order to describe the flux of this solute, we can start by writing the current density J in any point x of the solution as (no time dependence)

$$J_i(x) = -\tilde{D}_i z_i q \frac{dC_i(x)}{dx} + \tilde{\mu}_i z_i q\, C_i(x) E(x) \qquad (7.62)$$

where $C_i(x)$ is the *molecular* concentration of the solute i at x, z_i is its valence, q is the (positive) elementary charge, $\tilde{D}_i$ and $\tilde{\mu}_i$ are the molecule diffusion and mobility coefficients, respectively, and $E(x)$ is the electric field value at x. The two terms of the right side of Eq. (7.62) are clearly the diffusion and drift current density terms already introduced for electrons and holes in Chap. 3.

We can then transform Eq. (7.62) into a *molar* current density as follows:

$$J_{m,i}(x) = -D_i z_i \mathscr{F} \frac{dc_i(x)}{dx} + \mu_i z_i \mathscr{F} c_i(x) E(x) \qquad (7.63)$$

where $\mathscr{F}$ is the *Faraday constant* (see App. A) and $c_i(x)$ is the *molar* concentration of the solute i at x.

We can now introduce the *molar* flux F_{ci} of the solute i, defined as the moles of solute that cross the membrane (unit area) in the x direction per unit time

$$F_{ci}(x) = -D_i \frac{dc_i(x)}{dx} + \mu_i c_i(x) E(x) \qquad (7.64)$$

Let us now define the "generalized" mobility u_i as

$$u_i = \frac{\mu_i}{z_i \mathscr{F}} \qquad (7.65)$$

Then Eq. (7.64) can be written as, dropping (x) reference for simplicity,

$$F_{ci} = -D \frac{dc_i}{dx} + u_i c_i z_i \mathscr{F} E \qquad (7.66)$$

The Einstein relationship gives (for molar quantities)

$$D_i = RT u_i \qquad (7.67)$$

where R, the gas constant, is defined as

$$R = N_{AV} k \tag{7.68}$$

and N_{AV} and k are the *Avogadro* and *Boltzmann constants,* respectively. By making use of Eq. (7.67), we transform Eq. (7.66) to

$$F_{ci} = -u_i RT \frac{dc_i}{dx} + u_i c_i z_i \mathscr{F} E \tag{7.69}$$

which can be easily transformed into

$$F_{ci} = -u_i c_i \frac{d}{dx}[RT \ln c_i + z_i \mathscr{F} \phi(x)] \tag{7.70}$$

where
$$\frac{d\phi}{dx} = -E \tag{7.71}$$

Equation (7.70) is a simplified version of the *Nernst-Planck equation,* first introduced by Nernst (1888) and Planck (1890). The term inside the brackets can be recognized as a simplified expression of the *electrochemical potential energy* Π. (More general expressions of Π can be found in Refs. 6 and 7). Accordingly, we can write

$$F_{ci} = -u_i c_i \frac{d\Pi}{dx} \tag{7.72}$$

Moreover the gradient of an energy can be always related to a force f as follows:

$$f = -\frac{d\Pi}{dx} \tag{7.73}$$

In conclusion, we arrive at the *flux-force relationship*

$$F_{ci} = -u_i c_i f \tag{7.74}$$

In most books dealing with membrane transport, the Nernst-Planck Eq. (7.70) is deduced starting from Eq. (7.74). In concluding this section, two points can be stressed, namely:

1. Equation (7.62) is equivalent to Eqs. (3.46) and (3.47) derived for the current density of holes and electrons.
2. The generalized mobility u_i can be thought as a proportionality constant between the mean velocity v_i of particles i subject to a frictional drag and *any* force, acting on them, i.e.

$$v_i = u_i f \tag{7.75a}$$

For example, for a body freely moving in the atmosphere, Eq. (7.75a) becomes

$$v_i = u_i m_i g \tag{7.75b}$$

where m_i is the body mass and g the gravitational acceleration.

7.3.2 Solutions of the Nernst-Planck Equation

The Nernst-Planck equation is the starting point for several calculations, which are done by integration under appropriate boundary conditions and assumptions on charge and electric field. When applied to a concentration c_i of uncharged molecules (e.g., glucose), Eq. (7.70) reduces to

$$F_{ci} = -u_i c_i \frac{d}{dx}(RT \ln c_i) \tag{7.76}$$

or

$$F_{ci} = -u_i RT \frac{dc_i}{dx} \tag{7.77}$$

Let us assume that Eq. (7.77) describes the flux of the molecule i in the x direction at some point inside a homogeneous membrane of thickness Δx. Then, if the system under study is in a *steady state*, F_{ci} should have the same values at all points within the membrane. By assuming u_i also constant through the membrane, Eq. (7.77) can be easily integrated across the thickness Δx of the membrane separating two semi-infinite solutions of concentration c_a and c_b, respectively. Thus,

$$F_{ci} \int_0^{\Delta x} dx = -RTu_i \int_{\bar{c}_{i,a}}^{\bar{c}_{i,b}} dc \tag{7.78}$$

The quantity $\bar{c}_{i,a}$ represents the concentration of the solute "just inside" the membrane, on the side in contact with concentration $c_{i,a}$. Similarly, $\bar{c}_{i,b}$ represents the concentration of the solute "just inside" the membrane, on the side in contact with concentration $c_{i,b}$.

The meaning of the above definitions is that there is a discontinuity in matter when a particle is crossing a solution/membrane boundary. Concentrations "just inside" are linked[6] to concentrations "just outside" via *partition coefficients* $\beta(c_i)$.

The simplest hypothesis on the partition coefficients is to assume them independent of the actual concentration, i.e.,

$$\frac{\bar{c}_{i,b}}{c_{i,b}} = \beta_i = \frac{\bar{c}_{i,a}}{c_{i,a}} \tag{7.79}$$

Under conditions given by Eq. (7.79), Eq. (7.78) yields

$$F_{ci} = -RTu_i \beta_i \frac{\Delta c}{\Delta x} \tag{7.80a}$$

or, according to Eq. (7.67),

$$F_{ci} = -D_i \beta_i \frac{\Delta c}{\Delta x} \tag{7.80b}$$

Equation (7.80b) is easily recognized as the *Fick first law of diffusion* (see also Chaps. 3 and 5). By introducing the *permeability coefficient* P_i defined as

$$P_i = \frac{D_i \beta_i}{\Delta x} \tag{7.81}$$

Equation (7.80b) can be finally written as

$$F_{ci} = -P_i \Delta c_i \tag{7.82}$$

When the solute is charged (i.e., $z_i \neq 0$) and there is an electrical potential difference across the membrane, the situation is more complex, in consideration of the fact that Eq.

(7.70) *cannot* be integrated, unless the dependence of the electric potential ϕ on x is known.

Before making any choice on such a dependence, let us rewrite Eq. (7.70) as

$$F_{ci} = -D_i\left(\frac{dc_i}{dx} + \frac{c_i z_i \mathcal{F}}{RT}\frac{d\phi}{dx}\right) \tag{7.83}$$

Then, by multiplying both sides by exp $(z_i \mathcal{F}\phi/RT)$ and rearranging, we obtain

$$F_{ci} e^{z_i \mathcal{F}\phi/RT} dx = -D_i d(c_i e^{z_i \mathcal{F}\phi/RT}) \tag{7.84}$$

Assuming a steady state and that D_i is constant through the membrane, the integration of Eq. (7.84) across the membrane thickness gives

$$F_{ci}\int_0^{\Delta x} e^{z_i \mathcal{F}\phi/RT} dx = -D_i \int_{\bar{c}_{i,a},\bar{\phi}_a}^{\bar{c}_{i,b},\bar{\phi}_b} d(c_i e^{z_i \mathcal{F}\phi/RT}) \tag{7.85}$$

The right-hand side of Eq. (7.85) can be immediately integrated. On the other hand, in order to accomplish the integration of the left-hand side, the dependence of ϕ on x within the membrane must be known or assumed. A simple and most frequently used assumption, first proposed by Goldman in 1943, is that ϕ is a linear function of x, i.e. (with the overbar meaning the quantity "just inside" the membrane),

$$\bar{\phi} = \bar{\phi}_0 + \frac{\overline{\Delta\phi}}{\Delta x}x \tag{7.86}$$

where

$$\overline{\Delta\phi} = \bar{\phi}(\Delta x) - \bar{\phi}(0) \tag{7.87a}$$

or equivalently,

$$\overline{\Delta\phi} = \bar{\phi}(b) - \bar{\phi}(a) \tag{7.87b}$$

Equation (7.86) is known as the *constant field assumption*. By virtue of this assumption, Eq. (7.85) becomes

$$F_{ci}\int_0^{\Delta x} e^{z_i \mathcal{F}\bar{\phi}_a/RT} e^{z_i \mathcal{F}\overline{\Delta\phi}x/RT\Delta x} dx = -D_i(\bar{c}_{i,b} e^{z_i \mathcal{F}\bar{\phi}_b/RT} - \bar{c}_{i,a} e^{z_i \mathcal{F}\bar{\phi}_a/RT}) \tag{7.88}$$

Multiplying both sides by exp$(-z_i \mathcal{F}\bar{\phi}_a/RT)$, integrating and rearranging yields:

$$F_{ci} = -\frac{D_i z_i \mathcal{F}}{RT\,\Delta x}\overline{\Delta\phi}\left[\frac{\bar{c}_{i,b} e^{z_i \mathcal{F}\overline{\Delta\phi}/RT} - \bar{c}_{i,a}}{e^{z_i \mathcal{F}\overline{\Delta\phi}/RT} - 1}\right] \tag{7.89}$$

Equation (7.89) is usually referred as the *constant field flux equation* and it refers to *intramembrane* properties. However, by assuming again partition coefficients independent of the concentrations [Eq. (7.79)], it can be shown (see Sec. 7.3.4) that

$$\overline{\Delta\phi} = \Delta\phi \tag{7.90}$$

Under these conditions, and making use of Eq. (7.81), we finally arrive at

$$F_{ci} = -\frac{P_i z_i \mathcal{F}\,\Delta\phi}{RT}\left[\frac{c_{i,b} e^{z_i \mathcal{F}\Delta\phi/RT} - c_{i,a}}{e^{z_i \mathcal{F}\Delta\phi/RT} - 1}\right] \tag{7.91}$$

Equation (7.91) has been widely employed for the description of ion transport across biological membranes (a different approach will be shortly described in the next section).

Let us analyze some relevant features of Eq. (7.91). First of all, it should be appreciated that when $c_{i,a} \neq c_{i,b}$, the relation between F_{ci} and $\Delta\phi$ is clearly nonlinear. In other words, the membrane offers a different *resistance* to the flow of an ion depending on the direction of the flow. This asymmetric behavior is an example of *rectification*. The only exception to the rectifying behavior is given when $c_{i,a} = c_{i,b}$. The reader can easily verify that under this condition the relation between F_{ci} and $\Delta\phi$ is a straight line passing through the origin. The resulting equation is

$$F_{ci} = -\frac{P_i z_i \mathcal{F} \Delta\phi \, c_i}{RT} \tag{7.92}$$

By multiplying Eq. (7.92) by the ion charge $z_i \mathcal{F}$ and rearranging, we obtain a form of *Ohm's law*:

$$I_i = \left(-\frac{P_i z_i^2 \mathcal{F}^2 c_i}{RT}\right) \Delta\phi \tag{7.93}$$

Going back to the general case [Eq. (7.91)] when $c_{i,a} \neq c_{i,b}$, the condition of zero flux gives (see Sec. 6.2.1) the *Nernst equilibrium potential*:

$$\Delta\phi = \frac{RT}{z_i \mathcal{F}} \ln\left(\frac{c_{i,a}}{c_{i,b}}\right) \tag{7.94}$$

Finally, letting the potential drop $\Delta\phi$ go to zero, the resulting flux approaches the Fick law of diffusion (see Prob. 7.3),

$$F_{ci} = -P_i \, \Delta c_i = -D_i \beta_i \frac{\Delta c_i}{\Delta x} \tag{7.95}$$

The reader should be aware of the fact that Eq. (7.91) is obviously an incomplete picture of any real situation in the sense that it describes the flux of a *single* ion, not taking into account the fact that other ion species (at least one of opposite sign) should be present in the solution.

The general problem is faced at an elementary level when it is restricted to univalent anions and cations. Then, the flow of *each* cation c_+ is given by

$$F_{c+} = -\frac{P_+ \mathcal{F} \Delta\phi}{RT} \left[\frac{c_{+,b} e^{\mathcal{F}\Delta\phi/RT} - c_{+,a}}{e^{\mathcal{F}\Delta\phi/RT} - 1}\right] \tag{7.96a}$$

and the flow of *each* anion c_- is given by

$$F_{c-} = \frac{P_- \mathcal{F} \Delta\phi}{RT} \left[\frac{c_{-,b} e^{-\mathcal{F}\Delta\phi/RT} - c_{-,a}}{e^{-\mathcal{F}\Delta\phi/RT} - 1}\right] \tag{7.96b}$$

The steady-state condition implies the *zero current condition*, i.e.,

$$I = \mathcal{F}\left(\sum_c F_{c+} - \sum_a F_{c-}\right) = 0 \tag{7.97}$$

where $\sum_c F_{c+}$ is the sum of the flows of all cations and where $\sum_a F_{c-}$ is the sum of the flows of all anions.

Combining Eqs. (7.96) and (7.97) and solving for $\Delta\phi$, we obtain the *Goldman-Hodgkin-Katz (GHK) equation*

$$\Delta\phi = \frac{RT}{\mathscr{F}} \ln\left(\frac{\sum_c P_+ c_{+,o} + \sum_a P_- c_{-,i}}{\sum_c P_+ c_{+,i} + \sum_a P_- c_{-,o}}\right) \qquad (7.98)$$

where the subscript a has been replaced by o (outside) and b by i (inside).

In most biological membranes, where the predominant permeant ions are Na^+, K^+ and Cl^-, Eq. (7.98) reduces to

$$\Delta\phi = \frac{RT}{\mathscr{F}} \ln\left(\frac{P_{Na}c_{Na,o} + P_K c_{K,o} + P_{Cl} c_{Cl,i}}{P_{Na}c_{Na,i} + P_K c_{K,i} + P_{Cl} c_{Cl,o}}\right) \qquad (7.99)$$

Equation (7.99) is frequently used to estimate the potential drop across a biological membrane.

Example 7.1 Let us give an estimate of the membrane potential under conditions typical of an electrophysiology experiment (squid axon in seawater).

Answer Let us assume:

$$P_K : P_{Na} : P_{Cl} = 1 : 0.03 : 0.1$$

$c_{K,o} = 10$ mM $\qquad c_{Na,o} = 460$ mM $\qquad c_{Cl,o} = 540$ mM

$c_{K,i} = 400$ mM $\qquad c_{Na,i} = 50$ mM $\qquad c_{Cl,i} = 40$ mM

Then Eq. (7.99) gives

$$\Delta\phi \simeq -70 \text{ mV} \qquad (E7.1)$$

On the other hand, we can easily verify (see Prob. 7.4) that Cl is approximately at equilibrium. Therefore, Eq. (7.99) can be approximated with

$$\Delta\phi \simeq \frac{RT}{\mathscr{F}} \ln\left(\frac{P_{Na}c_{Na,o} + P_K c_{K,o}}{P_{Na}c_{Na,i} + P_K c_{K,i}}\right) \simeq -71 \text{ mV} \qquad (E7.2)$$

Before closing this section, let us finally consider a very simple case, where a homogeneous membrane separates two solutions a and b of a single $z{:}z$ salt. In this very special case the flux potential drop relation can be deduced by directly integrating the Nernst-Planck equation. According to this equation, we can write

$$F_{c+} = -c_+ u_+ \left[RT\left(\frac{d\ln c_+}{dx}\right) + z_+ \mathscr{F}\left(\frac{d\phi}{dx}\right) \right] \qquad (7.100)$$

and

$$F_{c-} = -c_- u_- \left[RT\left(\frac{d\ln c_+}{dx}\right) - z_+ \mathscr{F}\left(\frac{d\phi}{dx}\right) \right] \qquad (7.101)$$

Since bulk electroneutrality of each solution must be preserved, it follows that

$$c_+ = c_- = c \qquad (7.102)$$

and

$$F_{c+} = F_{c-} = F_c \qquad (7.103)$$

Equating Eqs. (7.100) and (7.101), making use of Eqs. (7.102) and (7.103), and rearranging, we obtain

$$\frac{d\phi}{dx} = -\left(\frac{u_+ - u_-}{u_+ + u_-}\right)\left(\frac{RT}{z\mathscr{F}}\right)\left(\frac{d}{dx}\ln c\right) \tag{7.104}$$

Assuming again constant partition coefficients, integration of Eq. (7.104) across the thickness of the membrane gives

$$\Delta\overline{\phi} = \Delta\phi = -\left(\frac{u_+ - u_-}{u_+ + u_-}\right)\left(\frac{RT}{z\mathscr{F}}\right)\ln\left(\frac{c_a}{c_b}\right) \tag{7.105}$$

Substituting Eq. (7.104) into Eq. (7.100) and combining terms yields

$$F_c = -\left(\frac{2RTu_+u_-}{u_+ + u_-}\right)\frac{dc}{dx} \tag{7.106}$$

By taking into account Eq. (7.103), we can finally write

$$F_c = -D_\pm\left(\frac{dc}{dx}\right) \tag{7.107}$$

where $D_\pm$, the diffusion coefficient of the *salt*, is

$$D_\pm = \frac{2RTu_+u_-}{u_+ + u_-} \tag{7.108}$$

This simple case allows us to discuss a relevant principle which is fundamental to the understanding of the origin of diffusion potentials in more complex systems. The principle is that, in the absence of an externally applied current, electroneutrality can be preserved only by the equivalent flow of anions and cations across a membrane. In the simple system just considered, this zero current condition (which does *not* imply equilibrium conditions) implies the constraint

$$F_{c+} = F_{c-} \tag{7.109}$$

and, therefore

$$I = z_+\mathscr{F}F_{c+} + z_-\mathscr{F}F_{c-} = 0 \tag{7.110}$$

The reader should note that, if $u_+ \neq u_-$, then Eq. (7.110) can be satisfied only if an electrical potential difference is generated with a magnitude proportional to the difference in mobilities and with an orientation appropriate to slow down the movement of the ion with the greater mobility and to speed up the movement of the ion with the lower mobility.

On the other hand, if

$$u_+ = u_- \tag{7.111}$$

then

$$\Delta\phi = 0 \tag{7.112}$$

In the limiting condition in which one of the mobilities is zero, e.g.,

$$u_+ = 0 \tag{7.113}$$

then, according to Eq. (7.105), we obtain

$$\Delta\phi = -\frac{RT}{z_+\mathscr{F}}\ln\left(\frac{c_b}{c_a}\right) \tag{7.114}$$

where $\Delta\phi$ is clearly the Nernst equilibrium potential. A similar result, with opposite orientation, would be obtained with $u_- = 0$.

Equation (7.114) means that, if an ion cannot cross the membrane, then *neither ion* is permitted to cross the membrane, otherwise electroneutrality would be violated. Thus, if

$$u_+ = 0 \tag{7.115a}$$

or

$$u_- = 0 \tag{7.115b}$$

then

$$F_{c+} = F_{c-} = 0 \tag{7.116a}$$

and

$$D_\pm = 0 \tag{7.116b}$$

In this *limiting* condition there is no ion flux and the system is in a state of equilibrium. We wish to underline that this equilibrium condition is never globally satisfied by living biological systems, which behave under *out-of-equilibrium* conditions.

7.3.3 Electrical Circuit Analogs

Let us rearrange the Nernst-Planck equation [Eq. (7.70)] of a *single ion i* in the form

$$F_{ci} = -u_i c_i z_i \mathscr{F} \left[\left(\frac{RT}{z_i \mathscr{F}} \right) \left(\frac{d}{dx} \ln c_i \right) + \left(\frac{d\phi}{dx} \right) \right] \tag{7.117}$$

The corresponding (molar) current density is then

$$J_{ci} = -u_i c_i z_i^2 \mathscr{F}^2 \left[\left(\frac{RT}{z_i \mathscr{F}} \right) \left(\frac{d \ln c_i}{dx} \right) + \left(\frac{d\phi}{dx} \right) \right] \tag{7.118}$$

and, rearranging, we obtain

$$\frac{J_{ci}}{u_i z_i^2 \mathscr{F}^2 c_i} = -\left(\frac{RT}{z_i \mathscr{F}} \right) \left(\frac{d \ln c_i}{dx} \right) - \left(\frac{d\phi}{dx} \right) \tag{7.119}$$

Assuming steady state (i.e., J_{ci} constant), Eq. (7.119) can be formally integrated over the thickness of the membrane Δx, yielding

$$J_{ci} \int_0^{\Delta x} \frac{dx}{u_i z_i^2 \mathscr{F}^2 c_i} = \frac{RT}{z_i \mathscr{F}} \ln \frac{c_{i,o}}{c_{i,i}} - \Delta\phi \tag{7.120}$$

Now we can define the integral resistance of the membrane to the ion i as

$$R_i = \int_0^{\Delta x} \frac{dx}{u_i z_i^2 \mathscr{F}^2 c_i} \tag{7.121}$$

and the voltage source of ion i as the Nernst potential

$$E_i = \frac{RT}{z_i \mathscr{F}} \ln \left(\frac{c_{i,o}}{c_{i,i}} \right) \tag{7.122}$$

By making use of Eqs. (7.121) and (7.122), we write Eq. (7.120) as

$$J_{ci} R_i = E_i - \Delta\phi \tag{7.123}$$

The circuital representation of Eq. (7.123) is given in Fig. 7.10.

Clearly, the equilibrium condition, when $J_{ci} = 0$, is given by

$$\Delta\phi = E_i = \frac{RT}{z_i \mathcal{F}} \ln\left(\frac{c_{i,o}}{c_{i,i}}\right) \quad (7.124)$$

The flow of *two ions* which, according to the Nernst-Planck equation, move independently of each other, can be represented as a two-branch parallel circuit (Fig. 7.11). Its current is given by

$$I = g_1(E_1 - \Delta\phi) + g_2(E_2 - \Delta\phi) \quad (7.125)$$

where

$$g_1 = \frac{1}{R_1} \quad (7.126a)$$

is the *conductance* of ion 1 and

$$g_2 = \frac{1}{R_2} \quad (7.126b)$$

is the conductance of ion 2; E_1 and E_2 are the two Nernst potentials.

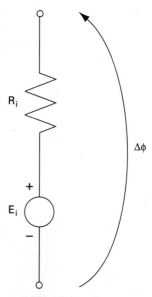

FIGURE 7.10 Equivalent circuit representation of the flow of a single ion.

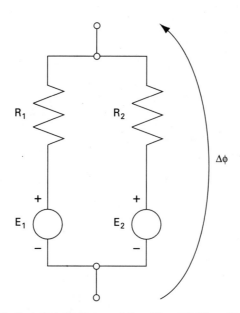

FIGURE 7.11 Equivalent circuit representation of the parallel flow of two ions.

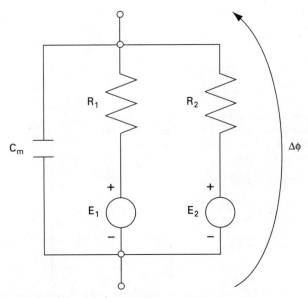

FIGURE 7.12 As Fig. 7.11, but now including the charging capability of the membrane.

If a charging process is also present on the membrane, a *capacitor* has to be inserted in parallel to the resistive branches (Fig. 7.12). The current density is now

$$I = C_m \frac{d}{dt}(\Delta\phi) + g_1(E_1 - \Delta\phi) + g_2(E_2 - \Delta\phi) \tag{7.127}$$

where C_m is the membrane capacitance per unit area.[9] These schemes will be further considered in Chap. 11 where the convention $(\Delta\phi - E_i)$ is chosen.

In closing this section, we wish to underline that electrical circuit rules, such as those routinely used in electrical engineering, deal with flux of *electrons* and *do not* recognize *different* ionic species. On the contrary, artificial and biological membranes *make these distinctions*. This consideration should be kept in mind when using the very valuable equivalent-circuit approach, to avoid misleading results.[6]

7.3.4 Modeling Transport through Structured Membranes

Fixed charges can be present on biological membranes (see Chap. 4). Moreover, particles crossing the membrane can interact with specific sites along their path, for example, inside a narrow channel (see Chap. 4). These situations, which have not been taken into account so far, will be considered in this section.

Donnan Equilibrium. Let us first consider the effect of *fixed charges* present on a membrane. In order to do so, let us first reconsider in greater detail the movement of an ion across a membrane. We can always imagine this motion as a three-step process: the first step is the crossing of the membrane's first border (i.e., the outer solution-membrane interface). Let us call this border the outside (*o*) border. The second step is the movement through the thickness of the membrane. The last step is the crossing of the inside (i) border (i.e., the membrane–inner solution interface). Consistent with this description, the potential drop across the membrane can be split into three parts, i.e.,

$$\Delta\phi = \Delta\phi_{o,\bar{o}} + \overline{\Delta\phi} + \Delta\phi_{\bar{i},i} \tag{7.128}$$

where $\Delta\phi_{o,\bar{o}}$ is the potential drop at the outside border, $\overline{\Delta\phi}$ is the potential drop through the interior of the membrane and $\Delta\phi_{\bar{i},i}$ is the potential drop at the inside border.

In contrast to the movement through the membrane thickness, the two border crossings can reasonably be assumed to be equilibrium phenomena.[6] Therefore we can indicate them as two Nernst equilibrium potentials, namely

$$\Delta\phi_{o,\bar{o}} = \frac{RT}{z\mathscr{F}} \ln\left(\frac{c_o}{\bar{c}_o}\right) \tag{7.129}$$

and

$$\Delta\phi_{\bar{i},i} = \frac{RT}{z\mathscr{F}} \ln\left(\frac{\bar{c}_i}{c_i}\right) \tag{7.130}$$

In accordance to Eq. (7.79), if the ratio at the borders is independent of the concentrations, we can then write

$$\frac{c_o}{\bar{c}_o} = \frac{1}{\beta} \tag{7.131a}$$

$$\frac{\bar{c}_i}{c_i} = \beta \tag{7.131b}$$

Therefore, under conditions expressed by Eqs. (7.131),

$$\Delta\phi = \overline{\Delta\phi} \tag{7.132}$$

We remind the reader that Eq. (7.132) was assumed without justification in the previous section [see Eq. (7.90)].

The situation becomes slightly more complex if we assume the presence of a concentration of a fixed molecular species in the membrane structure. As we already discussed in Chap. 4, this is a very reasonable assumption for biological membranes. In the presence of a fixed negative charge concentration, the electroneutrality condition, applied at any of the membrane borders to a monovalent salt of concentration c, reads

$$c_+ = c_- = c \tag{7.133}$$

and

$$c_{+,m} = c_{-,m} + |z|M \tag{7.134}$$

where c_+ and c_- represent the cation and anion concentration outside the membrane, the subscript m identifies ion concentrations in the membrane, and z is the valence of the negative chemical components fixed to the membrane, with concentration M. As compared to previous use of the electroneutrality condition, Eq. (7.134) is a *new* constraint, imposed by the presence of the fixed charged component of concentration M.

By assuming again an equilibrium condition at the solution-membrane interface, we can write

$$\Delta\phi = \frac{RT}{\mathscr{F}} \ln\left(\frac{c_{+,m}}{c_+}\right) = \frac{RT}{\mathscr{F}} \ln\left(\frac{c_-}{c_{-,m}}\right) \tag{7.135}$$

which implies

$$\beta = \frac{c_{+,m}}{c_+} = \frac{c_-}{c_{-,m}} \tag{7.136}$$

By using Eqs. (7.133) and (7.134), we transform Eq. (7.136) into

$$c\beta = \frac{c}{\beta} + |z|M \tag{7.137}$$

or

$$c\beta^2 - |z|M\beta - c = 0 \tag{7.138}$$

which gives

$$\beta = \frac{|z|M + (|z|^2M^2 + 4c^2)^{1/2}}{2c} \tag{7.139}$$

The equilibrium system just considered is referred to as the *Donnan equilibrium*. It depicts a situation where the partition coefficient β is a function of the salt concentration c. Therefore, indicating with β_i and β_o the partition coefficients at the two sides of the membrane, it is

$$\beta_i \neq \beta_o \tag{7.140}$$

Under these conditions, Eq. (7.128) becomes

$$\Delta\phi = -\frac{RT}{\mathscr{F}} \ln \beta_o + \overline{\Delta\phi} + \frac{RT}{\mathscr{F}} \ln \beta_i \tag{7.141}$$

and the first and third terms on the right-hand side of Eq. (7.141) do not cancel each other.

In the limiting case of $|z|M$ much greater than c, Eq. (7.141) reduces to

$$\Delta\phi = \frac{RT}{\mathscr{F}} \ln \beta_i - \frac{RT}{\mathscr{F}} \ln \beta_o = \frac{RT}{\mathscr{F}} \ln \frac{c_o}{c_i} \tag{7.142}$$

In this instance, $\Delta\phi$ is the sum of two equilibrium potentials that arise solely at the interfaces of the membrane through which no net flux occurs. This is the origin of the electrical potential differences generated in glass pH electrodes.

Obviously, the considerations made so far can be applied to *any* system made of two equilibrium compartments separated by a membrane, one of which contains charged species that cannot cross the membrane.

Discontinuous Flow. Biological (and also artificial) membranes often display saturation, transeffects, and competitive behavior that cannot be described by Nernst-Planck–type equations, where P_i, the permeability coefficient, is a constant and is not influenced by the concentration of the diffusing species or by the presence of other permeant ions.

In other words, the Nernst-Planck approach is not appropriate for describing very important entities such as ion channels. For these entities a quite different approach, based on the *absolute reaction rate theory* (proposed by Eyring in 1935), must be used. This approach will be illustrated with a simple example in the following. The reader can find further details in Refs. 6 and 8.

The fundamental assumption of this approach is that the movement of a particle inside a membrane is discontinuous, and the membrane can be viewed as a series of potential energy barriers, depicted as a series of peaks and valleys, that a particle must cross in order to pass from the outer (o) to the inner (i) solution. The key actor in the absolute rate theory is the rate constant K_{ij} which has the dimension of frequency [s^{-1}] and governs the movement of a particle from a minimum in energy (i.e., the "valley" i) to another minimum in energy (i.e., the "valley" j) over a maximum (i.e., the "peak" ij).

Without further justifications (which are provided in Ref. 6), let us assume that K_{ij} is given by the following Boltzmann (molar) distribution:

$$K_{ij} = \frac{kT}{h} e^{-\Delta G_{ij}/RT} \qquad (7.143)$$

where k is the Boltzmann constant, T the absolute temperature, h the Planck constant, ΔG_{ij} a molar energy indicating the height of the peak from the valley i, and R the gas constant. The reader can easily verify that K_{ij} has the dimension of frequency and that its numerical value increases with the absolute temperature T. Equation (7.143) originated from considerations of the average frequency ν of the vibrations (i.e., the movement) of a molecule at temperature T. This frequency ν can be estimated by equating

$$h\nu = kT \qquad (7.144)$$

or

$$\nu = \frac{kT}{h} \qquad (7.145)$$

At room temperature ν is in the order of 10^{12} s^{-1}.

The rate coefficient K_{ij} is obtained by weighting the frequency ν by a Boltzmann distribution.[10] The rate K_{ij} can then be related to a mean free path l in such a way that the mean velocity of a particle moving or rather "hopping" (the situation is similar to the one describing the electrode-solution interface; see Sec. 7.1.3) from valley i to valley j is given by

$$v_{ij} = K_{ij} \, l \qquad (7.146)$$

Consequently, the flux F_c of a particle is given by

$$F_c = K_{ij} \, l c \qquad (7.147)$$

Let us apply this approach to a simple two-barrier representation of a membrane as depicted in Fig. 7.13. Let us further assume that we are dealing with an uncharged molecule of concentrations c_0 (outside compartment) and c_i (inside compartment), so that no drift component has to be taken into account. Then, using the notation of Fig. 7.13, the flux from the outer compartment into the membrane (concentration c_m) is

$$F_{c,om} = K_{om} c_o l - K_{mo} c_m l \qquad (7.148)$$

Similarly, we can write the flux of exit from the membrane into solution i as

$$F_{c,mi} = K_{mi} c_m l - K_{im} c_i l \qquad (7.149)$$

When the diffusion of the molecules reaches a steady state, then we have

$$F_{c,om} = F_{c,mi} = F_c \qquad (7.150)$$

Also, if the barriers have the same amplitude, then

$$K_{om} = K_{im} \qquad (7.151)$$

and

$$K_{mo} = K_{mi} \qquad (7.152)$$

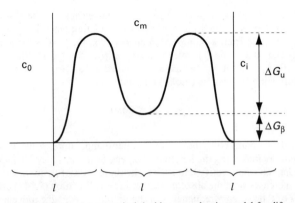

FIGURE 7.13 A symmetrical double energy barrier model for diffusion across a membrane.

Therefore, solving Eqs. (7.148) and (7.149), with constraints given by Eqs. (7.150) to (7.152), we obtain

$$F_c = \frac{K_{om}l}{2}(c_o - c_i) \tag{7.153}$$

The reader can note that Eq. (7.153) has the form of a diffusion equation, so that we may write

$$P = \frac{K_{om}l}{2} \tag{7.154}$$

where P is the permeability coefficient of the molecule.

Let us indicate with Δx the thickness of the membrane. Then, from Fig. 7.13, we can write

$$\Delta x = 2l \tag{7.155}$$

Therefore, since from Eq. (7.81)

$$P = \frac{D\beta}{\Delta x} \tag{7.156}$$

we can also write, taking into account Eq. (7.154),

$$D\beta = K_{om}l^2 \tag{7.157}$$

According to Eq. (7.143), the rate constant K_{om} can be written as

$$K_{om} = \frac{kT}{h}e^{-\Delta G/RT} \tag{7.158}$$

Moreover (see Fig. 7.13), we can split ΔG as

$$\Delta G = \Delta G_\beta + \Delta G_u \tag{7.159}$$

and, therefore we can write

$$K_{om} = \frac{kT}{h} e^{-\Delta G_\beta/RT} e^{-\Delta G_u/RT} \qquad (7.160)$$

and

$$D\beta = \frac{l^2 kT}{h} e^{-\Delta G_\beta/RT} e^{-\Delta G_u/RT} \qquad (7.161)$$

with the following identifications

$$\beta = e^{-\Delta G_\beta/RT} \qquad (7.162)$$

$$D = l^2 \frac{kT}{h} e^{-\Delta G_u/RT} \qquad (7.163)$$

Equation (7.163) could be further compared to the *Einstein-Smolucowsky relation*

$$D = \frac{\langle x^2 \rangle}{2} \frac{1}{t} \qquad (7.164)$$

The reader should appreciate that, unlike the approaches that involve integration of the Nernst-Planck equation, the rate theory allows, at least in principle, the parameters β and D to be expressed in terms of the physical and chemical structural details of the diffusion pathway.

The approach just introduced can be extended to (more realistic) multibarrier systems and also to charged species moving through channels containing binding sites.[6,8]

PROBLEMS

7.1 Discuss the frequency dependence of the double-layer equivalent impedance (Eq. 7.1).

7.2 Find out the expression for the minimum in the potential energy between two colloidal particles (see Sec. 7.1.5).

7.3 Deduce the Fick diffusion law from the constant field flux equation (Eq. 7.89).

7.4 By using the numerical values given in Example 7.1, find the numerical values of the Nernst equilibrium potential for Na^+, K^+ and Cl^-.

7.5 Find the expression for the membrane potential under steady-state conditions by making use of the equivalent circuit model (Eq. 7.127).

REFERENCES

1. O. M. Bockris and A. K. N. Reddy, *Modern electrochemistry,* Vol. 2, 3d ed., New York: Plenum Press, 1977.
2. A. J. Bard and L. R. Faulkner, *Electrochemical methods,* New York: John Wiley & Sons, 1980.
3. D. L. Harame, L. J. Bousse, J. D. Shott, and J. D. Meindl, "Ion-sensing devices with silicon nitride and borosilicate glass insulators," *IEEE Trans. Electron Devices,* ED-34: 1700–1707, 1987.

4. M. Grattarola, G. Massobrio, and S. Martinoia, "Modeling H^+-sensitive FET's with SPICE," *IEEE Trans. Electron Devices,* ED-39: 813–819, 1992.
5. J. Israelachvili, *Intermolecular and surface forces,* 2d ed., San Diego: Academic Press, 1991.
6. S. G. Schultz, *Basic principles of membrane transport,* New York: Cambridge University Press, 1980.
7. G. H. Wannier, *Statistical physics,* New York: Dover Publications, 1987.
8. B. Hille, *Ionic channels of excitable membranes,* 2d ed., Sunderland, Mass.: Sinauer Associates, 1992.
9. J. D. Bronzino, *The biomedical engineering handbook,* Boca Raton, Fla.: CRC Press–IEEE Press, 1995.
10. G. M. Barrow, *Physical chemistry,* 2d ed., New York: McGraw-Hill, 1966.

PART 4

DEVICES AND CAD

CHAPTER 8
METAL-OXIDE-SEMICONDUCTOR (MOS) STRUCTURE

The electronic processes and phenomena described in Chaps. 2, 3, and 6 will now be used to study an important electronic structure, the *MOS capacitor*, which represents the control element of the metal-oxide-semiconductor transistor. This transistor, which is the base for the most part of the semiconductor-based biosensors, belongs to a family of electronic devices called *unipolar* field-effect transistors, or simply *field-effect transistors* (FETs). The word *unipolar* derives from the fact that the current in these devices is carried mainly by one type of carrier, or one polarity of charge. The term *field effect* is used because the number of charges that can participate in the conduction process is controlled by an applied electric field. The MOS field-effect transistor is more usually referred to as a MOSFET. This is the name we will use throughout the book. The MOSFET will be described in Chap. 9.

8.1 MOS STRUCTURE

Fabrication of the MOS structure starts with a *p*- or *n*-type *semiconductor* material, uniformly doped, called the *substrate*, and assumed to be silicon from now on. An insulating layer is formed on top of the substrate. This layer is usually silicon dioxide (SiO_2) or silicon nitride (Si_3N_4) or alumina (Al_2O_3), simply called from now the *oxide* or *insulator*, and its thickness is usually 0.01 to 0.1 μm. A third layer, referred to as *gate*, is then formed on top of the insulator. The gate is made either of metal (usually aluminum) or of polycrystalline silicon. The described structure is often referred to by the acronym *MOS* (metal-oxide-semiconductor), regardless of whether the gate is made of metal or whether the insulator is silicon dioxide or other insulating material. In addition, since the gate and the semiconductor can be considered as the two plates of a capacitor, this structure is usually called the *MOS capacitor* (Fig. 8.1). The reader can find an electrochemical analog of this structure in Chap. 7, where the EIS (electrolyte-insulator-semiconductor) is considered.

The operating characteristics of the MOS capacitor are, however, quite different from those of a conventional parallel plate capacitor, the difference being due to the presence of a doped semiconductor as one of the capacitor's "plates." Indeed, unlike a conductor, the doped semiconductor has three distinct types of charge: electrons, holes, and the ionized impurity atoms. As a result, the MOS capacitor shows three distinct operating modes, which are called the *accumulation, depletion,* and *inversion* modes, and are distinguished

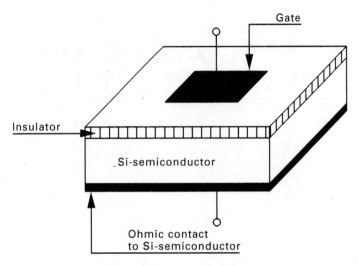

FIGURE 8.1 Structure of an MOS capacitor.

from each other in terms of the magnitude and polarity of the applied voltage.

We will describe these three modes of operation using, for purposes of illustration, a p-type silicon semiconductor and silicon dioxide as an insulator.

Many books, such as those indicated in Refs. 1 to 4, have been written about MOS structures (including MOSFET devices) because of the great importance of these structures in modern (bio)electronics. These books can be used as references for further readings.

8.2 ACCUMULATION OPERATING MODE

Let us apply a negative voltage V_G between the gate and the semiconductor of an MOS structure initially assumed to be neutral everywhere, as shown in Fig. 8.2a. Application of this negative voltage will induce a negative charge Q_G at the metal-oxide interface and will require a positive charge of equal magnitude to arise in the semiconductor to maintain the charge neutrality in the MOS system.

Because the semiconductor is p type, it has a lot of mobile positive charges, and thus it can supply the required positive charge. Then holes flow toward the semiconductor-oxide interface and *accumulate* in a thin region near the surface to provide the required charge.

This operating mode of the MOS capacitor is called the *accumulation* mode because the charge at the semiconductor-oxide interface is formed from the *accumulation* of *majority carriers*. In this mode of operation, the MOS capacitor behaves like a conventional parallel-plate capacitor. Therefore, the total charge of holes Q_p that accumulates for a fixed voltage V_G can be evaluated directly from the relation

$$Q_p = C_{ox}|V_G| \tag{8.1}$$

where C_{ox} is the capacitance of the oxide layer given by

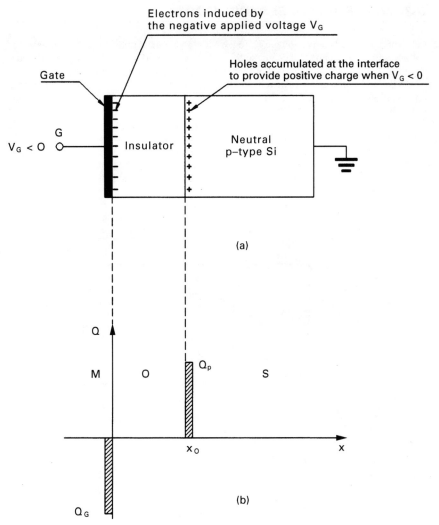

FIGURE 8.2 MOS capacitor in accumulation mode. (a) Effects of the applied negative bias voltage V_G and (b) corresponding charge balance.

$$C_{ox} = \frac{\varepsilon_{ox} A_G}{x_o} \qquad (8.2)$$

In Eq. (8.2), ε_{ox} is the dielectric constant of the oxide, A_G is the area of the gate, and x_o is the oxide thickness.

The absolute value of V_G has been introduced in Eq. (8.1) to point out that the charge Q_p is positive, since it must balance the negative charge Q_G induced by the negative applied voltage V_G at the metal gate of the MOS structure.

8.3 DEPLETION OPERATING MODE

Let us apply now a positive voltage V_G to the MOS structure. Application of this polarity of voltage will induce a positive charge at the metal-oxide interface and will require an equal negative charge in the semiconductor to maintain the charge neutrality in the MOS system.

As we will see later for relatively small values of applied voltages, we can neglect the effects of electrons (minority carriers) as a source of negative charge. Under this condition, the required negative charge in the semiconductor is obtained by depleting holes from a region near the semiconductor-oxide interface as shown in Fig. 8.3a. The electric field originated by the positive charge on the gate electrode pushes holes away in the semiconductor, leaving the bound acceptor atoms, which become negatively charged in the semiconductor, and therefore supplying the negative charge required.

Thus, a region depleted of mobile carriers, where the charge is due to the uncovered acceptor atoms (each of which contributes a negative charge $-q$), is formed at the semi-

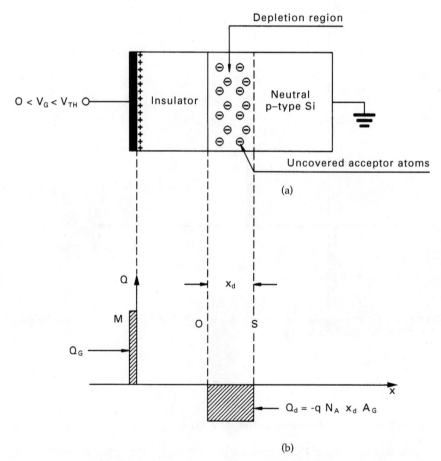

FIGURE 8.3 MOS capacitor in depletion mode. (*a*) Effects of the applied positive bias voltage V_G and (*b*) corresponding charge balance.

conductor-oxide interface. The width of this depletion region, indicated by x_d in Fig. 8.3a, is an important parameter in the operation of depletion-mode MOSFET devices and can be calculated once the applied voltage V_G is specified. The result is

$$V_G = \left(\frac{qN_A x_o}{\varepsilon_{\text{ox}}}\right) x_d + \left(\frac{qN_A}{2\varepsilon_s}\right) x_d^2 \tag{8.3}$$

Thus, if we know from the fabrication data sheets the values of the semiconductor doping concentration N_A and the oxide thickness x_o, the coefficients of x_d in Eq. (8.3) can be evaluated, and x_d can then be calculated for any given value of V_G. To derive Eq. (8.3), we proceed as follows.

Let us consider Fig. 8.4, which shows a view of an MOS capacitor made on a p-type substrate. The capacitor is assumed to be operating in the *depletion mode*. A charge Q_G then exists on the gate, matched by a charge $-Q_G$ in the depletion region. The width of the depletion region x_d can be obtained from charge balance considerations, i.e.,

$$qN_A x_d A_G = Q_G \tag{8.4}$$

To do this, we first compute the gate voltage V_G that results from the charge Q_G with the aid of *Gauss' law*.

The charges on the gate and in the semiconductor set up electric fields in the oxide and in the semiconductor: we indicate these fields by $E_{\text{ox}}(x)$ and $E_s(x)$, respectively. The corresponding voltage drops V_{ox} and V_s can be calculated from definition

$$\int_{V(0)}^{V(x_o)} dV = -\int_0^{x_o} E_{\text{ox}}(x)\, dx = V(x_o) - V(0) \tag{8.5a}$$

Hence,
$$V_{\text{ox}} = \int_0^{x_o} E_{\text{ox}}(x)\, dx \tag{8.5b}$$

where we have assumed V_{ox} to be positive when defined as $V_{\text{ox}} = V(0) - V(x_o)$, as indicated in Fig. 8.4.

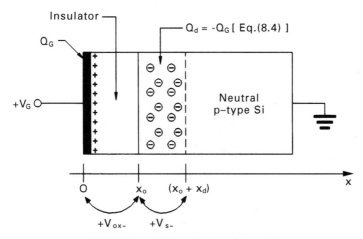

FIGURE 8.4 MOS capacitor in depletion mode.

In a similar way,

$$\int_{V(x_o)}^{V(x_o+x_d)} dV = -\int_{x_o}^{(x_o+x_d)} E_s(x)\, dx = V(x_o+x_d) - V(x_o) \tag{8.6a}$$

Hence,

$$V_s = \int_{x_o}^{(x_o+x_d)} E_s(x)\, dx \tag{8.6b}$$

where we have assumed V_s to be positive when defined as $V_s = V(x_o) - V(x_o + x_d)$, as indicated in Fig. 8.4.

Figure 8.4 shows that the sum of V_{ox} and V_s is just the applied positive voltage V_G. Thus the calculation of V_G can be carried out by evaluating E_{ox} and E_s, and then by using Eqs. (8.5) and (8.6) to obtain V_{ox} and V_s.

The electric fields $E_{ox}(x)$ and $E_s(x)$ can be calculated by using Gauss' law. For our purposes, this law is illustrated in Fig. 8.5a, where the gate with a charge $+Q_G$ and the depleted semiconductor with a negative charge $-Q_G = -qN_A A_G x_d$ are shown. Electric field lines originate on the positive charge and terminate on the negative charge, as shown.

Gauss' law, in its integral form, gives

$$\int \varepsilon \mathbf{E} \cdot d\mathbf{A} = Q \tag{8.7}$$

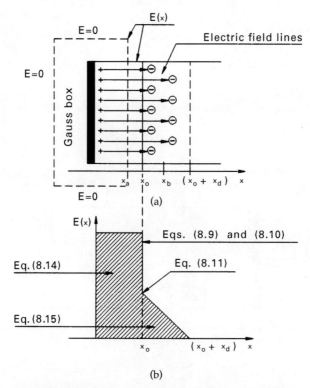

FIGURE 8.5 Calculation of electric fields and voltages in an MOS capacitor in depletion mode from Gauss' law. [*Adapted from lecture notes (Stanford University) by J. F. Gibbons.*]

where ε is the dielectric constant and $\mathbf{E}$ is the vector electric field flowing through a vector element of area $d\mathbf{A}$. The integral is to be calculated over a surface that encloses the charge Q within it.

For one-dimensional applications, such as the one shown in Fig. 8.5, it is convenient to consider a rectangular "box," whose right face is perpendicular to the electric field vector. The electric field then has a value only over an area A_G of the right face, and Eq. (8.7) yields

$$\varepsilon_x E(x) A_G = Q \tag{8.8}$$

where $E(x)$ is the electric field at the generic plane x, ε_x is the dielectric constant of the material at the plane x, and Q is the total charge enclosed in the box.

We can evaluate E_{ox} an E_s directly from Eq. (8.8). For E_{ox}, we consider the right face of the box to be at a point x_a within the oxide, as shown in Fig. 8.5a. The charge within the box is then Q_G, the charge on the gate, and ε_x is now ε_{ox}. Moreover, these values are independent of the position of the right face of the box as long as it is in the oxide. Therefore E_{ox} is *constant* through the oxide and has the value

$$E_{ox} = \frac{Q_G}{\varepsilon_{ox} A_G} \tag{8.9}$$

as shown in Fig. 8.5b. Equation (8.9) can also be rewritten as

$$E_{ox} = \frac{qN_A x_d}{\varepsilon_{ox}} \tag{8.10}$$

where we have used Eq. (8.4). Equation (8.10) shows that the electric field is independent of the gate area A_G.

In order to calculate E_s, we notice that, when the right face of the box is shifted just inside the semiconductor at $x = x_o$, the dielectric constant changes from ε_{ox} to ε_s. This change implies that the electric field at x_o will change abruptly from E_{ox} at the left side of the semiconductor-oxide interface to E_s on the right side, where

$$E_s(x_o) = \frac{Q_G}{\varepsilon_s A_G} = \frac{qN_A x_d}{\varepsilon_s} \tag{8.11}$$

Since ε_s (for Si) is about 1 pF/cm and ε_{ox} (for SiO$_2$) is about ⅓ pF/cm (see App. A for the exact values), the electric field will drop from E_{ox} to ⅓E_{ox} as we move across the semiconductor-oxide interface into the semiconductor. This situation is shown in Fig. 8.5b.

If we shift the right face of the box farther into the semiconductor, to a point such as x_b, then the charge within the box will be

$$Q(x_b) = Q_G - qN_A A_G(x_b - x_o) \tag{8.12}$$

The electric field at x_b will be then

$$E_s(x_b) = \frac{Q_G - qN_A A_G(x_b - x_o)}{\varepsilon_s A_G}$$

$$= E_s(x_o) - \frac{qN_A(x_b - x_o)}{\varepsilon_s} \tag{8.13}$$

Equation (8.13) states that the electric field decreases linearly as we proceed into the semiconductor.

The complete electric field distribution is then shown in Fig. 8.5b. Using this distribution we can calculate the voltage drops across the oxide layer and the depleted semiconductor region. Thus, from Eqs. (8.5) and (8.6),

$$V_{ox} = \int_0^{x_o} E_{ox}\, dx$$

$$= \frac{Q_G x_o}{\varepsilon_{ox} A_G} = \frac{qN_A x_d}{\varepsilon_{ox}} x_o = E_{ox} x_o \qquad (8.14)$$

and
$$V_s = \int_{x_o}^{(x_o+x_d)} E_s(x)\, dx = \int_{x_o}^{(x_o+x_d)} E_s(x_o)\, dx - \int_{x_o}^{(x_o+x_d)} \frac{qN_A}{\varepsilon_s}(x - x_o)\, dx$$

$$= \frac{qN_A x_d^2}{2\varepsilon_s} = \frac{x_d}{2} E_s(x_o) \qquad (8.15)$$

Equation (8.15) is similar to Eq. (6.30) applied to an n^+p junction when we replace W and ϕ_0 by x_d and V_s, respectively.

The voltage drops V_{ox} and V_s are represented by the areas under the proper parts of the field distribution shown in Fig. 8.5b.

We can summarize that, if we supply a positive charge Q_G to the gate of a p-type MOS capacitor driven into the *depletion mode*, we will

1. Develop a depletion of holes in the p-type semiconductor to a width x_d such that $Q_G = qN_A A_G x_d$.
2. Set up a constant electric field $E_{ox} = qN_A x_d/\varepsilon_{ox}$ in the oxide and an electric field in the semiconductor that decreases linearly from $qN_A x_d/\varepsilon_s$ at $x = x_o$ to zero at $x = (x_o + x_d)$.
3. Give rise to voltage drops V_{ox} and V_s across the oxide layer and the depleted semiconductor region.

The total positive voltage drop between the gate and the neutral region of the semiconductor is then

$$V_G = V_{ox} + V_s \qquad (8.16)$$

or
$$V_G = \left(\frac{qN_A x_o}{\varepsilon_{ox}}\right) x_d + \left(\frac{qN_A}{2\varepsilon_s}\right) x_d^2 = \frac{Q_G}{C_{ox}} + V_s \qquad (8.17)$$

which is the same as Eq. (8.3).

Conversely, if we apply such a voltage, given by Eq. (8.17), between the gate and the semiconductor, we will

1. Induce a positive gate charge Q_G.
2. Cause a depletion region of width x_d.
3. Create the electric field distribution shown in Fig. 8.5b.

In many applications (such as a depletion-mode MOSFET) we are interested in calculating the depletion width x_d when a voltage V_G is applied. Equation (8.17) shows that x_d can be evaluated when the semiconductor doping concentration N_A and the oxide thickness x_o are known, as pointed out by the following example.

Example 8.1 An MOS capacitor is made on silicon doped with $N_A = 10^{17}/\text{cm}^3$. Let the SiO_2 oxide thickness be $x_o = 100$ nm. Calculate V_{ox}, V_s, and x_d for $V_G = 1.0$ V. For the values of the quantities not specified, refer to App. A.

Answer Introducing in Eqs. (8.14) and (8.15) the specified values of N_A and x_o, we obtain

$$V_{ox} = 4.6 \times 10^5 \, x_d \tag{E8.1a}$$

$$V_s = 76 \times 10^9 \, x_d^2 \tag{E8.1b}$$

with x_d (in cm) to be calculated as follows.

If we apply $V_G = 1.0$ V to the MOS capacitor, we obtain from Eq. (8.17)

$$1.0 = 4.6 \times 10^5 \, x_d + 76 \times 10^9 \, x_d^2 \tag{E8.2}$$

which gives the depletion region width $x_d = 0.02$ μm. The corresponding values of V_{ox} and V_s are then 0.96 V and 0.03 V, respectively.

If N_A is reduced to $10^{15}/\text{cm}^3$, the depletion width obtained for the same applied voltage $V_G = 1.0$ V changes to $x_d = 0.88$ μm and the corresponding values of V_{ox} and V_s become 0.59 V and 0.41 V, respectively.

Equation (8.17) shows that the width of the depletion region x_d is determined by the applied voltage V_G. In particular, as the applied voltage is increased, the positive charge on the gate will increase, and the width of the depletion region x_d will also increase to a value such as to supply the necessary negative charge [see Eq. (8.4)]. In terms of the MOS capacitor, this implies that the distance between the capacitor "plates" increases as the applied voltage increases, and then the capacitance will decrease. This effect holds when the MOS structure is operating in the *depletion mode;* it is shown in Fig. 8.6.

The equivalent capacitance between the gate and the neutral semiconductor bulk is represented as the series of a fixed capacitor C_{ox} (modeling the oxide) and a variable ca-

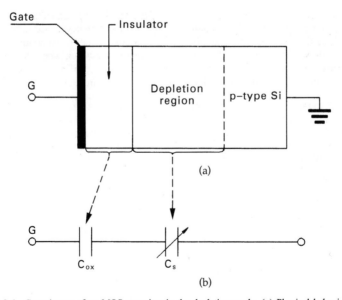

FIGURE 8.6 Capacitance of an MOS capacitor in the depletion mode. (*a*) Physical behavior and (*b*) equivalent circuit model of the total capacitance.

pacitance C_s (modeling the semiconductor). Both of the capacitances shown are defined as *incremental capacitances*, i.e.,

$$C_{ox} = \frac{dQ_G}{dV_{ox}} \qquad (8.18a)$$

$$C_s = \frac{dQ_G}{dV_s} \qquad (8.18b)$$

Introducing Eqs. (8.4), (8.14), and (8.15) into Eqs. (8.18), we can show that C_{ox} and C_s can both be evaluated from the *parallel-plate* relations

$$C_{ox} = \frac{\varepsilon_{ox} A_G}{x_o} \qquad (8.19a)$$

$$C_s = \frac{\varepsilon_s A_G}{x_d} \qquad (8.19b)$$

The equivalent capacitance between the gate and the neutral semiconductor bulk is then

$$C_{MOS} = \frac{C_{ox} C_s}{C_{ox} + C_s} = \frac{C_{ox}}{1 + \frac{\varepsilon_{ox} x_d}{\varepsilon_s x_o}} \qquad (8.20)$$

8.4 INVERSION OPERATING MODE

The analysis of the depletion-mode operation of the MOS *p*-type capacitor shows that, if electrons are neglected, all of the negative charge that must be developed in the semiconductor is a result of unneutralized doping acceptor atoms. Under these conditions, the depletion width x_d will increase uniformly as V_G increases. However, when the applied gate voltage reaches a *threshold* value V_{TH}, electrons, as a source of negative charge in the semiconductor, can no longer be neglected. In fact, we will prove that:

1. There is a gate voltage $V_G = V_{TH}$ applied to the MOS capacitor, such that the electron concentration at the semiconductor-oxide interface becomes equal to N_A. The semiconductor then has an *electron* concentration at its surface that is *equal* to what the hole concentration would be with no gate voltage applied. Therefore, the conductivity type of the semiconductor at its surface becomes *inverted* (i.e., from *p* type to *n* type).

2. For values of V_G greater than V_{TH}, all of the additional negative charge required in the semiconductor is provided by electrons accumulating at the semiconductor-oxide interface. The charge components for this condition are schematically shown in Fig. 8.7b, and consist of an amount of negative charge Q_a in the semiconductor contributed by acceptors (due to hole depletion) plus a thin layer of electrons at the semiconductor-oxide interface, forming a negative charge Q_n. The MOS capacitor is now said to be operating in the *inversion mode:* this term indicates that the surface conductivity is now *n* type (i.e., due to electrons) even if the semiconductor substrate is *p* type.

Electrons accumulating at the semiconductor surface of the *p*-type MOS capacitor can be considered to be caused by thermal generation within the depletion region. Electron-hole pairs that are generated in this region find themselves in an electric field that tends to

METAL-OXIDE-SEMICONDUCTOR (MOS) STRUCTURES

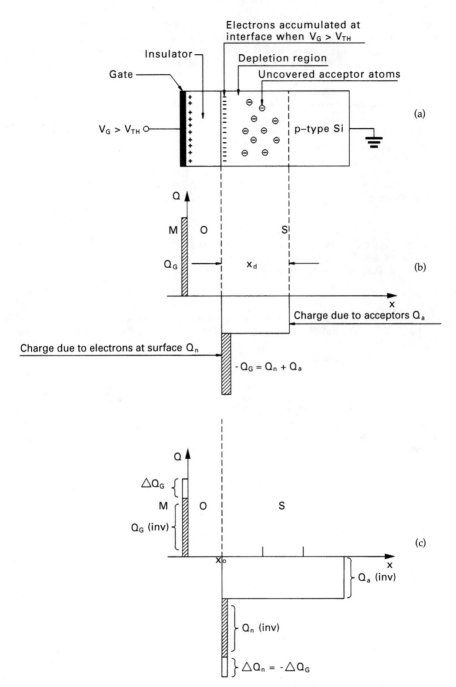

FIGURE 8.7 MOS capacitor in inversion mode. (*a*) Effects of the applied positive bias voltage $V_G > V_{TH}$, (*b*) corresponding charge balance, and (*c*) effects of the additional charge ΔQ_G.

transport the electrons to the semiconductor-oxide interface and the holes into the bulk of the semiconductor. This flow of electrons to the surface of the semiconductor increases the electron concentration at the semiconductor-oxide interface. The increased concentration, however, causes the *diffusion* of electrons from the semiconductor surface back to the bulk of the semiconductor, as shown in Fig. 8.8. Then we have two opposing processes; the electric field in the depletion region causes electrons to *drift* toward the semiconductor surface, and the resulting concentration gradient causes them to *diffuse back* toward the semiconductor bulk. Similar effects take place for holes, with the electric field now producing drift toward the bulk of the semiconductor and with the hole concentration gradient generating diffusion toward the surface.

An equilibrium condition will be reached when the oppositely directed drift and diffusion effects are in balance at every point for each carrier. When this condition is reached, the electron concentration at the semiconductor-oxide interface is given by Boltzmann relations [see Eqs. (6.10)], applied between semiconductor surface and bulk, i.e.,

$$n_s = n_{p0} e^{qV_s/kT} \tag{8.21}$$

where n_{p0} is the equilibrium electron concentration in the *p*-type semiconductor bulk and V_s is the voltage drop across the depletion region. In a similar way, the hole concentration p_s at the semiconductor-oxide interface is given by

$$p_s = p_{p0} e^{-qV_s/kT} \tag{8.22}$$

where p_{p0} is the equilibrium hole concentration in the *p*-type semiconductor bulk.

Equations (8.21) and (8.22) are very important relations for the operation of both *pn* junction and MOS devices (see Chaps. 6 and 9, respectively), and we will use them to discuss the concept of surface inversion.

To this purpose we first notice that according to Eqs. (8.21) and (8.22), n_s increases with V_s, while p_s decreases with V_s. Then, if we apply enough voltage, we can increase n_s and reduce p_s so that the electrons that accumulate at the surface are unlikely to recombine with holes. This is the first condition that must be achieved before an accumulation region of electrons can form. In addition, we need to ensure that n_s be greater than N_A so that the charge concentration contributed by electrons can exceed the charge concentration created by acceptor atoms. In other words, the condition for *inversion* requires that

$$n_s = N_A \tag{8.23}$$

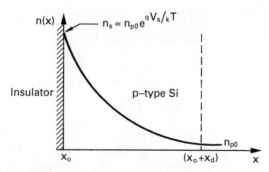

FIGURE 8.8 Electron concentration in the depletion region of an MOS capacitor made on a *p*-type silicon.

When this condition is reached, the surface is said to be *inverted* because the p-type semiconductor has a surface electron concentration equal to what the hole concentration would be with no applied bias voltage.

The voltage drop in the semiconductor which causes surface inversion can be evaluated by setting $n_s = N_A$ into Eq. (8.21) and by finding the corresponding value of V_s. Thus,

$$N_A = n_{p0} e^{qV_s/kT} \tag{8.24}$$

Let us indicate the value of V_s that meets this condition by V_s^*. Solving Eq. (8.24) for V_s^*, we obtain

$$V_s^* = \frac{kT}{q} \ln \frac{N_A}{n_{p0}} \tag{8.25}$$

Using Eq. (2.60), here rewritten for reader's convenience,

$$n_{p0} = \frac{n_i^2}{N_A} \tag{8.26}$$

we can rewrite Eq. (8.25) in the form

$$V_s^* = \frac{2kT}{q} \ln \frac{N_A}{n_i} = 2\phi_p \tag{8.27}$$

Equation (8.27) is important in the operation of p-type MOS devices. It states that the voltage drop in the semiconductor required to cause surface inversion depends only on the doping concentration N_A and the semiconductor material (through n_i). The voltage V_s^* is usually called the *surface inversion voltage* (see Nernst potential in Chap. 7).

The notation $2\phi_p$ in Eq. (8.27) is frequently encountered in literature and it comes from pn junction theory [see Eq. (6.8b)].

The value of V_G that is necessary to cause surface inversion, as already anticipated, is called the *threshold voltage* and is indicated as V_{TH}. This voltage is the sum of V_{ox} and V_s under the condition $V_s = V_s^*$.

To obtain an expression for V_{TH}, we notice that the depletion width required to produce V_s^* in the semiconductor is, from Eq. (8.15),

$$x_d^* = \sqrt{\frac{2\varepsilon_s V_s^*}{qN_A}} \tag{8.28}$$

The charge per unit area within this depletion region is, neglecting the electron contribution,

$$qN_A x_d^* = \sqrt{2\varepsilon_s qN_A V_s^*} \tag{8.29}$$

which is also the charge per unit area on the gate. The electric field in the oxide is then, from Eq. (8.10),

$$E_{ox}^* = \frac{qN_A x_d^*}{\varepsilon_{ox}} \tag{8.30}$$

and the voltage drop in the oxide is

$$V_{ox}^* = E_{ox}^* x_o = \frac{qN_A x_d^*}{\varepsilon_{ox}} x_o \tag{8.31}$$

The *threshold voltage* can now be obtained from

$$V_G^* = V_{TH} = V_{ox}^* + V_s^* = \frac{qN_A x_d^* x_o}{\varepsilon_{ox}} + \frac{2kT}{q} \ln \frac{N_A}{n_i}$$

$$= \frac{Q_G}{C_{ox}} + 2\phi_p \qquad (8.32)$$

The description and the analysis just given is valid as long as we do not consider nonideal effects that perturb the ideal situation described by Eq. (8.32). In particular, we have neglected the nonzero metal-semiconductor contact potential difference ϕ_{MS} and the presence of parasitic charges at the semiconductor-oxide interface and within the oxide. The analysis of these effects will be covered in Sec. 8.5, where a complete expression of V_{TH} will be obtained.

Example 8.2 An MOS capacitor is made on silicon material for which $N_A = 10^{16}/\text{cm}^3$. The oxide thickness is 100 nm. Calculate the depletion region width at surface inversion and the value of the threshold voltage. For the values of the quantities not specified, refer to App. A.

Answer We first calculate the surface inversion voltage. From Eq. (8.27) we obtain

$$V_s^* = \frac{2kT}{q} \ln \frac{N_A}{n_i} \simeq 0.7 \text{ V} \qquad (E8.3)$$

Using Eq. (8.28) for x_d^*, we find

$$x_d^* \simeq 0.295 \text{ }\mu\text{m} \qquad (E8.4)$$

Using $x_o = 0.1$ μm, we find from Eq. (8.31)

$$V_{ox}^* \simeq 1.4 \text{ V} \qquad (E8.5)$$

The threshold voltage is then, from Eq. (8.32),

$$V_{TH} = V_{ox}^* + V_s^* = 1.4 + 0.7 = 2.1 \text{ V} \qquad (E8.6)$$

When V_G is increased beyond V_{TH}, the charge distribution exhibits the shape shown in Fig. 8.7c. Under this operating condition, since no additional negative charge is produced by depletion of the semiconductor, the depletion region width x_d reaches its maximum value for $V_G = V_{TH}$, and it remains fixed at this value as V_G is increased beyond V_{TH}, and thus the additional negative charge comes from accumulation of electrons at the semiconductor-oxide interface. As Fig. 8.7c shows, the increase in Q_G above its inversion value is exactly matched by Q_n.

Thus, since all of the charge added on the gate is matched by the electrons accumulating at the semiconductor-oxide interface, the *incremental capacitance*

$$C = \frac{\Delta Q_G}{\Delta V_G} \qquad (8.33)$$

must be equal to the capacitance of the oxide layer C_{ox}. An equivalent way of expressing this is to write that the total charge of electrons in the inversion region is

$$|Q_n| = C_{ox}(V_G - V_{TH}) \qquad (8.34)$$

Equation (8.34) should be used with caution for V_G approximately equal to V_{TH}, because it is derived on the assumption that there are no electrons at the surface until the onset of strong inversion. This is, of course, an approximation that can be removed by considering the exact solution of Poisson's equation for the MOS capacitor (Sec. 8.7).

At this point, we summarize the main topics presented thus far regarding the operation of the MOS capacitor.

For a *p*-type MOS capacitor we can state:

1. If $V_G < 0$, the charge in the semiconductor is positive and forms from hole accumulation at the semiconductor-oxide interface. The capacitance of the structure is C_{ox}.
2. For $0 < V_G < V_{TH}$, the negative charge forms from hole depletion in a region of width x_d. The depletion region contains a fixed concentration of negative charge (acceptor atoms) and therefore the width of the depletion region x_d must increase as V_G increases. The capacitance of the MOS is the series combination of C_{ox} and C_s, where C_s is the capacitance of the semiconductor depletion region.
3. For $V_G > V_{TH}$, the depletion region width is fixed, and the additional negative charge arises from accumulation of electrons at the semiconductor-oxide interface. When the electron concentration at the surface exceeds the acceptor doping concentration, the surface is said to be *inverted*. The incremental capacitance in this situation is again C_{ox}, and the total charge of electrons in the inversion region is $|Q_n| = C_{ox}(V_G - V_{TH})$.

8.5 C-V PLOTS OF AN MOS STRUCTURE

The most important measurement in MOS work is the plot of the capacitance C_{MOS} as a function of the applied gate bias voltage V_G. This is known as a *C-V* (capacitance-voltage) plot. Its importance lies in the information that can be obtained from a single plot and from the fact that it can be performed on a relatively simple structure, allowing, for example, in-process quality control monitoring.

The ideal curve of incremental capacitance C_{MOS} versus V_G for an MOS structure made on a *p*-type semiconductor with $n_i = 10^{10}/cm^3$, $N_A = 10^{16}/cm^3$, and an oxide thickness of 200 nm, is shown in Fig. 8.9*a*. For all negative gate voltages, $C_{MOS} = C_{ox}$, or $C_{MOS}/C_{ox} = 1$. The structure is operating in the (majority carrier) *accumulation* mode.

For positive gate voltages in the range $0 < V_G < V_{TH} = 3.5$ V, the MOS structure is operating in the *depletion mode*, and the capacitance is given by Eq. (8.20). To find C_{MOS}/C_{ox} versus V_G, the depletion region width x_d is calculated versus V_G and Eq. (8.20) is then used to compute C_{MOS}/C_{ox}. At the boundary line between depletion and inversion, where $V_G = 3.5$ V, Eq. (8.20) provides $C_{MOS}/C_{ox} = 0.67$. The value of V_{TH} is obtained following the same procedure as in Example 8.2.

For gate voltages more positive than 3.5 V, the structure is operating in the *inversion* (or minority carrier accumulation) mode. The incremental capacitance C_{MOS} is C_{ox} in this mode of operation; then the ideal capacitance curve rises abruptly from 0.67 to 1.0 as V_G exceeds 3.5 V.

A realistic plot of C_{MOS}/C_{ox} versus V_G taken on an MOS capacitor for which $N_A = 10^{16}/cm^3$ and $x_o = 200$ nm is shown in Fig. 8.9*b*. The curve has the same general shape as the ideal one, with the following main differences:

> The abrupt changes in the slope of the *C-V* curve take place only in the simplified model. However, if we make the curves coincide by shifting one of them along the voltage axis, then the two curves are in quite satisfactory agreement with each other. In fact, the minimum capacitance is the same for both ideal and realistic, and the gate voltage change required to pass from $C_{MOS}/C_{ox} = 0.95$ to $C_{MOS}/C_{ox} = 0.67$ in the depletion mode is also the same.

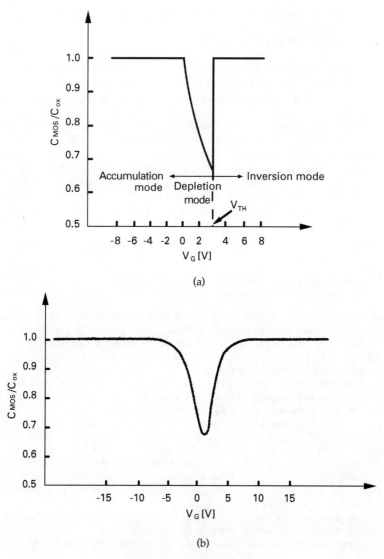

FIGURE 8.9 C-V plots for an MOS capacitor made on a *p*-type semiconductor under low frequency conditions. (*a*) Theoretical curve and (*b*) realistic curve.

The *voltage offset* has two main causes. First, the fact that the gate metal and the semiconductor are different materials means that a contact potential difference will develop between them. In fact, even though three different materials in series form the MOS capacitor (indeed, no matter how many materials are in the loop), the electrostatic potential difference between the two ends of the structure depends *only* on the first and last material because, except the metal contact potential ϕ_M and the semiconductor contact potential

ϕ_S, each of the other contact potentials in the loop (in particular ϕ_{ox}) appears twice in the sum (Kirchhoff's voltage law), once with a plus sign and once with a minus sign, then resulting in a cancellation. Thus, the contact potential developed by the MOS structure is given by[3]

$$\phi_{MS} = -(\phi_{\text{gate material}} - \phi_{\text{bulk material}}) \tag{8.35}$$

where ϕ_{MS} is a widely used symbol. The value of ϕ_{MS} can be calculated from Eq. (8.35) and Table 8.1. The contact potential can then be considered as a built-in bias voltage and must be included in the threshold voltage V_{TH}. It is, however, possible to make the gate from properly doped polysilicon, and then this contact potential difference disappears (or can be adjusted to some desired value).

A second cause of voltage offset is associated with the existence of an unneutralized layer of charge Q_o that is trapped at the semiconductor-oxide interface and within the oxide.

These charges are of four types[1,3]:

1. An *oxide fixed charge* exists near the semiconductor-oxide interface due to the mechanisms of oxide formation. This charge is found to be rather independent of oxide thickness, doping type (p or n), and doping concentration.
2. An *oxide trapped charge* can exist throughout the oxide, but usually close to either of its interfaces to the substrate or the gate. This charge can be acquired through radiation, photoemission, or the injection of high-energy carriers from the substrate.
3. A *mobile ionic charge* can exist within the oxide because of contamination by alkali ions (often Na$^+$) introduced by the environment. This charge can move within the oxide under the presence of an electric field.
4. An *interface trap charge* (also called *fast surface-state charge*) exists at the semiconductor-oxide interface. It is caused by defects at that interface, which give rise to charge traps; these can exchange mobile carriers with the semiconductor, acting as donors or acceptors.

Usually it is assumed that all parasitic charges are located at the semiconductor-oxide interface and that their value, denoted by Q_o is fixed. The charge Q_o, is called the *equivalent interface charge* and is always positive, for both p-type and n-type substrates. A voltage $Q_o C_{ox}$ is then associated to this charge.

Thus, these two nonideal effects (i.e., the metal-semiconductor potential difference

TABLE 8.1 Approximate Contact Potential of Materials to Intrinsic Silicon

Material (j)	ϕ_j, V
Ag	−0.4
Au	−0.3
Cu	0.0
Ni	+0.15
Al	+0.6
Mg	+1.35
Extrinsic Si	−ϕ_F
Intrinsic Si	0

and the equivalent interface charge) can be added by the principle of superposition into Eq. (8.32), giving

$$V_G^* = V_{TH} = \phi_{MS} - \frac{Q_o}{C_{ox}} + \frac{qN_A x_o}{\varepsilon_{ox}} x_d^* + \frac{2kT}{q} \ln \frac{N_A}{n_i}$$

$$= V_{FB} + 2\phi_p + \frac{Q_G}{C_{ox}} \tag{8.36}$$

where
$$V_{FB} = \phi_{MS} - \frac{Q_o}{C_{ox}} \tag{8.37}$$

is called the *flat-band voltage.*

The flat-band voltage can be considered as the value of the gate-substrate voltage V_G at which electron and hole concentrations are equal to their equilibrium values. Of the two terms on the right-hand side of Eq. (8.37), the first takes care of the contact potentials and the second takes care of the equivalent interface charge.

The results, illustrated in Fig. 8.9, are valid when the MOS structure operates at low frequency (10 to 100 Hz). In fact, for high-frequency (10-kHz) operations, the C-V plot looks more like Fig. 8.10.

The difference between high- and low-frequency C-V plotting is further explained by Fig. 8.11. It shows the fluctuation in charge distribution in the inversion condition in response to the fluctuation of the ac gate signal. Figure 8.11a is the high-frequency case. It shows that at inversion, there is indeed a layer of charge at the semiconductor-oxide interface, but the ac fluctuation of the gate charge takes place at a high enough rate that only the widening and narrowing of the depletion region can follow the charge fluctuation. This widening and narrowing occurs at the fast dielectric relaxation rate, which is on the order of 10^{-14} s. In Fig. 8.11b, the low-frequency case, the gate charge fluctuates slowly enough that the inversion charge can follow the variation directly. The frequency needs to be in the 10- to 100-Hz range before this can occur. In both cases, notice that the depletion region remains at its maximum width.[4]

The high-frequency capacitance-versus-voltage equations for an MOS capacitor are

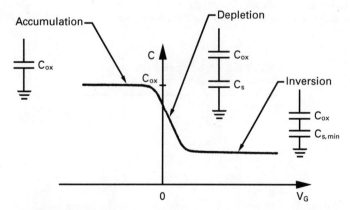

FIGURE 8.10 C-V plot for an MOS capacitor made on a p-type semiconductor under high frequency.

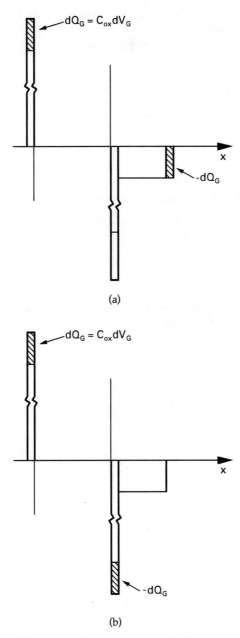

FIGURE 8.11 Fluctuations in charge distribution in an MOS capacitor at (a) high frequency and (b) low frequency.

now derived. All expressions for capacitances are per unit area. Furthermore, the semiconductor is assumed uniformly doped.

When the MOS capacitor is in *accumulation,* the measured total capacitance is

$$C_{\text{MOS}} = C_{\text{ox}} \tag{8.38}$$

where C_{ox} is the oxide capacitance.

In *depletion*, the capacitance drops with increasing gate voltage. To find the exact relationship, first call to mind that, from Eq. (8.17),

$$V_G = \frac{Q_G}{C_{\text{ox}}} + V_s = \frac{\sqrt{2\varepsilon_s q N_A V_s}}{C_{\text{ox}}} + V_s \tag{8.39}$$

Then

$$\frac{dV_G}{dQ_G} = \frac{1}{C_{\text{MOS}}} = \frac{1}{C_{\text{ox}}} + \frac{dV_s}{dQ_G} = \frac{1}{C_{\text{ox}}} + \frac{1}{C_s} \tag{8.40}$$

Equation (8.40) indicates that the total capacitance is the series combination of the oxide capacitance and depletion region capacitance. Another form of Eq. (8.40) is

$$\frac{C_{\text{MOS}}}{C_{\text{ox}}} = \frac{1}{1 + \frac{C_{\text{ox}}}{C_s}} \tag{8.41}$$

We now express C_{ox}/C_s as a function of V_G. To this purpose we refer again to Eq. (8.17), here rewritten for convenience as

$$V_G = \frac{qN_A x_d}{C_{\text{ox}}} + \frac{qN_A x_d^2}{2\varepsilon_s} \tag{8.42}$$

Since $C_s = \varepsilon_s/x_d$, then

$$V_G = \frac{qN_A \varepsilon_s}{C_{\text{ox}} C_s} + \frac{qN_A \varepsilon_s}{2C_s^2} \tag{8.43}$$

By defining

$$\frac{1}{V_{\text{ox}}} = \frac{2C_{\text{ox}}^2}{qN_A \varepsilon_s} \tag{8.44}$$

and multiplying Eq. (8.43) by it, and rearranging terms, we obtain

$$\left(\frac{C_{\text{ox}}}{C_s}\right)^2 + \frac{2C_{\text{ox}}}{C_s} - \frac{V_G}{V_{\text{ox}}} = 0 \tag{8.45}$$

Solving Eq. (8.45) for C_{ox}/C_s, and taking the meaningful root, we obtain

$$\frac{C_{\text{ox}}}{C_s} = -1 + \sqrt{1 + \frac{V_G}{V_{\text{ox}}}} \tag{8.46}$$

Substituting Eq. (8.46) into Eq. (8.41), we obtain

$$\frac{C_{\text{MOS}}}{C_{\text{ox}}} = \left(\sqrt{1 + \frac{V_G}{V_{\text{ox}}}}\right)^{-1} = \left(\sqrt{1 + \frac{2V_G C_{\text{ox}}^2}{qN_A \varepsilon_s}}\right)^{-1} \tag{8.47}$$

The above derivation assumes that $V_{\text{FB}} = 0$. If this does not hold true, we can simply replace V_G in Eq. (8.47) with $(V_G - V_{\text{FB}})$.

Once inversion is reached, the capacitance remains at a constant minimum value of

$$\frac{C_{\min}}{C_{\text{ox}}} = \left(\sqrt{1 + \frac{V_{\text{TH}}}{V_{\text{ox}}}}\right)^{-1} \tag{8.48}$$

where V_{TH} is the threshold voltage.

8.6 ION IMPLANTATION FOR THRESHOLD VOLTAGE CONTROL

In the previous section we have pointed out the causes of the difference between the theoretical and experimental values of V_{TH}. These changes can cause problems for the circuit designer and must therefore be minimized. Theoretically, we can choose oxide thickness and semiconductor doping concentrations that would cause V_{TH} to assume any value we like for each device. However, these choices can conflict with other device requirements.

Usually, we have a given oxide thickness and we want to adjust the threshold voltage under these conditions. Moreover, as already mentioned, the threshold voltage is also affected by an unneutralized layer of charge often *trapped* at the semiconductor-oxide interface.

The basic idea for using the technique of *ion implantation* to adjust the threshold voltage is pointed out in Fig. 8.12. Figure 8.12a shows the theoretical threshold condition for

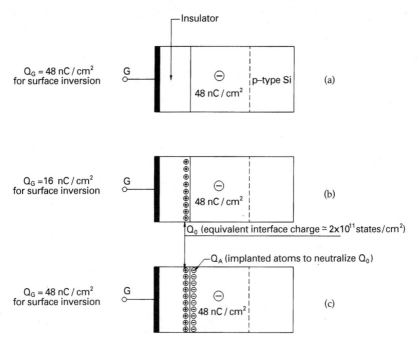

FIGURE 8.12 Ion implantation technique to control V_{TH} in an MOS capacitor. (*a*) Theoretical inversion condition, (*b*) inversion condition with a fixed $Q_o \simeq 2 \times 10^{11}/\text{cm}^2$, and (*c*) inversion condition with a fixed Q_o and ion implantation charge $Q_A = -Q_o$.

the MOS structure considered in the previous examples. The threshold voltage is about 2.1 V and the charge in the depletion region at surface inversion ($qN_Ax_d^*$) is about 48 nC/cm² of gate area.

Figure 8.12b shows the situation associated with the presence of a fixed positive surface charge Q_o (equivalent interface charge) introduced to represent the unneutralized layer of charge. This charge Q_o (usually positive) changes the electric field in the oxide, increasing the threshold voltage on p-channel devices and reducing it on n-channel devices.

Figure 8.12c shows the MOS structure with an appropriate charge of implanted atoms coming from group III of the periodic table, which resides just inside the semiconductor. As these atoms in the semiconductor, being acceptor impurities, are negative, we can cancel the effect of the surface charge Q_o by implanting a charge concentration Q_A which is exactly opposite to Q_o. This will then shift the actual threshold voltage back to the theoretical value as shown in Fig. 8.12a.

8.7 GENERAL ANALYSIS OF THE MOS STRUCTURE

In this section, we will derive the general relations between the surface potential V_s and the total charge per unit area in the semiconductor Q_s for any value of applied voltage V_G, be it in *accumulation, depletion,* or *inversion* including *weak inversion* neglected so far.[3]

The charge density at depth x in the semiconductor, is given by

$$\rho(x) = q[p(x) - n(x) + N_D - N_A] \tag{8.49}$$

From the Boltzmann equations, the electron and hole concentrations as a function of V and at x are given by

$$n_p(x) = n_{p0}\, e^{qV(x)/kT} \tag{8.50a}$$

$$p_p(x) = p_{p0}\, e^{-qV(x)/kT} \tag{8.50b}$$

where $V(x)$ is the potential with respect to the bulk of the semiconductor at x, taken to be zero far into the bulk of the semiconductor, and n_{p0} and p_{p0} are the equilibrium concentrations of electrons and holes, respectively, in the bulk of the semiconductor (p type, in the case here considered).

At the surface, Eqs. (8.50) reduce to

$$n_s(x) = n_{p0}\, e^{qV_s/kT} \tag{8.51a}$$

$$p_s(x) = p_{p0}\, e^{-qV_s/kT} \tag{8.51b}$$

where V_s is the surface potential [i.e., the value of $V(x)$ at the semiconductor surface].

In the bulk of the semiconductor, far from the surface, charge neutrality must exist; therefore $\rho(x) = 0$ and $V(x) = 0$. From these conditions, Eq. (8.49) yields

$$n_{p0} - p_{p0} = N_D - N_A \tag{8.52}$$

Using Eq. (8.52) and introducing Eqs. (8.50) into Eq. (8.49), and the result in Poisson's equation [Eq. (3.103)], we obtain (see also Sec. 7.2 for comparison)

$$\frac{d^2V}{dx^2} = -\frac{q}{\varepsilon_s}[p_{p0}(e^{-qV(x)/kT} - 1) - n_{p0}(e^{qV(x)/kT} - 1)] \tag{8.53}$$

Multiplying both sides of Eq. (8.53) by $2(dV/dx)$, we recognize the resulting left-hand side as $(d/dx)(dV/dx)^2$. Replacing x by a dummy variable, integrating from a point deep in the bulk (theoretically at infinity, where $V = 0$ and $dV/dx = 0$) to a point x, solving for dV/dx at a point x, and recalling that $E(x) = -dV/dx$, we obtain[2,3]

$$E(x) = -\frac{dV}{dx} = \pm\frac{2kT}{qL_D}F\left(\frac{qV}{kT}, \frac{n_{p0}}{p_{p0}}\right) \tag{8.54}$$

where
$$L_D \equiv \sqrt{\frac{2kT\varepsilon_s}{q^2 p_{p0}}} \tag{8.55}$$

is the extrinsic Debye length (indicated as χ in Sec. 7.2) for holes, and

$$F\left(\frac{qV}{kT}, \frac{n_{p0}}{p_{p0}}\right) \equiv \left[\left(e^{-qV/kT} - 1 + \frac{qV}{kT}\right) + \frac{n_{p0}}{p_{p0}}\left(e^{qV/kT} - 1 - \frac{qV}{kT}\right)\right]^{1/2} \geq 0 \tag{8.56}$$

Equation (8.54) is considered with positive sign for $V > 0$ and negative sign for $V < 0$.

The electric field at the surface (semiconductor-oxide interface) is determined from Eq. (8.54), by setting $V = V_s$. Therefore,

$$E_s = \pm\frac{2kT}{qL_D}F\left(\frac{qV_s}{kT}, \frac{n_{p0}}{p_{p0}}\right) \tag{8.57}$$

Using Gauss' law, we can relate the total charge for unit area Q_s in the semiconductor to the surface electric field by

$$Q_s = -\varepsilon_s E_s = \pm\frac{2\varepsilon_s kT}{qL_D}F\left(\frac{qV_s}{kT}, \frac{n_{p0}}{p_{p0}}\right) \tag{8.58}$$

where the negative sign in front of F must be used with $V_s > 0$ (*depletion* or *inversion*) and the positive sign with $V_s < 0$ (*accumulation*). This function behaves as shown in Fig. 8.13.

When $V_G = V_{FB}$, the surface charge is equal to zero. When $V_G < V_{FB}$, the surface charge is positive, corresponding to the accumulation regime, and when $V_G > V_{FB}$, the surface charge is negative, corresponding to the depletion and inversion regimes. The reader can show, as an exercise, that from Eqs. (8.56) to (8.58) the surface charge is proportional to $\exp(q|V_s|/2kT)$ in accumulation (when $|V_s|$ exceeds a few times kT/q) and in strong inversion. In depletion, the surface charge varies as $V_s^{1/2}$.

From Eq. (8.58), the differential capacitance per unit area of the semiconductor region is given by

$$C_s \equiv \frac{dQ_s}{dV_s} = \pm\frac{\varepsilon_s}{L_D}\frac{\left[1 - e^{-qV_s/kT} + \frac{n_{p0}}{p_{p0}}(e^{qV_s/kT} - 1)\right]}{F\left(\frac{qV_s}{kT}, \frac{n_{p0}}{p_{p0}}\right)} \tag{8.59}$$

Equations (8.58) and (8.59) are valid in all conditions (accumulation, depletion, and inversion).

The applied voltage can now be related to the surface potential V_s. The electric field at the semiconductor surface can be obtained from Eq. (8.57). Using the condition of continuity of the electric flux density

$$\varepsilon_s E_s = \varepsilon_{ox} E_{ox} \tag{8.60}$$

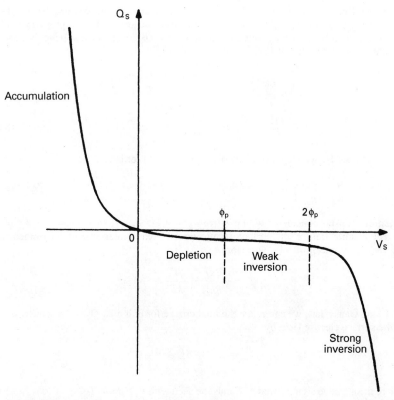

FIGURE 8.13 Variation of the charge density Q_s in the p-type semiconductor of an MOS capacitor as a function of the surface potential V_s.

we find the voltage drop $E_{ox}x_o$ across the oxide layer. Hence the applied voltage can be written as

$$V_G = V_{FB} + V_s + \varepsilon_s \frac{E_s}{C_{ox}} \tag{8.61}$$

where C_{ox} is the insulator capacitance per unit area.

Equation (8.61) allows us to calculate the threshold voltage V_{TH}, corresponding to the onset of inversion regime which occurs when the surface potential V_s is equal to $2\phi_p$. At this surface potential, the charge of the free carriers induced at the semiconductor-oxide interface is still small compared to the charge in the depletion region, which is given by

$$Q_s = \sqrt{2\varepsilon_s q N_A V_s} = \sqrt{2\varepsilon_s q N_A(2\phi_p)} \tag{8.62}$$

This charge causes the electric field at the semiconductor-oxide interface

$$E_s = \frac{Q_s}{\varepsilon_s} = \sqrt{\frac{2qN_A(2\phi_p)}{\varepsilon_s}} \tag{8.63}$$

Using Eq. (8.61), we find the following expression for the threshold voltage

$$V_{TH} = V_{FB} + 2\phi_p + \frac{\sqrt{2\varepsilon_s q N_A (2\phi_p)}}{C_{ox}} \tag{8.64}$$

which is the same as Eq. (8.36).

PROBLEMS

8.1 Calculate the flat band voltage V_{FB} for a p-type substrate with $N_A = 5 \times 10^3/\mu m^3$, an SiO_2 insulator with a thickness $x_o = 0.042$ µm, and an aluminum gate. The equivalent interface charge Q_o is 0.1 fC/µm².

8.2 An MOS capacitor is made on an n-type silicon substrate having a doping density $N_D = 4 \times 10^{15}/cm^3$. The oxide is grown under conditions for which the chemical rate constant $B = 2 \times 10^{-10} cm^2/h$.

  a. If the growth of the oxide is governed by the relation $x_o^2 = Bt$, where x_o is the oxide thickness, and t is the reaction time, how long should the growth be carried out to obtain an oxide capacitance $C_{ox} = 10^3$ pF/cm²?

  b. Using ε_{ox} (for SiO_2) and ε_s values given in App. A, calculate the threshold voltage V_{TH}.

  c. Calculate the depletion region width and the charge in this region (per square centimeter of capacitor area) when $V_G = V_{TH}$.

  d. Calculate the charge required to reach inversion if the dimensions of the gate are 10 µm × 100 µm.

8.3 Plot $(V_{TH} - V_{FB})$ versus oxide thickness x_o, from $x_o = 0.01$ µm to $x_o = 0.1$ µm, for $N_A = 10^2$, 10^3, and $10^4/\mu m^3$.

8.4 Find a relation which provides the ratio between the minimum capacitance C_{min} at high frequency of an MOS structure, and the oxide capacitance C_{ox}, as a function of the oxide thickness x_o and the substrate doping concentration.

8.5 Find the transition frequency between high- and low-frequency C-V characteristics, equating the generation rate in the depletion region and the charging current.

REFERENCES

1. A. S. Grove, *Physics and technology of semiconductor devices*, New York: John Wiley & Sons, 1967.
2. S. M. Sze, *Physics of semiconductor devices*, 2d ed., New York: John Wiley & Sons, 1981.
3. Y. P. Tsividis, *Operation and modeling of the MOS transistor*, New York: McGraw-Hill, 1987.
4. D. G. Ong, *Modern MOS technology: processes, devices and design*, New York: McGraw-Hill, 1984.

CHAPTER 9
METAL-OXIDE-SEMICONDUCTOR FIELD-EFFECT TRANSISTOR (MOSFET)

In Chap. 8 we presented the operating modes of the MOS structure and we anticipated that this structure could be considered as the control element of an important semiconductor device: the metal-oxide-semiconductor field-effect transistor (MOSFET). A large number of biosensors are based on the MOSFET, as will be described in Chap. 10. The goal of this chapter is to provide a description of MOSFET behavior, taking into account circuit analysis and design.

For circuit applications, the most important property of the MOSFET is the mathematical relation between the drain current I_{DS} and the drain-source and gate-source voltages, V_{DS} and V_{GS}, respectively.

We will then derive these equations and we will show how they can be used in circuit design, focusing on the circuit configurations mostly used in biological applications.

9.1 ENHANCEMENT-MODE MOSFET

In this section we will derive the drain characteristics of an *enhancement-mode n*-channel MOSFET, i.e., a relationship of the form

$$I_{DS} = I_{DS}(V_{GS}, V_{DS}) \tag{9.1}$$

The development follows directly from the results obtained in Chap. 8. In particular, An inversion layer can be formed at the semiconductor-insulator interface when we apply to an MOS capacitor a voltage V_G that is greater than its threshold voltage V_{TH} (see *strong inversion* in Sec. 8.3). The total mobile charge Q_n in the inversion layer is proportional to the difference between the applied voltage V_G and the threshold voltage V_{TH}, as follows from Eq. (8.34).

Subthreshold mode operation (*weak inversion*), based on diffusion rather than drift, will be briefly considered in Sec. 9.7.

The mobile charge that forms the inversion layer provides a conducting path between contacts placed on the surface of the semiconductor. This path, or inversion layer, is called a *channel*. Its width is defined to be, for the *p*-type semiconductor substrate assumed as an example, the depth in the semiconductor at which $n = N_A$. The width of the channel and the charge in the channel thus increase as V_{GS} increases. The resistance of this channel is inversely proportional to the total mobile charge in the inversion layer, and can therefore be controlled by varying the voltage applied to the MOS capacitor.

Figure 9.1 shows the basic structure of a MOSFET: it is an MOS capacitor that has been made between two n^+ (i.e., heavily doped n-type) contact regions in a p-type semiconductor. The *gate* (*G*) (or MOS capacitor electrode) has a length *L* and a width *Z*, as shown. The n^+ regions are called the *source* (*S*) and *drain* (*D*) regions. Their main function is to establish a low-resistance contact to the two ends of the n^+-type inversion layer, and a very high-resistance contact to the p-type semiconductor substrate. Indeed the n^+ regions can supply electrons to the inversion layer, as is required there for conduction, but they cannot supply holes to the p-type substrate, so they are a very poor ohmic contact to the substrate. Really they form *pn* junctions (see Chap. 6) with the substrate, with the polarity of bias between the source and drain regions and the substrate being such as to inhibit current flow. [Actually, a leakage current will flow across the n^+p junction (see Chap. 6) at the drain, but it is very small, so the source and drain are effectively isolated when there is no inversion layer connecting them.] In practice, only a negligible current can flow between the two n^+ regions unless a surface inversion layer is formed. This implies that current flows between source and drain only when we apply a voltage between gate and substrate that is greater than the threshold voltage for the MOS capacitor.

The device shown in Fig. 9.1 is called an *n-channel enhancement-mode* MOSFET; a *p-channel enhancement-mode* MOSFET can be made by using p^+ contact regions in an n-type semiconductor substrate. The array of possible MOSFETs is indicated in Fig. 9.2, with the arrow indicating the direction from p-type to n-type. The line connecting the source and drain is continuous for depletion-mode devices (channel exists for zero gate-source bias) and discontinuous for enhancement-mode devices (channel does not exist for zero gate-source bias). When MOSFETs are used as discrete devices, the substrate con-

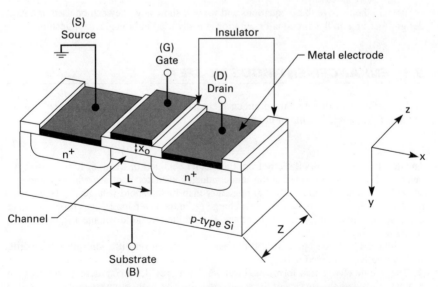

FIGURE 9.1 Structure of an *n*-channel enhancement-mode MOSFET.

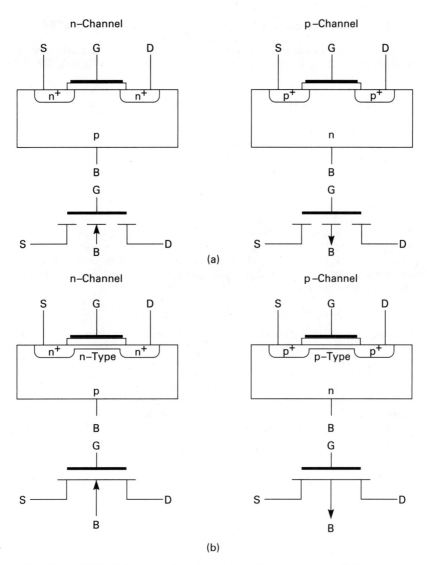

FIGURE 9.2 MOSFET structures and circuit symbols. (*a*) Enhancement mode; (*b*) depletion mode.

tact (labeled *B* for *body* in the figure) is usually connected to the source. In integrated circuits that use only one channel type, all of the transistors have a common substrate. Under these circumstances, it is common practice for a significant number of these devices to have their source terminals connected to other circuit components, not to the substrate. This connection results in complicated relationships between the gate-source voltage and drain current. For the circuits used in this book, we assume that the source and substrate are connected together.

9.1.1 Drain Characteristics for the Enhancement-Mode MOSFET

The basic electrical characteristics of a MOSFET can be described by a set of curves in which the drain current I_{DS} is plotted as a function of the drain-source voltage V_{DS} for given values of the gate-source voltage V_{GS}. These curves are called the *drain* (or *output*) *characteristics* of the MOSFET. Other curves, called *input characteristics*, relate the drain current I_{DS} to the gate-source voltage V_{GS} for given values of the drain-source voltage V_{DS}.

Typical output characteristics for an *n*-channel enhancement-mode MOSFET are shown in Fig. 9.3*a*. Figure 9.3*b* shows a typical input characteristic of a MOSFET, pointing out the regions of operation. Weak inversion characteristic will be considered in Sec. 9.7.

The drain nonsaturation characteristics, as shown later, are defined by the equation

$$I_{DS} = K\left[(V_{GS} - V_{TH})V_{DS} - \frac{V_{DS}^2}{2}\right] \tag{9.2}$$

The quantities K and V_{TH} are device parameters that must be either measured or calculated. The drain characteristics shown in Fig. 9.3*a* indicates that there are two basic regions of operation of the MOSFET, corresponding to very low $V_{DS} \ll (V_{GS} - V_{TH})$ (*nonsaturation region*) and very high $V_{DS} \gg (V_{GS} - V_{TH})$ (*saturation region*) values of the drain voltage. We will now derive the basic characteristics for these two regions. The locus of points dividing the saturation region from the nonsaturation region is defined by

$$V_{DS} = (V_{GS} - V_{TH}) \tag{9.3}$$

For digital applications, the MOSFET operates on both sides of the curve specified by Eq. (9.3). For analog applications, the MOSFET is usually operated in saturation.

We now consider what will occur when a drain voltage V_{DS} is applied to a MOSFET under the condition $V_{GS} > V_{TH}$, necessary to ensure that a conducting channel has been formed between the source and drain regions. To this purpose we consider Fig. 9.4, where x is the distance coordinate measured along the length of the channel from source to drain.

The application of V_{DS} will cause a voltage drop along the channel which we indicate by the function $V(x)$, and whose value varies from 0 to V_{DS} as we proceed from the source to the drain. We indicate the potential at a point x_{ch}, as shown in Fig. 9.4, by $V(x_{ch})$. The existence of such a potential implies that the voltage between the gate and a small section of channel placed at x_{ch} is

$$V'_{GS} = V_{GS} - V(x_{ch}) \tag{9.4}$$

The width and charge density at each point in the channel are then determined by V'_{GS} and will vary along the x direction as indicated in Fig. 9.4. In particular, the channel will show maximum width at the source end and minimum width at the drain end. The charge in the differential section Δx of channel at x_{ch} is then given by

$$\Delta Q_n = \Delta C_{ox}(V'_{GS} - V_{TH}) \tag{9.5}$$

where

$$\Delta C_{ox} = \varepsilon_{ox} Z \frac{\Delta x}{x_o} \tag{9.6}$$

In Eq. (9.6), C_{ox} is the capacitance of the oxide layer, ε_{ox} is the dielectric constant of the oxide, x_o is the oxide thickness, and Z is the width of the device (measured into the page).

The charge ΔQ_n can also be written in terms of V_{GS} and $V(x_{ch})$ by using Eq. (9.4), as

$$\Delta Q_n = \Delta C_{ox}[V_{GS} - V(x_{ch}) - V_{TH}] \tag{9.7}$$

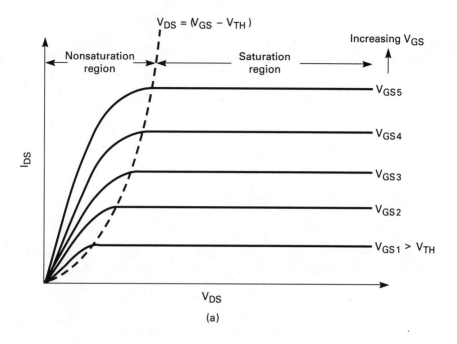

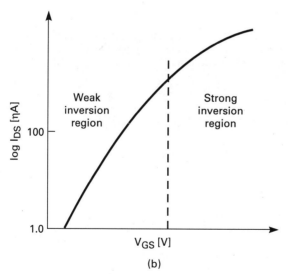

FIGURE 9.3 Characteristics of an *n*-channel enhancement-mode MOSFET. (*a*) Output characteristics; (*b*) input characteristics.

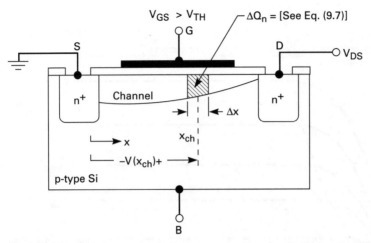

FIGURE 9.4 Shape of the channel width of an *n*-channel enhancement-mode MOSFET, caused by the application of a large drain-source voltage V_{DS}. [*Adapted from lecture notes (Stanford University) by J. F. Gibbons.*]

To calculate the drain current that flows under these conditions, we first notice that the time needed for carriers (electrons) to flow across the section Δx at x_{ch} is

$$\Delta t = \frac{\Delta x}{v_n} = -\frac{\Delta x}{\mu_n E(x_{ch})} \qquad (9.8)$$

where $E(x_{ch})$ is the field in the channel at x_{ch} induced by the application of V_{DS}. The minus sign in Eq. (9.8) derives from the definition of the positive direction of $E(x)$. Thus, the current flowing through the section Δx can be written as

$$I_{DS} = \frac{\Delta Q_n}{\Delta t} \qquad (9.9)$$

By using Eq. (9.7), we can rewrite Eq. (9.9) in the form

$$I_{DS} = \frac{\Delta C_{ox}[V_{GS} - V(x_{ch}) - V_{TH}]}{\Delta t} \qquad (9.10)$$

Using Eq. (9.6), we write Eq. (9.10) as

$$I_{DS} = -\frac{\mu_n \varepsilon_{ox} Z}{x_o}[V_{GS} - V(x_{ch}) - V_{TH}]E(x_{ch}) \qquad (9.11)$$

Since

$$E(x_{ch}) = -\frac{dV(x)}{dx}\bigg|_{x=x_{ch}} \qquad (9.12)$$

we obtain

$$I_{DS}\,dx = \frac{\mu_n \varepsilon_{ox} Z}{x_o}[V_{GS} - V(x_{ch}) - V_{TH}]\,dV(x) \qquad (9.13)$$

Equation (9.13) represents the current I_{DS} flowing through a section of width Δx centered at x. The channel, in its turn, is formed of a large number of such sections, all of which are connected in series going from the source to the drain. Moreover, the same current I_{DS} must flow through each section for a given V_{DS}. Therefore, we can integrate Eq. (9.13) as follows:

$$I_{DS}\int_0^L dx = \frac{\mu_n \varepsilon_{ox} Z}{x_o} \int_0^{V_{DS}} [(V_{GS} - V_{TH}) - V] \, dV \tag{9.14}$$

Carrying out the integration and dividing by the channel length L, we obtain[1-4]

$$I_{DS} = \frac{\mu_n \varepsilon_{ox} Z}{x_o L}\left[(V_{GS} - V_{TH})V_{DS} - \frac{V_{DS}^2}{2}\right]$$

$$= K\left[(V_{GS} - V_{TH})V_{DS} - \frac{V_{DS}^2}{2}\right] \tag{9.15}$$

which is exactly Eq. (9.2), when we set

$$K = \mu_n \frac{\varepsilon_{ox}}{x_o} \frac{Z}{L} \tag{9.16}$$

In Eq.(9.16), μ_n is constant up to sufficiently large values of V_{GS} and K begins to decrease when V_{GS} becomes large enough. Equation (9.16) states that, for a given type of channel (given μ), the parameter K can be determined directly once the gate width Z, the gate length L, and the oxide properties (x_o and ε_{ox}) are known.

At *low drain voltages*, i.e., for $V_{DS} \ll (V_{GS} - V_{TH})$, we can neglect the quadratic term in V_{DS} in Eq. (9.15). Thus, under this assumption, we can write

$$I_{DS} = K(V_{GS} - V_{TH})V_{DS} \tag{9.17}$$

Equation (9.17) states that the low drain voltage characteristics are a series of straight lines passing through the origin, with slopes that increase as V_{GS} increases. In general, each of these characteristics can be described by the equation

$$I_{DS} = G_{DS}V_{DS} \tag{9.18}$$

where G_{DS} is called the *drain-source conductance* of the MOSFET.

The variation in channel width with V_{DS} has given the physical key to obtaining the basic MOSFET equation [Eq. (9.15)]. However, this derivation is valid only as long as $V_{DS} < (V_{GS} - V_{TH})$, i.e., when the MOSFET is operated in the so-called *nonsaturation* region (see Fig. 9.3a), because when $V_{DS} = (V_{GS} - V_{TH})$ the channel will (in principle) be reduced to zero width at the drain end. The physical situation is shown in Fig. 9.5a: the channel is said to be *pinched off* at the drain. The drain current that flows under this condition can be calculated from Eq. (9.15) by substituting $V_{DS} = (V_{GS} - V_{TH})$. The result is

$$I_{DS} = \frac{K}{2}(V_{GS} - V_{TH})^2 \tag{9.19}$$

By differentiating Eq. (9.15), we find that the drain current has a horizontal tangent at $V_{DS} = (V_{GS} - V_{TH})$, as indicated in the drain characteristics of Fig. 9.3a.

For the current given by Eq. (9.19) to flow, it is necessary that some very high velocity electrons exist in a thin region near the drain. Then, the channel is in practice not pinched

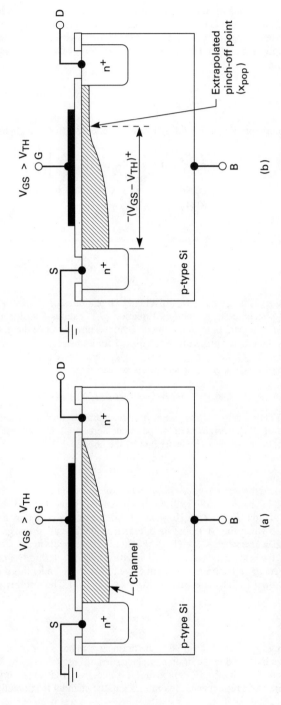

FIGURE 9.5 Shape of the channel width of an *n*-channel enhancement-mode MOSFET for (*a*) $V_{DS} = (V_{GS} - V_{TH})$ and (*b*) $V_{DS} > (V_{GS} - V_{TH})$.

down to exactly zero width, but rather to a point where there are just enough electrons to carry the drain current when each electron is traveling at its maximum velocity.

The presence of the thin pinched-off region is more evident when V_{DS} is increased beyond $(V_{GS} - V_{TH})$. When this happens, the actual channel pinch-off point, indicated by x_{pop} in Fig. 9.5b, will shift very slightly from the drain toward the source. The voltage drop from the source to the pinch-off point, $V(x_{pop})$, will be exactly $(V_{GS} - V_{TH})$, since this is the value needed to reduce the channel charge and width to zero. The remainder of the applied drain voltage, $[V_{DS} - (V_{GS} - V_{TH})]$, is dropped across the region from the extrapolated pinch-off point (x_{pop}) to the drain, and the charge in this region can be neglected.

Thus, it follows that the non-pinched-off portion of the channel behaves like a MOSFET that has a gate length $L = x_{pop}$ and is always operated at $V'_{DS} = (V_{GS} - V_{TH})$. The total charge in the non-pinched-off portion of the channel and the voltage drop across it are practically fixed at the values they have when $V_{DS} = (V_{GS} - V_{TH})$, so to a first approximation the current I_{DS} flowing in the channel will be independent of V_{DS} for all values of $V_{DS} > (V_{GS} - V_{TH})$. In other words, the drain current I_{DS} *saturates* at its pinch-off value. As a consequence we can extend the drain current characteristics horizontally from the pinch-off point to larger values of V_{DS} (see Fig. 9.3a). Thus, at *large drain voltages*, i.e., for $V_{DS} \gg (V_{GS} - V_{TH})$, the drain current I_{DS} in this region is given by Eq. (9.19).

The MOSFET is said to be operated in the *saturation mode*. From a circuit point of view, the fact that I_{DS} can be independent of V_{DS} implies that the output current I_{DS} is controlled entirely by the input voltage V_{GS}. Moreover, the output terminals of the MOSFET are therefore represented as a V_{GS}-controlled current source instead of a conductance as in the nonsaturated operation mode.

The physical phenomena just described and the corresponding electrical behavior of the MOSFET are summarized in Table 9.1.

In the preceding analysis, we considered the channel length L as a constant. In reality, L is determined by the distance between the depletion regions surrounding the source and drain. The widths of these regions are functions of the source-substrate and drain-substrate biases. The width of the drain depletion region increases with drain voltage, thereby decreasing the channel length, which increases the gradient of carrier concentration, and therefore increases the channel current. This increase in channel current due to *channel-length modulation* is called the *Early effect*. Because the dependence of I_{DS} on L is explicit (through the parameter K) in Eqs. (9.15) and (9.19), we can solve directly for the drain conductance G_{DS} of the MOSFET. Because the drain current is inversely proportional to

TABLE 9.1 Basic Properties of an *n*-Channel Enhancement-Mode MOSFET*

Gate and drain bias conditions	Channel conditions	Drain characteristics
$V_{GS} \leq V_{TH}$ and $V_{DS} \geq 0$	No inversion layer: $Q_n = 0$	$I_{DS} = 0$
$V_{GS} > V_{TH}$ and $0 < V_{DS} \ll (V_{GS} - V_{TH})$	Inversion layer exists: $Q_n = C_{ox}(V_{GS} - V_{TH})$	$I_{DS} = K(V_{GS} - V_{TH})V_{DS}$ voltage controlled conductance
$V_{GS} > V_{TH}$ and $0 < V_{DS} \leq (V_{GS} - V_{TH})$	Inversion layer exists: decreasing channel thickness near drain	$I_{DS} = K[(V_{GS} - V_{TH})V_{DS} - V_{DS}^2/2]$
$V_{GS} > V_{TH}$ and $V_{DS} \geq (V_{GS} - V_{TH})$	Inversion layer exists: pinch-off near drain	$I_{DS} = (K/2)(V_{GS} - V_{TH})^2$ voltage controlled current source

*Subthreshold mode is not considered.

L, the drain conductance (which manifests itself as a nonzero slope on the drain characteristics for large values of V_{DS}) is proportional to I_{DS} and inversely proportional to L. Usually, this conductance is approximated by a constant depending on the given process and device geometry. Because the conductance is proportional to I_{DS}, in this approximation the extrapolated drain curves all intersect the voltage axis at a single point, which is $V_{DS} = -V_0$. Thus, to take into account the channel-length modulation, we can simply add the term $G_{DS}V_{DS}$ to the expressions of the drain current just obtained.

9.1.2 Drain Voltage Breakdown

A breakdown process takes place at the drain region of a MOSFET when the drain-source voltage V_{DS} is increased beyond a certain value. The breakdown is caused by high electric fields generated in the vicinity of the n^+p junction that defines the boundary of the drain region. The source of this high electric field is indicated in Fig. 9.6a, where a positive drain voltage V_{DS} is applied between the drain and the source-substrate terminal of an n-channel enhancement-mode MOSFET. This voltage creates a depletion region at the drain region, and a corresponding electric field, shown in Fig. 9.6b, which exhibits its maximum value at the metallurgical boundary between the n^+ region and p substrate. The electric field can be obtained from Gauss' law, using an approach similar to that described in Sec. 8.3.

When a sufficient V_{DS} is applied to cause the maximum electric field to reach a value around 10 to 20 V/μm, electrons and holes that flow into (or are thermally generated within) the high field region can produce an *avalanche effect* (see Sec. 6.8.1), with the result that a very large current can flow from the substrate to the drain.

The drain breakdown process is not destructive if the drain current that flows is limited externally. However, the gate control over the drain current is lost, so, for circuit applications, the drain breakdown phenomenon fixes an upper limit on the drain voltage that can be applied. For typical MOSFETs this limit is in the range of 5 to 25 V.

9.2 DEPLETION-MODE MOSFET

In this section, we derive the drain characteristics of a *depletion-mode* MOSFET. The basic operating characteristics for the depletion-mode MOSFET can follow directly from the behavior of the MOS capacitor when it is operating in the depletion-mode (see Sec. 8.3).

We refer to an n-type substrate for the MOS capacitor so that the related depletion-mode MOSFET will be an n-channel device (and therefore comparable to the n-channel enhancement-mode device described in Sec. 9.1).

Application of a *negative* voltage on the gate of the MOS capacitor will deplete the n-type semiconductor. The width of the depletion region x_d can be calculated, if the semiconductor doping concentration N_D and the oxide thickness x_o are known, through the equation

$$\left(\frac{qN_D}{2\varepsilon_s}\right)x_d^2 + \left(\frac{qN_Dx_o}{\varepsilon_{ox}}\right)x_d = -V_{GS} \qquad (9.20)$$

Equation (9.20) can be compared to Eq. (8.17). Notice that for the n-type substrate we must apply negative voltage on the gate to produce the desired depletion of electrons.

The depletion width x_d increases as V_{GS} is made more negative, and, if the semicon-

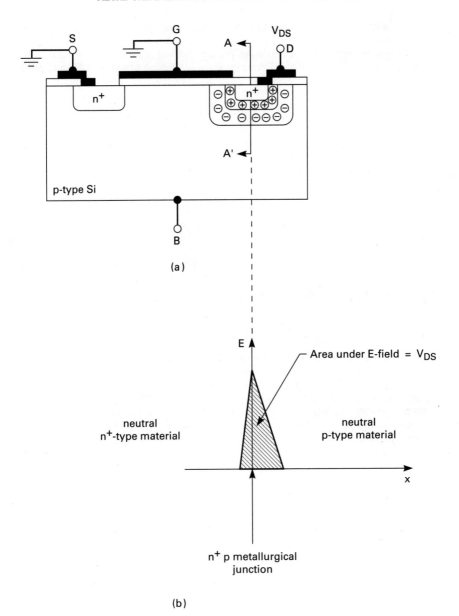

FIGURE 9.6 MOSFET breakdown. (*a*) Depletion region produced by an applied V_{DS} at the drain and (*b*) electric field along A-A'. [*From lecture notes (Stanford University) by J. F. Gibbons.*]

ductor substrate is thin enough, the entire substrate can be depleted before the surface inversion condition is reached.

The negative voltage that is needed to produce full substrate depletion is called the *pinch-off voltage* and is usually indicated by V_p. It can be calculated by substituting the thickness of the semiconductor region W for x_d in Eq. (9.20), i.e.,

$$-V_p = \left(\frac{qN_D}{2\varepsilon_s}\right)W^2 + \left(\frac{qN_D x_o}{\varepsilon_{ox}}\right)W \qquad (9.21)$$

For many depletion-mode MOSFETs, the oxide is sufficiently thick that the voltage drop across the oxide, V_{ox}, is much larger than the voltage drop across the depleted semiconductor region, V_s, for all operating conditions. When this approximation holds, the term in x_d^2 can be neglected and Eqs. (9.20) and (9.21) simplify to

$$x_d \simeq -\frac{\varepsilon_{ox} V_{GS}}{qN_D x_o} \qquad (9.22)$$

and

$$V_p \simeq -\left(\frac{qN_D x_o}{\varepsilon_{ox}}\right)W \qquad (9.23)$$

It is also worth noticing that, under this approximation,

$$\frac{x_d}{W} \simeq \frac{V_{GS}}{V_p} \qquad (9.24)$$

These properties of the MOS capacitor can be used to describe an *n-channel depletion-mode* MOSFET where, unlike the enhancement-mode MOSFET, the *n* region is obtained by acting on the technology processes of the device.[3,4] A drain characteristic for an *n*-channel depletion-mode MOSFET is shown in Fig. 9.7. The drain characteristic is similar in form to an enhancement-mode characteristic. However, the behavior of the MOSFET with gate voltage V_{GS} is completely different. In particular, the depletion-mode MOSFET delivers maximum current for $V_{GS} = 0$ and pinches off completely ($I_{DS} = 0$) for $V_{GS} = V_p$.

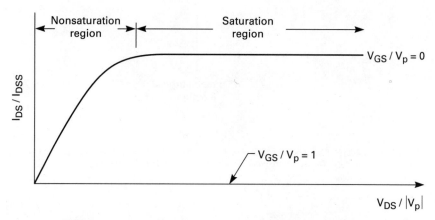

FIGURE 9.7 I_{DS}-V_{DS} characteristics of an *n*-channel depletion-mode MOSFET.

9.2.1 Drain Characteristics for the Depletion-Mode MOSFET

In analogy with the enhancement-mode MOSFET, we choose conditions so that the drain current will saturate when V_{DS} becomes sufficiently large. To this purpose, the gate and drain voltages must cooperate to reduce the thickness of the conducting channel at the drain end below the value it has at the source end. This goal is reached in the enhancement-mode MOSFET by making V_{DS} and V_{GS} be of the *same polarity* (see Sec. 9.1). This causes the gate-channel voltage to be smaller at the drain end than at the source end, with the result that the channel is thinner at the drain than it is at the source.

To get a similar result for the depletion-mode MOSFET, we need V_{DS} and V_{GS} to be of *opposite polarity*. This will increase the voltage difference between the gate and the channel at the drain end, thus increasing the width of the depletion region and decreasing the thickness of the conducting channel. In fact, the channel becomes pinched off at the drain end by a drain voltage of $V_{DS} = (V_{GS} - V_p)$. If $V_{GS} = 0$, the channel pinches off when V_{DS} reaches $-V_p$.

The channel profile at pinch-off for the condition where $V_{GS} = 0$ is shown in Fig. 9.8a. To calculate the drain characteristics for this case, we take a small section of the conducting channel placed at a distance x from the source and we consider Eq. (3.17), here rewritten for reader's convenience in the one-dimensional form,

$$J = \sigma E \quad (9.25)$$

With appropriate substitutions, Eq. (9.25) yields

$$\sigma \frac{dV}{dx} = \frac{I_{DS}}{Z(W - x_d)} \quad (9.26)$$

and then

$$I_{DS}\, dx = \sigma Z W \left(1 - \frac{x_d}{W}\right) dV \quad (9.27)$$

For the case under consideration, where $V_{GS} = 0$, we have [see Eq. (9.24)]

$$\frac{x_d}{W} = -\frac{V(x')}{V_p} \quad (9.28)$$

where $V(x)$ is the voltage at x measured with respect to the source. Using Eq. (9.28) in Eq. (9.27) and integrating, we obtain

$$I_{DS} \int_0^L dx = \sigma Z W \int_0^{V_{DS}} \left(1 + \frac{V}{V_p}\right) dV \quad (9.29)$$

and then[1-4]

$$I_{DS} = \frac{\sigma Z W}{L}\left(V_{DS} + \frac{V_{DS}^2}{2V_p}\right) \quad (9.30)$$

The drain current saturates at $V_{DS} = -V_p$, having a value of

$$I_{DSS} = -\frac{\sigma Z W}{L} \frac{V_p}{2} \quad (9.31)$$

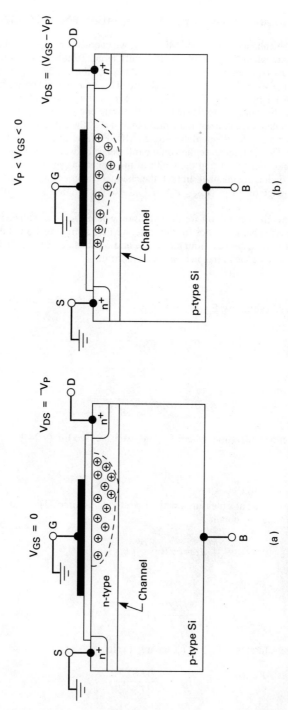

FIGURE 9.8 Profile of the channel width at pinch-off of an *n*-channel depletion-mode MOSFET. (*a*) For $V_{GS} = 0$ and $V_{DS} = -V_P$; (*b*) Application of V_{GS} reduces channel thickness at source. [*Adapted from lecture notes (Stanford University) by J. F. Gibbons.*]

This is the current that flows into the drain when the channel is just pinched off at the drain end, as shown in Fig. 9.8a.

If V_{DS} is increased further, the pinch-off point will move slightly toward the source; however, the voltage drop along the non-pinched-off channel will remain at V_p. A thin region in which the electrons move with their maximum velocity will extend from the drain to the actual pinch-off point.

If V_{GS} becomes more negative, the thickness of the channel at the source end will be reduced, and the drain voltage required to pinch off the channel at the drain end will decrease, as shown in Fig. 9.8b. The mathematical theory for the drain characteristics then leads to[1-4]

$$I_{DS} = -I_{DSS}\left[2\left(1 - \frac{V_{GS}}{V_p}\right)\frac{V_{DS}}{V_p} + \left(\frac{V_{DS}}{V_p}\right)^2\right]$$

$$= \frac{\sigma Z W}{L}\left[\left(1 - \frac{V_{GS}}{V_p}\right)V_{DS} + \frac{V_{DS}^2}{2V_p}\right] \quad (9.32)$$

for a *non-pinched-off* channel, under the assumption that x_d is directly proportional to the gate-channel voltage drop at each point along the channel. Though this assumption is not exactly true, it is a reasonable approximation for many practical depletion-mode MOSFETs.

At *low drain voltages,* we can neglect the quadratic term in V_{DS} in Eq. (9.32). Thus, under this assumption we can write[1-4]

$$I_{DS} = \frac{\sigma Z W}{L}\left(1 - \frac{V_{GS}}{V_p}\right)V_{DS} \quad (9.33)$$

which states that the low drain voltage characteristics are a series of straight lines passing through the I_{DS} versus V_{DS} origin. The slopes decrease as V_{GS} decreases (i.e., becomes more negative) until the channel is fully pinched-off, at which point $I_{DS} = 0$ for any V_{DS}.

The pinch-off condition for the channel is

$$V_{DS} = V_{GS} - V_p \quad (9.34)$$

where V_p and V_{GS} are negative numbers. Equation (9.34) can be used with Eq. (9.32) to evaluate the drain current at the pinch-off point in terms of V_{GS}. The result is[1-4]

$$I_{DS} = I_{DSS}\left(1 - \frac{V_{GS}}{V_p}\right)^2 \quad (9.35)$$

The drain current I_{DS} is constant at the value given in Eq. (9.35) for all $V_{DS} > (V_{GS} - V_p)$.

The two parameters I_{DSS} and V_p that were used in the equations for the drain characteristics play roles that are equivalent to those of K and V_{TH} for the enhancement-mode MOSFET.

9.3 MOSFET AMPLIFIER

We will now present the basic ideas necessary to design discrete MOSFET circuits with particular emphasis on those circuit configurations used in biological applications. MOSFETs are electronic circuit devices which, under certain conditions, can deliver more signal power to a load than they absorb at their inputs (because of this property, they are usually called *amplifiers*). They are then *active* circuit devices, as distinguished from *passive* circuits [resistance-inductance-capacitance (*RLC*)], which cannot deliver more signal power to a load than they absorb from a source. However, MOSFETs are not active devices at all frequencies, but only up to some maximum frequency; at sufficiently high frequencies, they also behave like passive circuits (i.e., three-terminal *RLC* circuits).

The signal power delivered to the load arises from a power supply (e.g., a battery). The MOSFET converts some of the dc power available from this supply into ac signal power and makes this signal power available at its terminal.

It is convenient to begin the study of the applications of MOSFETs with some graphical considerations.

9.3.1 Graphical Analysis

Let us consider the circuit of a basic MOSFET amplifier using an *n*-channel enhancement-mode device as shown in Fig. 9.9. In this circuit, the signal source is represented by a signal voltage $v_s(t)$ in series with a source resistor R_S. The battery V_{BB} and the resistor R_B give the gate-source bias, while the capacitor C_S is used to prevent the flow of dc current through the signal source. The battery V_{DD} and the resistor R_L provide the drain-source bias, in a way that will be described later.

One method of studying the signal-amplifying properties of the circuit of Fig. 9.9 is to solve *Kirchhoff's laws* graphically on a set of drain characteristics.[5-8] This method has some quantitative limitations, though it provides a convenient way to point out the basic circuit operation.

Before going on, a remark on notation must be made. The following notation[5,6] will be used to specify quantities that have both dc and *time-varying* components. The *total value* of a variable is expressed as a lowercase letter with uppercase subscripts: for example, i_{DS} or v_{DS}. The dc *value* of a variable is expressed as a uppercase letter with uppercase subscripts: for example, I_{DS} or V_{DS}. The *time-varying* value of a variable is expressed as a lowercase letter with lowercase subscripts: for example, v_{ds} or i_{ds}. In some cases the time variation can also be explicitly indicated: $i_{ds}(t)$ or $v_{ds}(t)$. Thus, $i_{DS} = I_{DS} + i_{ds}(t)$ and $v_{DS} = V_{DS} + v_{ds}(t)$.

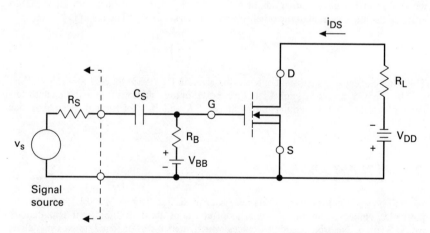

FIGURE 9.9 Basic *n*-channel enhancement-mode MOSFET amplifier.

To study the signal transmission properties of the circuit of Fig. 9.9, we first evaluate the values of v_{DS} and i_{DS} that are allowed in the drain circuit. To do this we notice that the values (v_{DS}, i_{DS}) must have to undergo the following constraints:

1. Satisfy the *load equation*

$$v_{DS} = V_{DD} - R_L i_{DS} \qquad (9.36)$$

2. Be compatible with the drain characteristics of the MOSFET.

These two conditions are met by tracing a *load line* [i.e., the equivalent of Eq. (9.36) for given values of V_{DD} and R_L] on the drain characteristics of the MOSFET, as shown in Fig. 9.10, where $V_{DD} = 30$ V and $R_L = 5$ kΩ. The drain characteristics are those of a typical MOSFET.

The choice of the values for V_{DD} and R_L usually depends on two factors: (1) the application to which the amplifier is devoted and (2) the MOSFET ratings, which indicate the allowed area of the i_{DS} versus v_{DS} plane in which a load line can lie. The MOSFET manufacturer usually specifies a maximum drain current $I_{DS,\text{max}}$, a maximum drain-source voltage $v_{DS,\text{max}}$, and a maximum power dissipation P_{max}. For the MOSFET characteristics shown in Fig. 9.10, these values are $I_{DS,\text{max}} = 100$ mA, $v_{DS,\text{max}} = 50$ V, and $P_{\text{max}} = 300$ mW. Thus, the constraints

$$v_{DS} < v_{DS,\text{max}} \qquad i_{DS} < I_{DS,\text{max}} \qquad v_{DS} i_{DS} < P_{\text{max}} \qquad (9.37)$$

must be all verified simultaneously within the shaded area shown in Fig. 9.10, which determines the i_{DS} versus v_{DS} plane in which the load line can lie. In general it is convenient

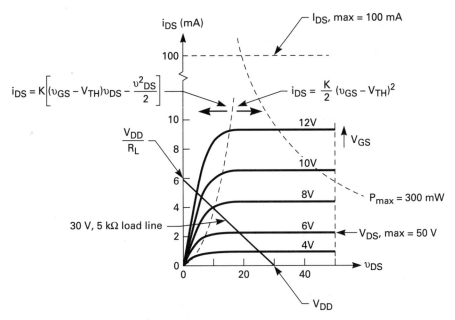

FIGURE 9.10 Graphical analysis of the basic MOSFET amplifier of Fig. 9.9. Drain characteristics (i_{DS}-v_{DS}) of the *n*-channel enhancement-mode MOSFET showing allowed operating region and load line.

to reduce the maximum ratings by about 10 percent and then redefine the allowed area accordingly.

9.3.2 Signal Transfer Characteristics

Once the load line has been chosen, the signal-transfer properties of the circuit can be developed. For the circuit of Fig. 9.9, the signal transfer characteristics are represented by either of the two transfer characteristics shown in Fig. 9.11, each obtained by considering different values for v_{GS} and reading the corresponding value of i_{DS} (or v_{DS}) from the load line.

The drain current versus gate-source voltage transfer characteristic is usually used because of its independence of R_L so long as the operating point remains in the saturated region of the drain characteristics. Under this condition, Eq. (9.19), here rewritten for reader's convenience as

$$i_{DS} = \frac{K}{2}(v_{GS} - V_{TH})^2 \tag{9.38}$$

holds until v_{GS} becomes so large that the MOSFET is driven out of the saturation region of its drain characteristics.

9.3.3 Small-Signal Voltage Gain

The transfer characteristic defined by Eq. (9.38) and shown in Fig. 9.11a (saturation region) can be used to evaluate the small-signal voltage gain. To this purpose, we notice that the total gate-source voltage for the circuit of Fig. 9.9 can be written in the form

$$v_{GS} = V_{BB} + v_s(t) \tag{9.39}$$

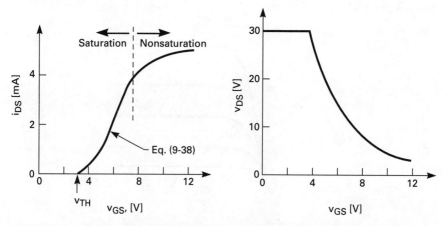

FIGURE 9.11 Graphical analysis of the basic MOSFET amplifier of Fig. 9.9. Transfer characteristics obtained from load line of Fig. 9.10: (a) i_{DS}-v_{GS} and (b) v_{DS}-v_{GS}.

Equation (9.39) is justified because, since the gate is insulated from the source and drain by the gate oxide, no dc current can flow in it. Hence the dc gate voltage for the circuit shown in Fig. 9.9 must be V_{BB}. Introducing Eq. (9.39) into Eq. (9.38), we get

$$i_{DS} = \frac{K}{2}[(V_{BB} - V_{TH}) + v_s(t)]^2 \tag{9.40a}$$

or

$$i_{DS} = \frac{K}{2}(V_{BB} - V_{TH})^2 + K(V_{BB} - V_{TH})v_s(t) + \frac{K}{2}v_s^2(t) \tag{9.40b}$$

The amplifier is said to be operating under *small-signal* conditions when

$$Kv_s^2(t) \ll 2K(V_{BB} - V_{TH})v_s(t) \tag{9.41a}$$

or

$$v_s(t) \ll 2(V_{BB} - V_{TH}) \tag{9.41b}$$

When Eqs. (9.41) are valid, the total drain current becomes

$$i_{DS} = \frac{K}{2}(V_{BB} - V_{TH})^2 + K(V_{BB} - V_{TH})v_s(t)$$

$$= I_{DS} + i_{ds}(t) \tag{9.42}$$

The time-dependent component of the drain current, $i_{ds}(t)$, is thus proportional to $v_s(t)$, i.e.,

$$i_{ds}(t) = K(V_{BB} - V_{TH})v_s(t) \tag{9.43}$$

Now, if we introduce Eq. (9.42) into the load equation expressed by Eq. (9.36), we find

$$v_{DS} = (V_{DD} - I_{DS}R_L) - i_{ds}(t)R_L \tag{9.44}$$

Thus, the time-dependent component of the output voltage, $v_{ds}(t)$, is

$$v_{ds}(t) = -i_{ds}(t)R_L \tag{9.45}$$

Combining Eq. (9.43) and Eq. (9.45), yields

$$v_{ds}(t) = -K(V_{BB} - V_{TH})R_L v_s(t) \tag{9.46}$$

Equation (9.46) indicates that the small-signal output voltage is proportional to the signal voltage. Thus, we define a *small-signal voltage gain* for the circuit as

$$A_v = \frac{v_{ds}(t)}{v_s(t)} \tag{9.47}$$

For the circuit of Fig. 9.9, Eq. (9.47) gives

$$A_v = -K(V_{BB} - V_{TH})R_L \tag{9.48}$$

Using Eq. (9.42), we can express the voltage gain in terms of the dc drain current as

$$A_v = -2\frac{I_{DS}}{(V_{BB} - V_{TH})}R_L \tag{9.49}$$

Equation (9.49) is more suitable than Eq. (9.48), because V_{TH} can be evaluated by inspection of the drain characteristics, V_{BB} and I_{DS} are evaluated directly from the bias point, and it does not contain the K parameter.

The minus sign in Eqs. (9.48) and (9.49) means that the time-dependent part of the output voltage will be an inverted, magnified image of the input signal. If the input signal is sinusoidal, the minus sign implies a phase shift of 180° between input and output.

Example 9.1 Evaluate the gain in the MOSFET amplifier shown in Fig. 9.9 for a bias voltage $V_{BB} = 6$ V.

Answer Using the drain characteristics and load line given in Fig. 9.10, we find $I_{DS} \simeq 2$ mA, $V_{TH} \simeq 3.5$ V, and $R_L = 5$ kΩ. From Eq. (9.49), we obtain therefore

$$|A_v| = 2\left(\frac{2 \times 10^{-3}}{6 - 3.5}\right) 5 \times 10^3 = 8 \tag{E9.1}$$

If the input signal is so large (or the bias voltage V_{BB} is so small) that the small-signal condition [Eqs. (9.41)] is not satisfied, the output of the amplifier cannot be a magnified image of the input signal, and the concept of the small signal voltage gain loses its meaning.

When this happens, the amplifier is said to *distort* the input signal (i.e., the output is a distorted image of the input). This form of distortion takes place when the transfer characteristic of a network is nonlinear, and then it is always present, to some degree, in the basic MOSFET amplifier, even when it is operating under small-signal conditions.

When the circuit is driven, for example, with a sinusoidal input signal, the dc drain current changes, and there are frequencies present in the output current that do not appear in the input signal. When these effects take place, we say that the amplifier has *distorted* the signal. This particular form of distortion is called *harmonic distortion* because the new frequencies are harmonics of the input frequency f_s (i.e., multiples of f_s). However, we can reduce distortion by choosing operating conditions to ensure that the total harmonic distortion remains below a certain value. This criterion of limited distortion sets a limit on the maximum input signal that the amplifier can handle, but at the same time, it quantifies the concept of small-signal conditions. Then, when the distortion is adequately limited, the output signal will look like the input signal, and we can define a voltage gain as above. A treatment of distortion is outside the aim of the book; the reader interested in a quantitative analysis can, for example, refer to Ref. 6.

9.4 BIASING CIRCUITS FOR THE MOSFET

The previous analysis has shown that the gate-source bias voltage V_{BB} determines the small-signal voltage gain of the basic MOSFET amplifier [see Eq. (9.49)]. Thus, it is fundamental to set the bias voltage at some desired value and to ensure that it will remain at that chosen point under changing operation conditions.

Circuits that achieve these results are called *biasing circuits*.[5–10] In this section, we will describe the most widely used biasing circuits for enhancement-mode MOSFETs.

In an enhancement-mode MOSFET, the gate and drain terminals both have the same polarity of voltage. Thus, for an *n*-channel MOSFET, they must both be positive with respect to the source. This constraint allows us to bias the gate with the same power supply (e.g., a battery) that is used to bias the drain.

A circuit that achieves this purpose is the *voltage divider* shown in Fig. 9.12. Because

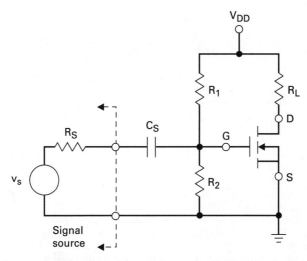

FIGURE 9.12 Voltage divider biasing circuit for an *n*-channel enhancement-mode MOSFET.

the gate is insulated from the underlying substrate material by the gate oxide, no dc current can flow into the gate terminal of the MOSFET. Thus, the gate bias provided by the biasing circuit is

$$V_{GS} = V_{DD} \frac{R_2}{R_1 + R_2} \qquad (9.50)$$

For example, if we want to set V_{GS} at 6 V and we are using a bias source $V_{DD} = 30$ V, then Eq. (9.50) states that the ratio R_1/R_2 must be 4. Then we are free to choose any values of R_1 and R_2 which meet the condition $R_1/R_2 = 4$.

Another requirement usually has to be satisfied: the resistance between the gate terminal and ground must be large compared to R_s, so that essentially all of the signal voltage $v_s(t)$ will appear between the gate and source. If this requirement is not satisfied, a fraction of $v_s(t)$ can be dropped across the internal resistance of the signal source, with the signal voltage delivered to the amplifier will then be smaller than it could be.

By considering the power supply to have zero internal resistance, the resistance between the gate terminal and ground is

$$R_{GS} = \frac{R_1 R_2}{R_1 + R_2} \qquad (9.51)$$

Therefore, the above-mentioned requirement is mathematically expressed by the relation

$$\frac{R_1 R_2}{R_1 + R_2} \gg R_S \qquad (9.52)$$

In practice, it is convenient to choose the left side at least 100 times the right side.

Example 9.2 Design a voltage-divider biasing circuit assuming that $V_{DD} = 30$ V, $R_L = 5$ kΩ, and $R_S = 500\Omega$. Moreover, let us assume operation at $V_{GS} = 6$ V.

Answer Equation (9.50) yields

$$\frac{R_2}{R_1+R_2} = \frac{V_{GS}}{V_{DD}} = \frac{6}{30} = 0.2 \tag{E9.2}$$

and

$$\frac{R_1 R_2}{R_1 + R_2} \geq 100 \text{ k}\Omega \tag{E9.3}$$

where we have imposed that the left side of Eq. (9.52) be 200 times the right side. The minimum values of R_1 and R_2 that satisfy these two equations under the above constraints are $R_1 = 500$ kΩ, $R_2 = 125$ kΩ. From Fig. 9.10 we can find also the drain current $I_{DS} = 2$ mA.

Another biasing configuration is the so-called *single-resistor biasing circuit* shown in Fig. 9.13. Here the gate is fed back to the drain terminal through a large resistor R_F. The main advantage of this biasing circuit configuration is that it provides *self-regulation* of the bias point. In fact, if the supply voltage V_{DD} increases slightly, both V_{DS} and V_{GS} will tend to rise with it. However, a rise in V_{GS} will cause an increase in I_{DS}, which will in turn reduce the originally assumed change in V_{GS} (because of an increased voltage drop in R_L). Then, the bias point tends to be stabilized against random changes in the supply voltage.

The operating point obtained from this circuit configuration can be determined from Fig. 9.14, where a curve representing the condition $V_{GS} = V_{DS}$ has been drawn. The point at which the load line meets this curve defines the *operating point*. The values are those of Fig. 9.10.

The self-regulating process of the biasing circuit can be investigated by calculating the change in V_{GS} that is caused by a change in V_{DD}. Thus, combining the equations

$$V_{GS} = V_{DS} \tag{9.53a}$$

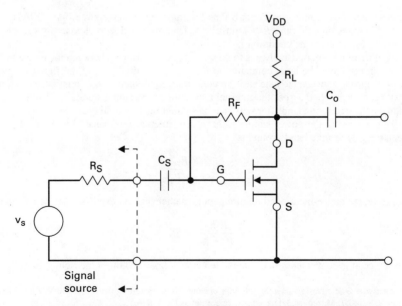

FIGURE 9.13 Feedback biasing circuit for an *n*-channel enhancement-mode MOSFET.

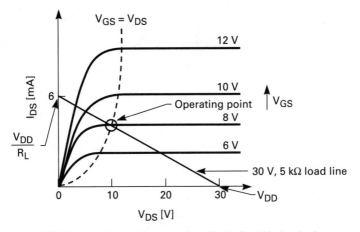

FIGURE 9.14 Graphical representation of the feedback biasing circuit.

$$I_{DS} = \frac{K}{2}(V_{GS} - V_{TH})^2 \tag{9.53b}$$

$$V_{DS} = V_{DD} - I_{DS}R_L \tag{9.53c}$$

we obtain

$$V_{GS} = V_{DD} - \frac{K}{2}(V_{GS} - V_{TH})^2 R_L \tag{9.54}$$

Differentiating Eq. (9.54), we obtain

$$\frac{dV_{GS}}{dV_{DD}} = 1 - [K(V_{GS} - V_{TH})R_L]\frac{dV_{GS}}{dV_{DD}} \tag{9.55}$$

Taking into account Eq. (9.48), we find the magnitude of the voltage gain of the amplifier is

$$|A_v| = K(V_{GS} - V_{TH})R_L \tag{9.56}$$

and using this result in Eq. (9.55), we have

$$\frac{dV_{GS}}{dV_{DD}} = \frac{1}{1 + |A_v|} \tag{9.57}$$

Therefore, a change of ΔV_{DD} in V_{DD} produces a change

$$\Delta V_{GS} = \frac{\Delta V_{DD}}{1 + |A_v|} \tag{9.58}$$

in V_{GS}. If we assume, for example, $|A_v|$ is 10, a 1-V drift in V_{DD} will produce a change of only 0.9 V in V_{GS}, which is negligible for most practical purposes.

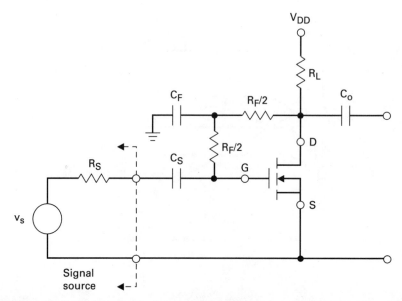

FIGURE 9.15 Feedback biasing circuit for an *n*-channel enhancement-mode MOSFET. This configuration eliminates the signal path feedback effects. Typical values: $R_F = 10$ MΩ, $C_F = 1$ nF.

The resistor R_F also determines a signal path from input to output that is in parallel with the MOSFET. Then the small-signal voltage gain of the circuit will be reduced even if the gain reduction can be minimized by choosing R_F large (a value in the 1- to 10-MΩ range is usually chosen). Moreover, the gain reduction can be eliminated by splitting R_F in half and connecting a small capacitor to ground, as shown in Fig. 9.15.

9.5 SMALL-SIGNAL MODELS FOR THE MOSFET

In this section we will show that when a MOSFET is operating under conditions of limited distortion, it can be represented by a *small-signal circuit model*.

The importance of such a model arises mainly from the consideration that small-signal models are *linear*. Then, if we know that small-signal operating conditions hold, we can substitute each (nonlinear) MOSFET in the circuit with its linear model, applied at the given operating point. We can then use the theory of linear circuit analysis to compute the circuit characteristics.

The development of small-signal models is based on an application of the Taylor approximation. To verify this methodology, we consider the MOSFET as a black box with a pair of input and output terminals, as shown in Fig. 9.16a, where reference polarities for voltages and currents are also shown.

To develop a model for the drain circuit, we notice that i_{DS} is function of the two variables v_{GS} and v_{DS}, i.e.,

$$i_{DS} = i_{DS}(v_{GS}, v_{DS}) \tag{9.59}$$

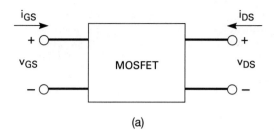

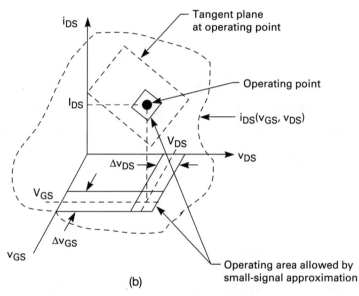

FIGURE 9.16 Graphical aid for the development of a small-signal model for the MOSFET. (*a*) Dipole representation of the MOSFET and (*b*) $i_{DS} = f(v_{GS}, v_{DS})$ representation. (*Adapted from J. F. Gibbons*[8].)

We represent i_{DS} graphically as a surface over the v_{GS} versus v_{DS} plane, as shown in Fig. 9.16*b*. Moreover, the MOSFET is biased by the application of a dc gate voltage V_{GS} and drain voltage V_{DS}, which define an operating dc current I_{DS} given formally by

$$I_{DS} = I_{DS}(V_{GS}, V_{DS}) \tag{9.60}$$

and represented in Fig. 9.16*b* as a point *P* on the i_{DS} surface.

If we assume that the surface has a tangent plane at *P*, and that the gate voltage is made to vary by a small amount Δv_{GS}, then estimating Δi_{DS} from the tangent plane will provide approximately the same result as estimating Δi_{DS} from the actual surface.[8]

In general, the tangent plane is defined by the two slopes $\partial i_{DS}/\partial v_{GS}$ and $\partial i_{DS}/\partial v_{DS}$, each calculated at the bias point *P*. Thus, the change in i_{DS}, for given changes in v_{GS} and v_{DS}, will be

$$\Delta i_{DS} = \Delta v_{GS} \left.\frac{\partial i_{DS}}{\partial v_{GS}}\right|_{V_{GS}, V_{DS}} + \Delta v_{DS} \left.\frac{\partial i_{DS}}{\partial v_{DS}}\right|_{V_{GS}, V_{DS}} \tag{9.61}$$

The reader must notice that Δv_{GS} and Δv_{DS} need not be static: they can vary with time so long as they meet the small-signal condition. Therefore, for instance, we can identify Δv_{GS} with $v_{gs}(t)$, the time-variable signal voltage applied to the gate; of course, similar considerations hold for $v_{ds}(t)$. These considerations allow us to rewrite Eq. (9.61) in the form

$$i_{ds} = D_1 v_{gs} + D_2 v_{ds} \quad (9.62)$$

where i_{ds}, v_{gs}, and v_{ds} are the *small-signal* voltages and current, and the D's have been introduced to indicate the derivatives.

We can now develop a small-signal circuit model for the MOSFET by redrawing the box of Fig. 9.16a in a form that is consistent with Eq. (9.62). This is shown in Fig. 9.17a, where only the small-signal voltages and currents are indicated. To understand the circuit

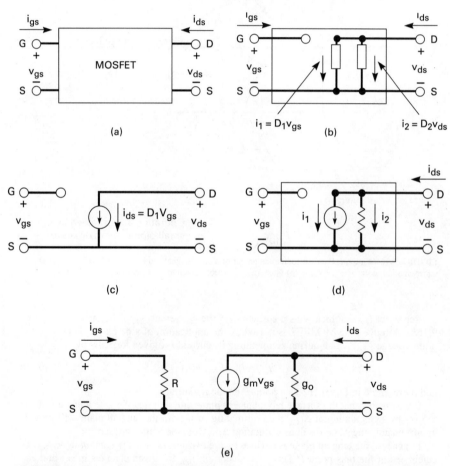

FIGURE 9.17 Circuit representation of Eq. (9.62). (*a*) MOSFET as a black box, (*b*) representation of the drain current as the sum of two components, one due to v_{gs} and one due to v_{ds}, (*c*) representation of a voltage-controlled current source, (*d*) representation of the drain characteristics by the Norton equivalent circuit, and (*e*) small-signal model at low-frequency operation for the MOSFET.

element configuration in the model, we notice that Eq. (9.62) states that i_{ds} is the sum of two components, one due to v_{gs} and one due to v_{ds}. Therefore, in Fig. 9.17b, we use two rectangular boxes connected in parallel, so that i_{ds} will be equal to the sum of the currents flowing in each box (Kirchhoff's current law). The current which flows in the left box is $D_1 v_{gs}$, and the current which flows in the right box is $D_2 v_{ds}$.

We have now to define the boxes in terms of the circuit elements (resistances, inductances, capacitances, voltage sources, and current sources) of the linear circuit theory. In particular, we need a current source, or better a voltage-controlled current source, to define the left box. A model (or circuit representation) for the voltage-controlled current source is shown in Fig. 9.17c. Here the current source provides a current $D_1 v_{gs}$ to the output terminals regardless of the value of v_{ds}. The current is controlled by the value of a "remote" circuit voltage, v_{gs}, called the *control variable*. The controlled source uses the value of v_{gs} and delivers a current that is proportional to it at every time.

The second term in Eq. (9.62), or the right box in Fig. 9.17b, implies a contribution to i_{ds} that is proportional to v_{ds}. This contribution can be defined by inserting a conductance $g_o = D_2$ across the output terminals. The complete model for the output terminals is then shown in Fig. 9.17d. This circuit configuration meets *Norton's theorem*, which states that a linear network can be represented at a pair of terminals by a (remotely controlled) current source and a shunt impedance. In the low-frequency case, the shunt impedance is simply a conductance.

Because both D_1 and D_2 multiply voltages to provide currents, they must both have the dimensions of mhos. They are called the *transconductance* g_m and the *output conductance* g_o, respectively. In terms of g_m and g_o, Eq. (9.62) becomes

$$i_{ds} = g_m v_{gs} + g_o v_{ds} \tag{9.63}$$

Values for g_m and g_o are obtained either by measurements or, if the functional form given in Eq. (9.59) is known, by analysis.

Let us consider an enhancement-mode MOSFET operating in the saturated region, where we know that

$$i_{DS} = \frac{K}{2}(v_{GS} - V_{TH})^2 \tag{9.64}$$

Under these assumptions, we obtain

$$D_1 = g_m = \left.\frac{\partial i_{DS}}{\partial v_{GS}}\right|_{V_{GS}, V_{DS}} = K(V_{GS} - V_{TH}) \tag{9.65a}$$

$$D_2 = g_o = \left.\frac{\partial i_{DS}}{\partial v_{DS}}\right|_{V_{GS}, V_{DS}} = 0 \tag{9.65b}$$

Since the dc drain current component of i_{DS} can be written as

$$I_{DS} = \frac{K}{2}(V_{GS} - V_{TH})^2 \tag{9.66}$$

then g_m can be written in the form

$$g_m = \frac{2 I_{DS}}{(V_{GS} - V_{TH})} \tag{9.67}$$

Equation (9.67) is important because it specifies g_m as a function of the bias point values, I_{DS} and V_{GS}.

To the sake of completeness, a similar analysis for the gate circuit should be performed to obtain a Norton representation also for the gate terminals. However, this is not necessary since we have assumed that i_{gs} is zero at low frequencies, i.e., the gate oxide is a perfect insulator. We can therefore represent the gate-source connection in the model by an open circuit. The complete *small-signal model* for the MOSFET at low-frequency operation is then shown in Fig. 9.17e.

9.5.1 Amplifier Analysis under Low-Frequency Operation

Let us consider again the basic MOSFET amplifier as shown in Fig. 9.18a, with the corresponding small-signal model in Fig. 9.18b. In the small-signal model, the power supply line is considered as a ground terminal, since no time-variable voltages exist on this line (assuming the power supply has no internal impedance).

We can evaluate the voltage gain of the amplifier by finding v_{gs} in terms of v_s and then taking into account that the small-signal output voltage is

$$v_{ds} = -g_m v_{gs}\left(R_L \,\Big\|\, \frac{1}{g_o}\right) \tag{9.68}$$

Since C_S is assumed to be a short circuit at signal frequencies, and $(R_1 \| R_2) \gg R_S$ (by design of the bias circuit), all of v_s appears at the gate-source terminals. Therefore, we can write

$$v_{gs} = v_s \tag{9.69}$$

Moreover, if we can neglect $1/g_o$ in comparison with R_L, we obtain

$$v_{ds} = -g_m v_{gs} R_L = -g_m v_s R_L \tag{9.70}$$

where Eq. (9.69) has been used.

Using Eq. (9.70), we obtain the voltage gain

$$A_v = \frac{v_{ds}}{v_s} = -g_m R_L \tag{9.71}$$

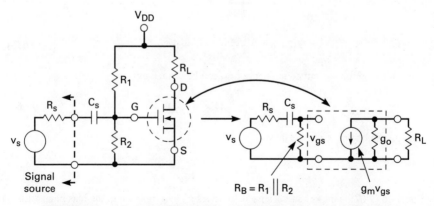

FIGURE 9.18 (a) Basic MOSFET amplifier and (b) corresponding small-signal model.

Introducing Eq. (9.67) into Eq. (9.71), we find

$$A_v = -\frac{2I_{DS}R_L}{(V_{GS} - V_{TH})} \quad (9.72)$$

which is the same as Eq. (9.49).

If the output conductance g_o cannot be neglected, the gain can be written in a straightforward way, using Eqs. (9.68), (9.69), and (9.71), as

$$A_v = -g_m\left(R_L \left\| \frac{1}{g_o} \right.\right) \quad (9.73)$$

showing one of the advantages of using a small-signal model.

9.5.2 Amplifier Analysis under High-Frequency Operation

In this section we will consider the MOSFET operating at high frequency, and we will show how the small-signal model can be used to understand the circuit operation of the device. For this purpose, we have to modify the low-frequency model to include the MOSFET capacitances, which set an upper limit on the signal frequency at which the analysis just performed is valid.

To do this, we recall that the gate and channel form an MOS capacitor whose value is approximately

$$C_{ox} = \frac{\varepsilon_{ox} A_G}{x_o} \quad (9.74)$$

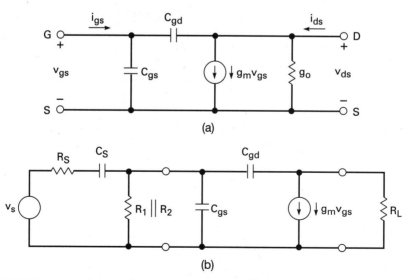

FIGURE 9.19 (a) High-frequency small-signal model for a MOSFET and (b) circuit model used to analyze the high-frequency behavior of the basic MOSFET amplifier of Fig. 9.18a.

where A_G is the gate area and x_o is the oxide thickness. In reality, this total capacitance is distributed along the channel, though for a first approximation, we can split it into two components, C_{gs} and C_{gd}, that are connected between gate, source, and drain terminals as shown in Fig. 9.19a. As a first approximation, we expect that C_{gs} and C_{gd} are approximately equal; this approximation is true for low values of V_{DS}. However, they are also bias-dependent, and it is possible to choose operating conditions such that $C_{gs} \gg C_{gd}$. For a typical MOSFET biased to operate as an amplifier, C_{gs} will be in the range 2 to 10 pF and C_{gd} is in the range 0.2 to 0.5 pF.

The effect of these capacitances on the behavior of the amplifier can be understood by analyzing the small-signal model shown in Fig. 9.19b. Using linear circuit analysis techniques, we find that the circuit provides the low-frequency voltage gain A_o to an upper frequency

$$f_u = \frac{1}{2R_S[C_{gs} + C_{gd}(1 + A_o)]} \tag{9.75}$$

9.6 MOSFET-BASED OPERATIONAL AMPLIFIER

Before we examine some MOSFET-based circuits used in biological applications, it is worth mentioning briefly a class of amplifiers in which the output signal is precisely proportional to the input signal, even when the amplifying components themselves tend to distort the signal and, in addition, have properties that vary with temperature and possibly over time. This type of amplifier is called an *operational amplifier* or *op-amp* for short.

The basic block diagram of an operational amplifier is shown in Fig. 9.20a. It consists of three main components: a *differential input* stage, a *high-gain amplifier* stage, and a *level-shifting output* stage.

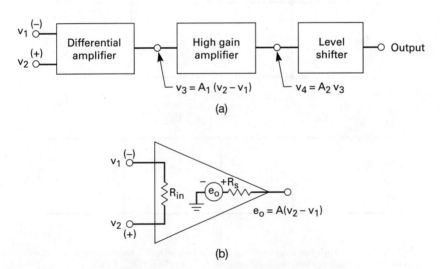

FIGURE 9.20 (a) Stages of an operational amplifier and (b) circuit symbol of the operational amplifier.

The *differential amplifier input stage* provides an output signal (v_3 in Fig. 9.20a) that is proportional to the difference between the input signals applied to the two input terminals. The stage provides a very high input resistance between the input terminals, so that the op-amp does not draw any appreciable signal current from the signal sources that are driving it.

The second block of the op-amp shown in Fig. 9.20a is a very *high gain amplifier*. Its output (v_4 in Fig. 9.20a) should be 10^4 to 10^5 times its input (v_3 in Fig. 9.20a). This will ensure that a relevant output signal can be generated even when the difference between the two input signals is very small.

The last block in Fig. 9.20a is called a *level shifter*. It ensures that the final output of the op-amp is at 0 V dc when there is no difference between the two input signals. The level shifter is also operated under conditions such that the output impedance of the op-amp is very low.

A circuit symbol for the op-amp that summarizes the functions of the three blocks is shown in Fig. 9.20b. It consists of a triangle pointing in the direction of high gain. There are two inputs and one output. The resistance between the two input terminals is indicated with R_{in}, and it is usually assumed to be infinite. The output is represented by a *Thevenin-equivalent* voltage source e_o in series with the output resistance R_s. The output signal voltage is then

$$e_o = A(v_2 - v_1) \tag{9.76}$$

where A is the gain of the amplifier, v_2 is the signal voltage applied to the *noninverting* (+) input terminal and v_1 is the voltage applied to the *inverting* (−) input terminal. The terms *inverting* and *noninverting* indicate the phase relationship that exists between v_1 or v_2 and e_o; e.g., a positive signal applied to the inverting input terminal provides a negative output signal.

A full description of each of the op-amp stages and of the possible applications of this particular type of amplifier would require a book in itself. Thus, here we have preferred to provide only a "definition" of the op-amp, and then to suggest that the reader interested in a deeper treatment of this subject refer, for example, to Refs. 6 and 9.

9.7 SUBTHRESHOLD OPERATION OF THE MOSFET

The simplified model discussed so far assumes that $I_{DS} = 0$ for $V_{GS} \leq V_{TH}$. In reality, this is not the case, and MOFSET can also operate in the so-called *subthreshold* region. This operating condition is characterized by the gate voltage V_{GS} that is well below the threshold voltage V_{TH}. As V_{GS} decreases to V_{TH}, the I_{DS}-V_{GS} characteristics change from square-law to exponential. Whereas the region where V_{GS} is above the threshold is called the *strong inversion* region, the region below is called the *subthreshold*, or *weak inversion* region. This is illustrated in Fig. 9.21a, where a MOSFET input characteristic is shown with the square root of current plotted as a function of the gate-source voltage. When the gate-source voltage decreases to the value designated as V_{ON} (a function of both V_{TH} and kT/q[11]), the current changes from square-law to an exponential-law behavior.

In the subthreshold condition, the current I_{DS} flows through the channel by *diffusion*, from the region of high carrier concentration to that of low concentration, and can be expressed in the form[12]

$$I_{DS} = I_o e^{+qV_{GS}/nkT}(1 - e^{-qV_{DS}/kT}) \tag{9.77}$$

where I_0 is a constant and n is the *threshold slope factor*.[11]

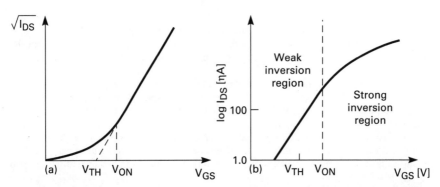

FIGURE 9.21 Subthreshold or weak inversion characteristics of the MOSFET. (*a*) Square-root axis for I_{DS}; (*b*) logarithmic axis for I_{DS}.

The main advantages of subthreshold operation are

1. Power dissipation is very low, from 10^{-12} to 10^{-6} W for a typical circuit.
2. The drain current saturates in a few kT/q, allowing the MOSFET to operate as a current source over most of the voltage range from near ground to V_{DD}.
3. The exponential nonlinearity is an ideal computation primitive for many applications.

This operation condition is very useful when low power is needed, such as for human implantable biomedical devices or for chips emulating networks of real neurons. Circuits, operating under standard strong inversion conditions are proposed to the reader as problems at the end of the chapter.

9.7.1 Current Flow in the Subthreshold Condition

In the subthreshold region, the mobile charge in the channel is much smaller than is the depletion charge in the substrate. Therefore we can neglect the effect of mobile charge on the electrostatics of the MOSFET.

Because the substrate and gate potentials are constant along the channel, the fact that the mobile charge can be neglected compared with the gate and substrate charges implies that the surface potential is constant along the channel. This self-consistent set of conditions defines the *subthreshold* regime of operation. Because there is no electric field along the channel, current cannot flow by drift; hence, diffusion must be the dominant current-flow mechanism. Thus, the only current that flows is the diffusion current [see Eq. (3.40)], here rewritten for convenience as

$$J_{n,\text{diff}} = qD_n \frac{dn}{dx} \tag{9.78}$$

where D_n is the electron diffusion coefficient [see Eq. (3.39)]. Current continuity equation for electrons [see Eq. (3.99)] in a steady state without generation-recombination processes requires

$$\frac{\partial J_n}{\partial x} = 0 \tag{9.79}$$

and this then forces the electron concentration to be a linear function of distance x, decreasing from the source toward the drain. Therefore,

$$n(x) = \left(\frac{n(0) - n(L)}{L}\right)x - n(0) \tag{9.80}$$

where L is the channel length, and the electron concentrations $n(0)$ at the source side of the channel and $n(L)$ at the drain side of the channel are generated by the Boltzmann distribution (neglecting the threshold slope factor n)

$$n(0) = n_0 e^{-q(\phi_0 - V_{GS})/kT} \tag{9.81a}$$

$$n(L) = n_0 e^{-q(\phi_0 - V_{GD})/kT} = n_0 e^{-q(\phi_0 - V_{GS} + V_{DS})/kT} \tag{9.81b}$$

where ϕ_0 is the built-in voltage between source and channel and n_0 is the carrier concentration at the Fermi level.

From Eqs. (9.78) and (9.80), we obtain

$$J_{n,\text{diff}} = J_{DS,\text{subth}} = qD_n\left(\frac{n(0) - n(L)}{L}\right) \tag{9.82}$$

Applying Eqs. (9.81) to Eq. (9.82), we obtain the *subthreshold current*

$$I_{DS,\text{subth}} = \frac{qAD_n n_0}{L} e^{-q\phi_0/kT} e^{+qV_{GS}/kT}(1 - e^{-qV_{DS}/kT})$$

$$= I_0 e^{+qV_{GS}/kT}(1 - e^{-qV_{DS}/kT}) \tag{9.83}$$

where A is the cross-sectional area, and we have absorbed the preexponential factors into a constant I_0. Notice that in Eq. (9.82), the roles of the source and drain are symmetrical; then if we interchange them, the magnitude of the current given by Eq. (9.83) is identical, with the current flowing in the opposite direction.

In some practical situations, it is convenient to rewrite Eq. (9.83) in the form

$$I_{DS,\text{subth}} = I_{\text{sat}}(1 - e^{-qV_{DS}/kT}) \tag{9.84}$$

where

$$I_{\text{sat}} = I_0 e^{+qV_{GS}/kT} \tag{9.85}$$

is the *saturated drain current*, exponential in the gate-source voltage, which represents the flat part of the curve in the I_{DS}-V_{DS} plane.

9.7.2 Diode-Connected MOSFETs

An often-used circuit configuration is shown in Fig. 9.22, where the MOSFET is *diode-connected;* that is, its gate is connected to its drain. This connection provides a voltage that is proportional to the logarithm of the input current. This voltage can be used to control the output currents of other MOSFETs, but it is below the range of usable inputs for circuits such as transconductance amplifiers (see Sec. 9.7.5).

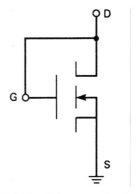

FIGURE 9.22 Diode-connected n-channel enhancement-mode MOSFET.

9.7.3 Current-Mirror-Connected MOSFETs

It is common to have a current of a certain sign and an input that requires an equal but opposite current. A circuit that performs this inversion of current polarity is shown in Fig. 9.23, where the MOSFETs are n-channel enhancement-mode devices. The input current I_{in} biases a diode-connected MOSFET M_1. The resulting V_{GS} is just sufficient to bias the second MOSFET M_2 to a saturation current I_{out} equal to I_{in}. The value of I_{out} will be nearly independent of the drain voltage of M_2 as long as M_2 is in saturation. The n-channel current mirror *reflects* a current from V_{DD} into a current to ground, so it is called a *current mirror*.[9,10,12] A similar arrangement holds for currents of the opposite sign using p-channel MOSFETs. The p-channel circuit reflects a current to ground into a current from V_{DD}.

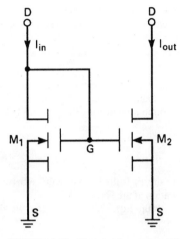

FIGURE 9.23 Current-mirror-connected n-channel enhancement-mode MOSFETs.

9.7.4 Differential Pair Cell

Many circuits have an input signal represented as a difference between two voltages. These circuits use the so-called *differential pair* circuit shown in Fig. 9.24 as an input stage. We will analyse its characteristics in the following.

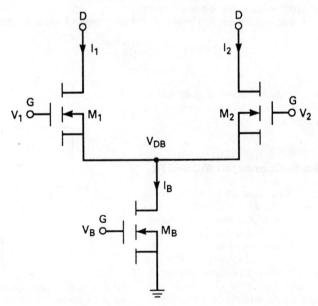

FIGURE 9.24 MOSFET differential pair.

In Fig. 9.24, the MOSFET M_B is used as a current source; under normal conditions, its drain voltage V_{DB} is large enough that the drain current I_B is saturated at a value set by the gate voltage V_B. The way in which I_B is shared between M_1 and M_2 is a function of the difference between V_1 and V_2, which is the purpose of the operation of the stage.

Applying Eq. (9.84) to M_1 and M_2, we obtain

$$I_1 = I_0 e^{q(V_1-V_{DB})/kT} \tag{9.86a}$$

$$I_2 = I_0 e^{q(V_2-V_{DB})/kT} \tag{9.86b}$$

The sum of the two drain currents must equal I_B. Then,

$$I_B = I_1 + I_2 = I_0 e^{-qV_{DB}/kT}(e^{qV_1/kT} + e^{qV_2/kT}) \tag{9.87}$$

By solving Eq. (9.87) for the voltage V_{DB}, we obtain

$$e^{-qV_{DB}/kT} = \frac{I_B}{I_0} \cdot \frac{1}{e^{qV_1/kT} + e^{qV_2/kT}} \tag{9.88}$$

Substituting Eq. (9.88) into Eq. (9.86), we obtain expressions for the drain currents of M_1 and M_2. Therefore,

$$I_1 = I_B \frac{e^{qV_1/kT}}{e^{qV_1/kT} + e^{qV_2/kT}} \tag{9.89a}$$

$$I_2 = I_B \frac{e^{qV_2/kT}}{e^{qV_1/kT} + e^{qV_2/kT}} \tag{9.89b}$$

If V_1 is more positive than V_2 by many kT/q, MOSFET M_2 gets turned off, so essentially all the current flows through M_1, I_1 is approximately equal to I_B, and I_2 is approximately equal to 0. On the other hand, if V_2 is more positive than V_1 by many kT/q, M_1 gets turned off, I_2 is approximately equal to I_B, and I_1 is approximately equal to 0 (Refs. 9, 10, 12).

9.7.5 Transconductance Amplifier

The current mirror stage decribed in Sec. 9.7.3 is used, for example, in the *differential transconductance amplifier*[9,10,12] to generate an output current that is proportional to the difference between the drain currents of M_1 and M_2, i.e.,

$$I_1 - I_2 = I_B \frac{e^{qV_1/kT} - e^{qV_2/kT}}{e^{qV_1/kT} + e^{qV_2/kT}} \tag{9.90}$$

Multiplying both the numerator and denominator of Eq. (9.90) by $e^{-(qV_1/kT+qV_2/kT)/2}$ we can express every exponent in terms of voltage differences. The result is

$$I_1 - I_2 = I_B \tanh\left(\frac{q}{2kT}(V_1 - V_2)\right) \tag{9.91}$$

The circuit of the differential transconductance amplifier is shown in Fig. 9.25. It consists of a differential pair and a single current mirror, like the one shown in Fig. 9.23, which is used to subtract the drain currents I_1 and I_2. The current I_1 drawn out of M_3 is reflected as an equal current out of M_4; the output current is thus equal to $(I_1 - I_2)$, and is given by Eq. (9.91). The current out of the amplifier, as a function of $(V_1 - V_2)$, is then a curve very close to a hyperbolic tangent.

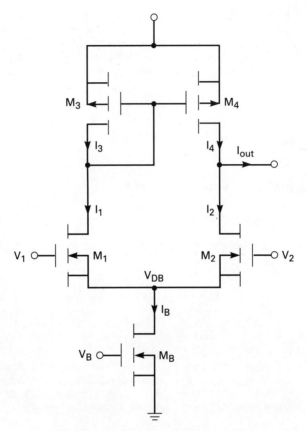

FIGURE 9.25 Basic transconductance amplifier.

9.7.6 Exponential and Logarithmic Function Cells

We saw in Sec. 9.7.2 that a diode-connected MOSFET provides a voltage that is proportional to the logarithm of the input current. However, better results can be obtained by using the circuit configuration of Fig. 9.26. Thus, a voltage can be generated by two diode-connected MOSFETs in series, as shown in Fig. 9.26a. The inverse operation (i.e., creating a current proportional to the exponential of a voltage) is obtained by the circuit of Fig. 9.26b.

Applying Eq. (9.85) to M_1 and M_2, we obtain

$$I = I_0 e^{qV_1/kT} = I_0 e^{q(V_2 - V_1)/kT} \tag{9.92}$$

Taking logarithms of the last two terms of Eq. (9.92), we obtain

$$V_2 = 2V_1 \tag{9.93}$$

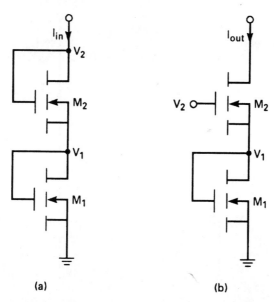

FIGURE 9.26 Two circuits showing the logarithmic voltage-current characteristic of the MOSFET. (*a*) The current-input configuration provides an output voltage proportional to the logarithm of the input current and (*b*) the voltage-input configuration provides an output current exponentially related to the input voltage.

and then

$$\ln \frac{I}{I_0} = V_1 = \frac{V_2}{2} \qquad (9.94)$$

These cells can be used, for example, in modeling the Boltzmann relations.

9.7.7 Square Root Function Cell

A variant of the circuit of Fig. 9.26 is the one shown in Fig. 9.27. The voltage across M_1 can be written

$$V_3 = V_1 + \ln \frac{I_{in}}{I_0} \qquad (9.95)$$

From Eq. (9.94), we have

$$\ln \frac{I_{out}}{I_0} = \frac{V_3}{2} \qquad (9.96)$$

Substituting Eq. (9.95) into Eq. (9.96), we find the dependence of I_{out} on I_{in}:

$$\frac{I_{out}}{I_0} = \left(\frac{I_{in}}{I_0} e^{qV_1/kT} \right)^{1/2} \qquad (9.97)$$

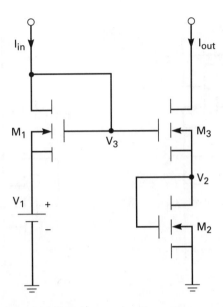

FIGURE 9.27 Circuit that generates an output current proportional to the square root of the input current.

9.8 CONTRIBUTIONS OF ORGANIC CHEMISTRY TO THE DEVELOPMENT OF ELECTRONIC DEVICES

Starting from the 1960s, the dimensions of silicon devices have been continuously reduced, and this process has been witnessed first by the introduction of the expression *microelectronics* in the '80s and then *nanoelectronics* in the '90s. The expression *nanoelectronics* underlines the fact that technological tools are available which allow the designer to fabricate MOSFET devices with lateral dimensions below the micrometer limit (e.g., a channel with a length of 500 nm or less). It can be easily foreseen that this miniaturization (or *top-down*) process will not go on forever, because of both technological limits and intrinsic physical limits. Just to give a very simplistic feeling of it, it is worth observing that the expression "silicon doped with 10^{18} atoms/cm^3," applied to a silicon cube with a 10^{-2}-μm linear dimension, means the introduction of just one atom into the cube.

Switching now from nanoelectronics to organic chemistry and molecular biology, we immediately realize that the nanometer is the natural scale of dimension for polymers and biological macromolecules, which can be defined as devices, originated by a *bottom-up* process, to operate in the nanometer range. In conclusion, nanoelectronics and polymer chemistry (including molecular biology) show an increasing overlap in the scale of integration they are dealing with, and this has resulted, starting from the beginning of the '90s, in the proposal of *hybrid silicon-organic devices* and even of *single-molecule transistors*. This topic is an impetuously growing one and its systematic treatment is out of the scope of this book. On the other hand, this topic is based on a truly bioelectronic approach, and we think it is appropriate to close this chapter by at least indicating a few examples of unconventional electronic devices based on this approach.

9.8.1 Organic Field-Effect Transistors

The interest in field-effect transistors made from easily processable thin organic films originates from the possibility of using such transistors in applications such as smart cards and flexible displays. Toward this goal, FETs incorporating *oligothiophenes* have been proposed,[13] resulting in thin-film transistors (TFTs). The TFT structure includes a conductive Si substrate with an Au contact, which functions as a gate with either SiO_2 (thickness in the order of 300 nm) or MgF_2 (thickness in the order of 100 nm) as gate insulator. Source and drain Au pads are photolithographically defined on top of the insulator. A thin organic film of oligothiophenes is then evaporated over the gate insulator. The organic molecules are oriented perpendicular to the substrate, and the electron transport direction is transverse to the chains (see Fig. 9.28).

Mobilities μ at room temperature in the order of 0.02 to 0.05 cm^2/V-s have been reported for different classes of oligothiophenes.[13]

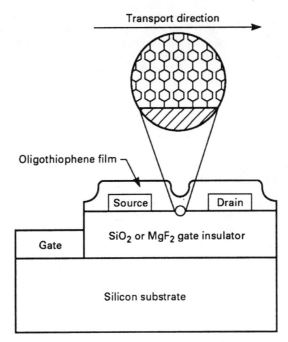

FIGURE 9.28 TFT structure: a conductive Si substrate, with an Au contact, functions as a gate (G), and either SiO_2 or MgF_2 is the gate insulator. Source (S) and drain (D) Au pads are photolithographically defined on top of the insulator. A thin (50-nm) organic film is thermally evaporated over the gate insulator. The inset shows that the organic molecules are oriented almost perpendicular to the substrate and the transport direction is transverse to the chains. (*From L. Torsi et al.*[13])

9.8.2 Proteins as Computer Memories

Computer card memories incorporating the *protein bacteriorhodopsin* as a component have been proposed.[14] The idea is to optically write and read data within a three-dimensional volume containing the protein in a rigid polymer matrix. Such a three-dimensional system would store 3.5 gigabytes of data with access time comparable to that of semiconductor memory. Overall, the memory should be able to store about 100 times more data than presently (year 1997) available semiconductor memory.

The protein bacteriorhodopsin seems to be an ideal photoactive material. It is the light-harvesting protein in the purple membrane of a microorganism, *Halobacterium salinarium*, also known as *Halobacterium halobium*. When absorbing light, the protein pumps protons across the bacterium membrane (the purple membrane), generating a chemical and osmotic potential that serves as a source of energy for metabolic processes. As a result of its interaction with light, bacteriorhodopsin can switch from one conformational state to another, and this property is at the basis of its use as an optical memory element. The protein, after absorbing a photon of green light, undergoes a photocycle that creates a series of colored states and eventually returns the protein to its resting state in about 10 milliseconds. However, if a red photon is directed at the protein 2 milliseconds after the green photon has activated the photocycle, a branching reaction occurs to form a highly stable blue-absorbing product.

As a result, by using orthogonal laser beams, data can be written at any location within a three-dimensional solid containing the protein, which can be addressed as a kind of optical organic AND. Reading can be also carried out, through a more difficult process.[14] The speed of the write/read process is not advantageous over more traditional systems. The advantage of using proteins stems from the fact that bacteriorhodopsin can be manipulated atom by atom using genetic and protein engineering in a way that is not achievable by standard lithographic techniques applied to semiconductor materials. It is worth underlining that this protein disproves the common perception regarding the fragility of biological materials: it is not denatured up to 80°C, and it can be switched optically from one state to another for at least 10 million times.

9.8.3 Single-Molecule Transistors

As already indicated, organic chemistry and condensed-matter physics have converged in the development of ultrasmall devices with nanometer-length scales. A nanometer-scale transistor has been proposed, consisting of only a single, carbon-based, molecule.[15] It is based on a carbon nanotube, which is a single molecule belonging to the *fullerene complexes*, whose discoverers were awarded the 1996 Nobel Prize in chemistry.[16] A single carbon nanotube can be made to contact two microscopic metallic strips, a few hundreds of nanometers apart, which are used as source and drain contacts. The current then flows from source to drain in the nanotube, according to quantum mechanics rules, giving rise to a *molecular transistor*. Further details can be found in the quoted papers.

PROBLEMS

9.1 Consider the inverter circuit with a passive load resistor R_L shown in Fig. 9.29. Assume that the MOSFET is an *n*-channel enhancement-mode device with a threshold voltage $V_{TH} = 1$ V, $K = 1$ mA/V^2, $R_L = 2$ kΩ, and $V_{DD} = 5$ V. For the MOSFET operating in the cut-off, saturation, and nonsaturation regions, find the value of V_{out} and draw the transfer characteristics of V_{out} versus V_{in} and I_{DS} versus V_{in}.

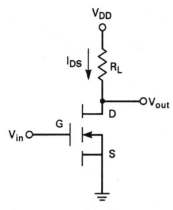

FIGURE 9.29 An *n*-channel enhancement-mode MOSFET inverter with a passive load resistor.

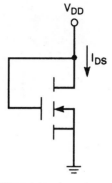

FIGURE 9.30 An *n*-channel enhancement-mode MOSFET with $V_{GS} = V_{DS}$.

9.2 Verify mathematically that the enhancement-mode *n*-channel MOSFET of Fig. 9.30, in which the gate is connected to the drain, acts like a nonlinear resistor. Draw also the I_{DS} versus V_{DS} plot, qualitatively. (For the reader: this circuit configuration usually replaces the R_L resistor in the inverter circuit of Fig. 9.29, because in integrated circuit fabrication, R_L consumes large areas on chip compared to MOSFETs).

9.3 Consider the inverter circuit with the saturated MOSFET M_1 as a load shown in Fig. 9.31. Assume that the MOSFETs are *n*-channel enhancement-mode devices, and be V_{TH1} and V_{TH2} the threshold voltages of the MOSFETs M_1 and M_2, respectively. For the MOSFETs operating in the cut-off, saturation, and nonsaturation regions, find the value of V_{out} and draw the transfer characteristics of V_{out} versus V_{in} and I_{DS} versus V_{in}. Make a quantitative analysis assuming $V_{DD} = 5$ V, $V_{TH1} = V_{TH2} = 1$ V and $V_{in} = 5$ V. Consider two cases: (a) $K_{M2}/K_{M1} = 1$ and (b) $K_{M2}/K_{M1} = 4$.

9.4 Consider the inverter circuit with an *n*-channel depletion-mode load MOSFET M_1 and an *n*-channel enhancement-mode driver MOSFET M_2 shown in Fig. 9.32. For the MOSFETs operating in the cut-off, saturation, and nonsaturation regions, find the value of V_{out} and draw the transfer characteristics of V_{out} versus V_{in} and I_{DS} versus V_{in}. Make a quantitative analysis assuming $V_{DD} = 5$ V, $V_{TH2} = 1$ V, $V_{TH1} = -2$ V and $V_{in} = 5$ V. Consider two cases: (a) $K_{M2}/K_{M1} = 1$ and $K_{M2}/K_{M1} = 4$.

9.5 Consider the *n*-channel, enhancement-mode MOSFET current source shown in Fig. 9.33, and assume $V_{DD} = 10$ V, $V_{TH} = 1$ V, and $\mu_n C_{ox} = 3 \times 10^{-5}$ s/V-cm². If $I_k = 25$ μA, $I_h = 100$ μA, and

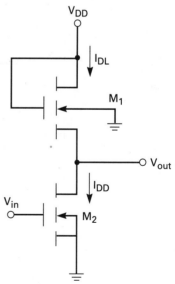

FIGURE 9.31 An *n*-channel enhancement-mode MOSFET inverter with a saturated *n*-channel enhancement-mode MOSFET as a load.

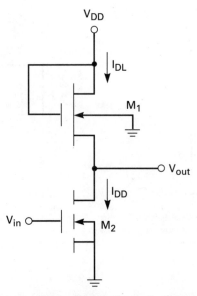

FIGURE 9.32 An *n*-channel enhancement-mode MOSFET inverter with an *n*-channel depletion-mode MOSFET as a load.

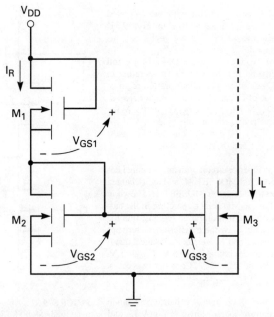

FIGURE 9.33 An *n*-channel enhancement-mode MOSFET current source.

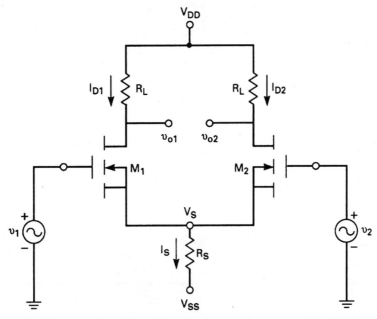

FIGURE 9.34 Differential amplifier with n-channel enhancement-mode MOSFETs.

V_{G3} has to be kept at a level of 1.5 V, find the width and length dimensions for each MOSFET.

9.6 Consider the differential amplifier with n-channel enhancement-mode MOSFETs shown in Fig. 9.34. Assume the MOSFET circuit of Fig. 9.34 is designed to operate with $V_{DD} = V_{SS} = 10$ V and $I_{D1} = I_{D2} = 0.25$ mA, when $v_1 = v_2 = 0$ V. Assuming also values of $K = 0.05$ mA/V² and $V_{TH} = 1$ V for the threshold voltage, find the values of V_{GS}, R_S and R_L when the MOSFETs are assumed to be in the saturation region. Also draw the dc load line in the I_{DS} versus V_{DS} plane.

9.7 Figure 9.35 shows a two-stage MOSFET amplifier. The MOSFETs have the parameters $V_{TH} = 1$ V and $K = 0.1$ mA/V². The circuit operates in the midband frequency, and the MOSFET small-signal output impedances are $r_{d1} = r_{d2} = \infty$. Find the dc operating point of the MOSFETs, the small-signal transconductances of the two MOSFETs, the small-signal voltage gain $A_d = V_{out2}/V_{in1}$ and the input and output resistance.

REFERENCES

1. R. S. Muller and T. I. Kamins, *Device electronics for integrated circuits,* 2nd ed., New York: John Wiley & Sons, 1986.
2. M. Shur, *Physics of semiconductor devices,* Englewood Cliffs, N.J.: Prentice Hall, 1990.
3. DeWitt G. Ong, *Modern MOS technology: processes, devices, and design,* New York: McGraw-Hill, 1984.
4. Y. P. Tsividis, *Operation and modeling of the MOS transistor,* New York: McGraw-Hill, 1987.
5. P. E. Gray and C. L. Searle, *Electronic principles,* New York: John Wiley & Sons, 1969.

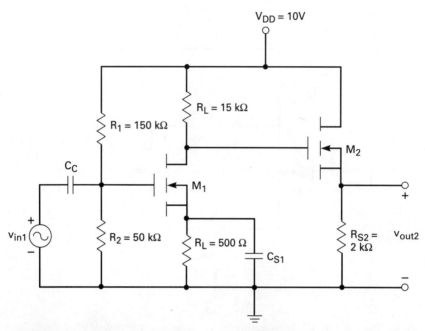

FIGURE 9.35 Two-stage *n*-channel enhancement-mode MOSFET amplifier with a common-source/common-drain configuration.

6. J. Millman, *Microelectronics,* New York: McGraw-Hill, 1979.
7. W. G. Oldham and S. E. Schwarz, *An introduction to electronics,* New York: Holt, Rinehart and Winston, 1972.
8. J. F. Gibbons, *Semiconductor electronics,* New York: McGraw-Hill, 1966.
9. P. R. Gray and R. G. Meyer, *Analysis and design of analog integrated circuits,* New York: John Wiley & Sons, 1977.
10. P. E. Allen and D. R. Holberg, *CMOS analog circuit design,* New York: Holt, Rinehart and Winston, 1972.
11. G. Massobrio and P. Antognetti, *Semiconductor device modeling with SPICE,* 2d ed., New York: McGraw-Hill, 1993.
12. C. Mead, *Analog VLSI and neural systems,* Reading, Mass.: Addison-Wesley, 1989.
13. L. Torsi, A. Dodabalapur, L. J. Rothberg, A. W. P. Fung, and H. E. Katz, "Intrinsic transport properties and performance limits of organic field-effect transistors," *Science,* 272: 1462–1464, 1996.
14. J. A. McDonald, "Bonafide bioelectronic memory on the horizon," *Biosensors & Bioelectronics,* 10(6/7): xvii–xxii, 1995.
15. L. Kouwenhoven, "Single-molecule transistors," *Science,* 275: 1896–1897, 1997.
16. R. F. Service, "A captivating carbon form," *Science,* 274: 345–346, 1996.

CHAPTER 10
MOSFET-BASED BIOELECTRONIC DEVICES: BIOSENSORS

The MOSFET (see Chap. 9) is the basis for the development of a series of sensors for the measurement of physical and chemical parameters. The equations of the MOSFET drain current exhibit a number of parameters that can be directly influenced by external quantities, and small technological variations of the original MOSFET configuration also give rise to a large number of sensing properties. All the devices exhibit the common property that a surface charge is measured on a silicon chip, depending on an electric field in the adjacent insulator.

MOSFET-based sensors such as the GASFET, OGFET, ADFET, SAFET, CFT, PRESSFET, ISFET, CHEMFET, REFET, ENFET, IMFET, BIOFET, and others are widely discussed in Refs. 1 to 3.

In this chapter we are not trying to review the literature in the area of biosensors, since that would constitute a book in itself. The purpose is to provide an introduction to this field, with emphasis on MOSFET-based sensors. The reader should bear in mind that the field of biosensors includes many kinds of devices which are not based on MOSFET technology, such as those based on screen printing technology or fiber-optic technology. A short overview of different kinds of biosensors is given in the first section of the chapter. The interested reader can find comprehensive descriptions of the topic in Refs. 2 to 5.

Then, we will describe, as an example, the operational mechanisms of three MOS technology-based potentiometric biosensors: the ISFET (ion-sensitive field-effect transistor), the ENFET (enzyme field-effect transistor), and the LAPS (light-addressable potentiometric sensor).

10.1 BIOSENSOR OVERVIEW

The *biosensor* was first described by Clark and Lyons in 1962,[6] when the term *enzyme electrode* was used. In this first enzyme electrode, an oxido-reductase enzyme was held next to a platinum electrode in a membrane sandwich as shown in Fig. 10.1. The platinum anode polarized at +0.6 V responded to the peroxide produced by the enzyme reaction with the substrate. The primary target substrate for this system was glucose, and it led to the development of the first glucose analyzer for the measurement of glucose in whole blood. The same technique has since then been used for many other oxygen-mediated oxido-reductase enzyme systems.

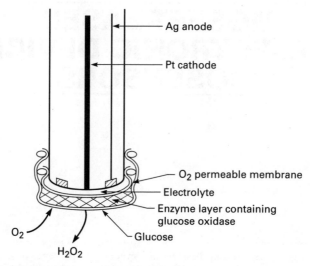

FIGURE 10.1 Modification of the Clark oxygen electrode to give an enzyme electrode.

Biosensors are small analytical bioelectronic devices that combine a *transducer* with a *sensing biological component* (biologically active substance). The transducer, which is in intimate contact with the biologically sensitive material, can be one for measuring weight, electrical charge, potential, current, temperature, or optical activity. The biologically active species can be an enzyme, a multienzyme system, an antibody or an antigen, a receptor, a population of bacterial or even eucaryotic cells, or whole slices of mammalian or plant tissue, to name a few. Substances such as sugars, amino acids, alcohols, lipids, nucleotides, and others can be specifically identified and their concentration measured by these devices.

A schematic functional representation of a biosensor is shown in Fig. 10.2. It consists

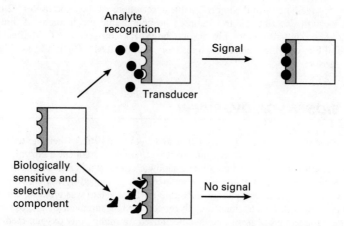

FIGURE 10.2 Schematic functional representation of a biosensor.

of a biological sensing element in close proximity or integrated with a signal transducer, to produce a reagentless sensing system specific for the target analyte.

Each component of the biosensor, its characteristics and applications will be investigated in this section.

10.1.1 Biological Component

The *biological component* of the biosensor, which represents the molecular detecting element, is made of highly specialized macromolecules (antibodies, antigens, enzymes, receptors) or of complex systems (cells, tissues) with the appropriate specificity and sensitivity. Biosensors can be classified according to the biocomponent used for detection and sensing. Examples are given for the following biological components:

Enzymes. Enzymes (see Sec. 5.4.2) are proteins that lower the energy threshold at which a given reaction takes place (catalytic function). The molecule being modified (e.g., oxidized, reduced, hydrolyzed) by the reaction is called an *enzyme substrate*, and represents the species to be measured by the biosensor.

Widely used *enzymatic biosensors*[4,5] are being produced with multiple enzyme systems, to increase the number of measurable quantities (as the biosensor for the measurement of polyamine, using amino oxidase and peroxidase), as well as to amplify the reaction (as in the case of glucose oxidation by glucose-oxidase in the presence of glucose-reductase). The enzymatic biosensors provide a linear response for a wide range of substrate concentrations.

Antibodies. The antibodies (*Ab*) are glycoproteins produced by the immune system against specific external substances (antigens *Ag*). Theoretically it is possible to produce antibodies able to identify any antigen. Biosensors using antibodies as sensing elements are called *immunobiosensors*.[4,5]

Receptors. The regulation of biological processes at the molecular level is based upon specialized protein structures, called *receptors* (see Sec. 4.7.2), able to recognize a number of physiological signals. As an example, the neurotransmitter action is mediated by receptors present in the plasma membrane of the target cells; in this case the biologically active site can be an ionic channel. The acetylcholine receptor is the best known receptor in the field of neurotransmission (see Chaps. 4 and 11).

Cells and Tissues. The measurement of a molecular species, in some cases, requires not only its interaction with the biosensor molecule, but also its transformation (via biochemical reactions) into a measurable product. This could be done by using several enzymes with the appropriate cofactors; it is, however, simpler to operate with populations of cells where the metabolic pathway is naturally present. A relevant example of the complexity of the reactions at the surface of a biosensor caused by a population of bacteria is given by the *L-arginine* biosensor. In this case, bacterial cells of *Streptococcus faecium* are used in combination with an ammonia gas–sensing electrode. Arginine is assayed via the metabolism of arginine by the microorganisms, as shown in Fig. 10.3. It is difficult to obtain such complex reactions outside a cellular structure.

Instead of using a cell population as sensing element, it is possible to utilize slices of tissue (plant or animal); in this case there is the advantage of maintaining the cells in a natural environment, thus eliminating the danger of damaging them by trying to isolate them. For example, an *adenosine* biosensor,[4,5] has been proposed in which the biosensing

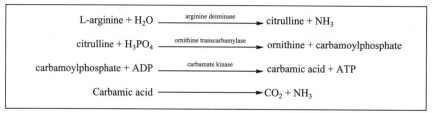

FIGURE 10.3 Metabolism of arginine by the microorganism sequence.

element is made of tissues obtained from mouse small-intestinal mucosal cells. Even the biocatalytic properties of the banana pulp have been used for *dopamine* sensing.[4,5]

10.1.2 Transducer

The transducer transforms the detection-induced physicochemical variations occurring in the biosensing element into a signal (usually electric), which is then amplified by an *ad hoc* designed electronic circuit and used in the control of external devices (such as an insulin pump). The transducer can be *electrochemical* (amperometric, potentiometric, conductometric), *optical, piezoelectric,* or *calorimetric.* Very often this classification is used to identify the type of biosensor.

Amperometric Biosensor. This is an electrochemical biosensor measuring the flux of electrons (current), generated by oxidation-reduction reactions induced by the biological component, toward a fixed-voltage electrode.[2,3] In these conditions the current is linearly dependent on the analyte concentration. The performance of these biosensors is directly related to the efficiency of electron transfer. This can be obtained with the use of redox molecules (*mediators*), such as *ferrocene,* placed on the electrode surface. The use of mediators reduces the possibility of interference on the oxidation-reduction processes caused by other analytes. The amperometric biosensor is characterized by its simplicity and its low implementation cost. This important advantage has lead to the introduction of disposable sensors, such as those for the measurement of blood glucose for diabetic patients ("reactive stripes"). The reactions on which this biosensor is based are schematically shown in Fig 10.4.

Potentiometric Biosensor. This is an electrochemical biosensor, operating at constant current (usually zero). It measures the charge-density variations on the electrode surface following a catalytic process or a surface modification due to the selective binding of a molecule.

A type of potentiometric biosensor consists of a modified MOSFET. Its operation is based on the interaction between H^+ ions (present in the solution) and the surface of the

$$\text{Glucose} + O_2 \xrightarrow{\text{glucose oxidase}} \text{gluconic acid} + H_2O_2$$

$$H_2O_2 \xrightarrow{+0.65\ \text{V}} 2H^+ + O_2 + 2e^-$$

FIGURE 10.4 Reactions in an amperometric biosensor for the measurement of glucose concentration.

insulating layer (Si_3N_4, Al_2O_3, Ta_2O_5); such interaction induces a channel modulation in the transistor substrate.[1,3] On this principle, a simple pH-measuring device is based (see Sec. 7.1.4). Because of its robustness, miniaturization, and modularity, it can be employed where other more traditional pH meters cannot be used. This structure is the base of several enzymatic devices (ENFETs) designed to measure the pH variation caused by a reaction catalyzed by an enzyme, and consequently the substrate concentration.[7] Figure 10.5 shows the cross-section of the ENFET device, compared to the cross section of the traditional MOSFET.

Advantages of potentiometric biosensors based on MOS technology are related to their miniaturization, to their capability of measuring different species on the same silicon chip, and finally to the large-scale production capabilities of the microelectronic industry. Their major drawback consists in packaging difficulties.

Conductometric Biosensor. This is an electrochemical biosensor able to measure the electric conductivity of a solution, following the application (between two electrodes) of an electric field at a frequency of about 1 kHz. With these biosensors[4] it is possible to monitor the concentration changes of an ionic species or its migration velocity. An example is the measurement of urea via the urease enzyme. This enzyme decomposes the neutral substrate into ionic species, thus increasing the conductivity of the solution.

Optical Biosensor. This device is based on the emission, absorption, or scattering of light[5] and makes large use of optical fibers. Optical biosensors are comparable in dimensions to those based on MOS technology, but they do not require electrical contacts. The necessary instrumentation is, however, rather expensive. A relevant biosensor is the optical immunobiosensor, based on the evanescent wave phenomenon.[4]

Piezoelectric Biosensor. This contains a vibrating crystal, where the frequency of vibration is a function of the crystal mass.[4,5] The bond formation (direct or indirect, physical or chemical) between molecules and the crystal surface (e.g., due to their absorption or to the interaction with antibodies present on the crystal surface) produces a mass variation, which then induces a shift of the crystal resonance frequency. It can be used to measure gaseous pollutants with concentrations in the order of a few parts per billion.

Calorimetric Biosensor. The conductivity of this device has a high temperature coefficient, which allows the measurement of temperature variations in the order of millide-

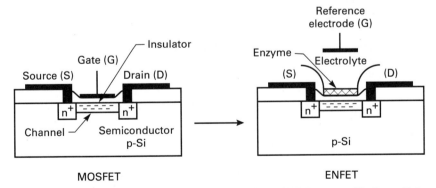

FIGURE 10.5 ENFET structure obtained by modification of the MOSFET structure. (*Kindly provided by Andrea Massobrio.*)

grees Celsius.[4] It is used to monitor fermentation processes, and can be coupled with enzymes, cells, and tissues.

10.1.3 Characteristics of Biosensors

The immobilization of the biosensing material represents a critical step in the production of biosensors, especially in dealing with enzymes and antibodies. A biosensor is specified by the following characteristics.

Selectivity. This is the capability to select and to measure only the desired biochemical species, with minimal interference from the many other species present in the environment. Just to give an example, it is not sufficient to use an antibody for which the desired species represents an antigen, because many antibodies can bind with other elements having on their surface the same functional group. It is very difficult to obtain total selectivity, not only of antibodies, but also of enzymes, microbic populations, and other biosensing elements.

Detection Limit. This defines the minimum concentration of the analyte to be measured, and determines the choice of certain types of biosensors when the analyte concentration values are very low.

Reversibility. To use a biosensor continously, it must be based on a reversible reaction. When the sensing element is an enzyme, its catalytic function makes it possible to return to the native form after the end of the chemical reaction. When using antibodies, the kinetics of the Ab-Ag binding is described by the affinity constant $K_a = [AgAb]/[Ag][Ab]$, where [AgAb] is the concentration of the complex, and [Ag] and [Ab] are the concentrations of free antigen and antibody, respectively. Since the K_a value is usually very high (typical range 10^4 to 10^{12} M^{-1}), the Ab-Ag reaction is practically irreversible, and this puts some constraints to the design of immunobiosensors.

Lifetime. This parameter is related to the fact that the biosensing element can, with time, undergo degenerative processes, which cause the loss of the original biological function. A parameter which strongly influences the lifetime of the biosensor is related to the method of immobilization of the biosensing element, since some molecules may, with time, detach from the biosensor surface.

Response Time. The enzymatic or immunological reactions are characterized by fast kinetics; thus the response time of the biosensor is often not reaction-limited, but is rather diffusion-limited.

Biocompatibility. Some biosensors are designed to be used *in vivo;* since the introduction of a foreign element causes the body to start a response reaction, these biosensors have to use *biocompatible materials.* A widely used technique is to treat the surfaces with *albumin* (to reduce the contact between body and foreign material), and with *heparin* (to reduce the possibility of blood clot formation near the biosensor). As of today, there are no biosensors with good performances *in vivo,* and many efforts are devoted to solving this problem. In the case of glucose measurement (very important for diabetic people) promising results have been obtained with *ex vivo* systems.[8]

10.1.4 Applications of Biosensors

A way of classifying biosensor applications is according to the field in which the device is used.

1. *Clinical applications.* The diagnostic procedures are usually based on the measurements of species such as ions, gases, and hormones. Implantable biosensors can provide real-time measurements in critical conditions and drive feedback systems, in, for example, surgical operations.[9,10] It is thus possible to imagine that in the future the laboratory could be directly at bedside, in the home of the patient. Commercially available sensors include, for example, biosensors for (quick) tests for pregnancy, cholesterol, HDL, urine and blood glucose, and *Helicobacter* in the gastric mucose.
2. *Agriculture and horticulture.* In this area, biosensors show promising results in the measurement, in the field, of the degree of ripeness of several products.
3. *Industry.* Biosensors can be used as on-line monitors by canning and fermentation industries.[11]
4. *Environmental pollution.* In this field, biosensors can be used to monitor the presence of pesticides, herbicides, chemical fertilizers, and several water pollutants. Since on-line measurements are very important, the use of biosensors will probably represent a valuable early warning system.[12]
5. *Food quality.* In this field, biosensors are being used to monitor the presence of heavy metallic ions (e.g., Hg).[13]
6. *Defense.* Biosensors can be used in defense applications, to monitor the presence, for example, of nerve gases in the environment.

10.2 ION-SENSITIVE FIELD-EFFECT TRANSISTOR (ISFET)

As stated in the introduction of the chapter, and in accordance with the aim of the book, we will focus our attention now on a MOSFET-based biosensor.

The ion-sensitive field-effect transistor (ISFET), a device which measures pH or ions in solution, was reported first by Bergveld.[14] The structure of the ISFET, as shown in Fig. 10.6, is fundamentally a MOSFET, the essentials of which are given in Chaps. 8 and 9, which can be rendered H^+-sensitive by eliminating its metal gate electrode in order to expose the gate insulator to the solution. The introduction of an ion-selective membrane can make it sensitive to other specific ions.

The ISFET, in its basic form, is a potentiometric pH electrode. Therefore, strictly speaking, it is not a biosensor, but rather a chemical sensor, which can be compared with the traditional pH-glass membrane electrode, based on the Nernst potential. The ISFET advantages are planar construction, small dimensions, low impedance, fast response, large-scale production, ease of multisensor (differential) realization, and immediate use after dry storage. A drawback is the larger drift rate and the necessary stringent encapsulation of the chip edges and bonding leads. In the next section, the ISFET will be used as the basis of potentiometric enzyme sensors (ENFETs).[7]

The gate insulator of the ISFET senses the H^+ ion concentration, generating an interface potential on the gate; the corresponding drain-source current change in the semiconductor channel is observed. From a comparison of the MOSFET and ISFET structures, it is apparent that the only difference between the electrical circuits is the replacement of the

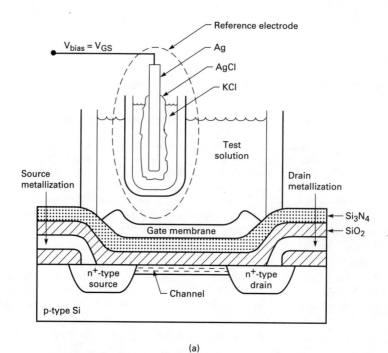

FIGURE 10.6 (*a*) Schematic representation of an ISFET; where a gate membrane is also shown; (*b*) expanded view of semiconductor-insulator-electrolyte interface (SiO_2 indicated for simplicity).

metal gate of the MOSFET by the series combination of a reference electrode, electrolyte solution, and a chemically sensitive insulator. The reference electrode provides a stable potential (see Sec. 7.1.1) in the solution independent of changes in the dissolved species or in the pH of the solution. The gate voltage is applied to the reference electrode to produce the channel in the semiconductor.

As already stated, the operational mechanism of the ISFET originates from the pH sensitivity of the inorganic gate oxide, such as SiO_2, Al_2O_3, Si_3N_4, or Ta_2O_5. This mechanism is a surface phenomenon which can be explained by the site dissociation model (see Sec. 7.1.4).

Surface hydroxyl groups react with the analyte in an acidic or a basic way, resulting in a corresponding surface charge and potential. Depending on the specific properties of the gate oxide, this surface potential is in the order of 25 mV/pH (SiO_2) to 59 mV/pH (Ta_2O_5). The response is determined by the kinetics of the surface reactions, and response times in the order of milliseconds are typical. It should be pointed out that the potential drop at the solid-liquid junction is influenced by the properties of the solid material, which can change over time, giving rise to apparently "nonnernstian" responses.

The temperature sensitivity of the ISFET is mainly determined by the semiconductor component of the sensor. A method to compensate for this temperature sensitivity is to use a differential pair of ISFETs or an ISFET and a MOSFET, as in electronic systems. Since the ISFET is an open transistor, it is also sensitive to light, but this can also be compensated for by the application of a differential pair.

The requirement of the reference electrode is difficult to meet by using integrated circuit technology to build it on a silicon chip. A reference electrode uses a chemical reaction to move ions into solution from an electrode. For example, with the Ag/AgCl reference electrode, the motion of chloride ions from or to the solid Ag/AgCl electrode carries the current

$$e^- + AgCl \rightleftharpoons Ag + Cl^- \tag{10.1}$$

The Nernst equation for the potential with such a reaction is

$$V = V_o - (RT/\mathscr{F}) \ln [Cl^-] \tag{10.2}$$

where V is the potential of the Ag electrode, V_o a constant, R the gas constant (see App. A), $\mathscr{F}$ the Faraday constant (see App. A), T the absolute temperature, and $[Cl^-]$ the concentration of chloride ions. If there is a constant concentration of chloride ions, the potential V is stable, as desired. Such a stable concentration can be easily obtained with a typical reference electrode where the Ag/AgCl is isolated from the solution to be tested by a frit (or a salt bridge) and KCl is provided on the Ag/AgCl side of the frit. This is not so easy to obtain if the reference electrode is, for example, a silver film deposited on a silicon chip for compatability with FET technology.[3]

In this section we have given a qualitative overview of the ISFET; in the next sections we will present a quantitative description of the device operation mechanisms.

10.2.1 ISFET Operation

All chemical sensors based on the field-effect operation share a common feature. Their measurable properties can be described in terms of the flat-band potential V_{FB}. As we have seen, the threshold voltage, V_{TH} is directly related to V_{FB} [Eq. (8.36)]. In the following, we show how we can make V_{FB}, and consequently V_{TH} and I_{DS}, sensitive to different chemical entities in the medium contacting the sensor.

Then let us consider a system consisting of the following components: a reference electrode, which contains a stable solid-liquid interface, an electrolyte, an insulating layer (eventually consisting of layers of different insulators), a silicon substrate, and a metal back contact. In the usual electrochemical representation, this system is written as

M | Si | Insulator | Electrolyte | Reference electrode | M'

The most important measurable parameter of this system is its flat-band voltage V_{FB}, defined as the voltage applied to M' which makes the silicon surface potential zero. In Ref. 15 it is shown that

$$V_{FB} = (E_{ref} + \varphi_{lj}) - (\varphi_{eo} - \chi_e) - \frac{\phi_{Si}}{q} - \frac{Q_o}{C_{ox}} \qquad (10.3)$$

where E_{ref} is the reference electrode potential relative to vacuum, which is obtained by adding 4.7 V to the potential relative to the normalized hydrogen electrode; φ_{lj} is the liquid-junction potential difference between the reference solution and the electrolyte; φ_{eo} is the potential drop in the electrolyte at the insulator-electrolyte interface; χ_e is the surface dipole potential of the solution; ϕ_{Si} is the work function of Si; C_{ox} and Q_0 are the insulator capacitance and effective charge per unit area (see Sec. 8.6).

In the expression of V_{FB} of the MOSFET, the terms in parenthesis are replaced by the work function of the contacting metal [ϕ_{MS}; see Eq. (8.36)]. Thus, the flat-band voltage, and therefore the threshold voltage, is the quantity measured in chemically sensitive electronic devices based on the field-effect principle. Taking into account Eqs. (8.36), (8.37), and (10.3), we can write

$$V_{TH}(ISFET) = V_{TH}(MOSFET) + E_{ref} + \varphi_{lj} + \chi_e - \varphi_{eo} - \frac{\phi_M}{q} \qquad (10.4)$$

where ϕ_M is the work function of the metal back contact of the semiconductor relative to vacuum. Equation (10.4) replaces the MOSFET threshold voltage in the I_{DS}-V_{DS} equations [see Eqs. (9.15) and (9.19)]. The main source of the pH sensitivity of the ISFET, among the terms in Eq. (10.3), is the potential drop φ_{eo} in the electrolyte at the insulator-electrolyte interface.

On the basis of site-binding theory,[16] applied to the amphoteric metal oxide gate materials used in ISFETs, the sensitivity of this device is described in terms of the intrinsic buffer capacity of the oxide surface, β_s, and the electrical surface differential capacitance C_s. The ISFET sensitivity toward changes in the bulk pH is described by the ratio β_s/C_s.

An approach, originally proposed by Bergveld and coworkers,[17] to the description of the acid-base properties of ISFETs is analogous to the classical description of the acid-base properties of protein molecules (see Chap. 4). The acid-base titration of proteins is also determined by the ratio between the intrinsic buffer capacity and the electrical double-layer capacitance.

The conclusion that ISFET surfaces and protein molecules behave in a similar way with respect to their acid-base properties is an alternative description of the operational mechanism of the ISFET, and it is simpler and more easily understood than the original theory,[18] which will be used in Chap. 12.

10.2.2 pH Dependency of φ_{eo}: A Measure of the pH Sensitivity of the ISFET

The origin of the potential φ_{eo} is the interaction of the insulator surface with ions present in the electrolyte. The main feature of all theories for this interaction is that the presence of discrete surface sites is assumed. The model used can have one or more different types of sites, each with acidic, basic, or amphoteric character. For the Al_2O_3 oxide which we will take into account (the use of insulator Si_3N_4 is considered in Chaps. 7 and 12), it is usual to consider that only one type of site is present, of the type A-OH, where A represents Al.[18] To account for the fact that both signs of charge have been experimentally observed in colloid chemical studies, the site considered must be amphoteric, which means it can act as a proton donor or acceptor. We then assume that the oxide surface contains sites in three possible forms: A-O$^-$, A-OH, and A-OH$_2^+$. The acidic and basic character of the neutral site A-OH can be characterized by two equilibrium constants K_a and K_b (acidity and basicity constants, respectively):

$$\text{A-OH} \rightleftharpoons \text{A-O}^- + \text{H}_s^+ \qquad \text{with } K_a = \frac{[\text{A-O}^-][\text{H}^+]_s}{[\text{A-OH}]} \qquad (10.5)$$

$$\text{A-OH} + \text{H}_s^+ \rightleftharpoons \text{A-OH}_2^+ \qquad \text{with } K_b = \frac{[\text{A-OH}_2^+]}{[\text{A-OH}][\text{H}^+]_s} \qquad (10.6)$$

In Eqs. (10.5) and (10.6), [A-OH] is the surface concentration of neutral sites, and [A-O$^-$] and [A-OH$_2^+$] the surface concentration of negative and positive surface sites respectively, the values of which are given by the respective equilibrium constants and the pH of the bulk solution, pH_b. The quantity $[H^+]_s$ indicates the volume concentration of protonated water molecules, $[H_3O^+]$, in contact with the oxide surface, the value of which can be obtained by combining Eqs. (10.5) and (10.6), i.e.,

$$[H^+]_s = \sqrt{\frac{K_a}{K_b} \frac{[\text{A-OH}_2^+]}{[\text{A-O}^-]}} \qquad (10.7)$$

Equation (10.7), for the neutral surface—that is, when $[\text{A-OH}_2^+] = [\text{A-O}^-]$—reduces to

$$[H^+]_s = \sqrt{\frac{K_a}{K_b}} \qquad (10.8)$$

or, equivalently, we can write

$$pH_s = -\log[H^+]_s = -\log\sqrt{\frac{K_a}{K_b}} \qquad (10.9)$$

This neutral equilibrium situation is established for a value where $pH_b = pH_s$, which is generally known as the pH at the *point of zero charge* (pH_{pzc}), which is characteristic of a certain type of oxide (see Sec. 7.1.4). When the surface is in contact with an electrolyte with a value of $pH_b \neq pH_{pzc}$, the effect on pH_x can be expressed in terms of the *intrinsic buffer capacity of the surface*, β_s (see Sec. 4.2.2), which is by definition the ratio between a small amount of strong base $d[B]$ (or acid) concentration added to the solution and the resulting change in pH, indicated as dpH_s, i.e.,

$$\beta_s = \frac{d[B]}{dpH_s} \qquad (10.10)$$

The sign of Eq. (10.10) is correct if OH$^-$ ions are considered, otherwise, if H$^+$ ions are considered, the sign must be changed. Therefore,

$$[B] = [\text{A-O}^-] - [\text{A-OH}_2^+] = -\frac{\sigma_s}{q} \qquad (10.11)$$

where σ_s is the net surface concentration charge of the titrated groups and q is the electronic charge.

Considering the total concentration of charged and neutral surface sites

$$N_s = [\text{A-O}^-] + [\text{A-OH}_2^+] + [\text{A-OH}] \tag{10.12}$$

we can calculate, from Eqs. (10.5), (10.6), and (10.12), the concentration of the negative and positive surface concentration sites, as

$$[\text{A-OH}^+] = N_s \frac{K_a}{K_a + [\text{H}^+]_s + K_b[\text{H}^+]_s^2} \tag{10.13}$$

and

$$[\text{A-OH}_2^+] = N_s \frac{K_b[\text{H}^+]_s^2}{K_a + [\text{H}^+]_s + K_b[\text{H}^+]_s^2} \tag{10.14}$$

Thus, Eqs. (10.10) and (10.11) give[17]

$$\beta_s = \frac{d[\text{B}]}{d\text{pH}_s} = \frac{d[\text{B}]}{d[\text{H}^+]_s} \cdot \frac{d[\text{H}^+]_s}{d\text{pH}_s}$$

$$= N_s \frac{K_a + 4K_aK_b[\text{H}^+]_s + K_b[\text{H}^+]_s^2}{(K_a + [\text{H}^+]_s + K_b[\text{H}^+]_s^2)^2} \cdot 2.3[\text{H}^+]_s \tag{10.15}$$

Equation (10.15) indicates that the buffer capacity of an amphoteric surface becomes larger with a larger value of N_s, as well as with a larger value of the product K_aK_b, which is equivalent to considering a small value of $\Delta pK = (pK_a + pK_b)$.

Because of the intrinsic buffer capacity of the surface, the value of pH_s does not follow the value of pH_b during titration, and this gives rise to a surface potential φ_{eo}, according to the Boltzmann distribution

$$[\text{H}^+]_s = [\text{H}^+]_b \, e^{-q\varphi_{eo}/kT} \tag{10.16}$$

or, equivalently,

$$\varphi_{eo} = 2.3 \frac{kT}{q} (\text{pH}_s - \text{pH}_b) \tag{10.17}$$

According to the derivation given in Sec. 7.2.1, the charge in the diffuse layer can be written as

$$\sigma_d = -(8kT\varepsilon_0\varepsilon_r c_0)^{1/2} \sinh\left(\frac{zq\varphi_{eo}}{2kT}\right) = -\sigma_s \tag{10.18}$$

where c_0 is the ion concentration in the bulk and z is the ion valence. Equation (10.18) indicates the integral capacitance of the system which will be further considered in Sec. 10.3.4 when the response of the ISFET to stepwise changes in the ionic strength, will be considered.

The ability of the electrolyte to store charge in response to a change in the electrostatic potential is the differential capacitance

$$\frac{d\sigma_d}{d\varphi_{eo}} = -\frac{d\sigma_s}{d\varphi_{eo}} = -C_s \tag{10.19}$$

A complete expression of the charge balance should include the charge in the semiconductor component. In this treatment, this contribution will be neglected; it will be taken into account in Chap. 12.

A change with the bulk pH_b induces a change in pH_s, which in turn results in a change in the surface potential φ_{eo}.

According to Eqs. (10.10), (10.11), and (10.19), we can write

$$\frac{d\varphi_{eo}}{dpH_s} = \frac{d\sigma_s}{dpH_s}\frac{d\varphi_{eo}}{d\sigma_s} = -\frac{q\beta_s}{C_s} \tag{10.20}$$

under the condition that C_s is independent of φ_{eo}, which is in most cases a reasonable assumption,[17] and taking into account the convention on the sign for β_s.

Equation (10.20) indicates that for a small value of dpH_s, a large value of $d\varphi_{eo}$ will take place when the surface has a large intrinsic buffer capacity β_s, especially in combination with a small electrical surface capacitance C_s.

Since an ISFET is a device that measures the gate insulator-electrolyte interface potential φ_{eo}, as a function of the pH of the bulk electrolyte pH_b, Eq. (10.17) must be differentiated to determine the pH sensitivity of the ISFET, i.e.,

$$\frac{d\varphi_{eo}}{dpH_b} = 2.3\frac{kT}{q}\left(\frac{dpH_s}{dpH_b} - 1\right)$$

$$= 2.3\frac{kT}{q}\left(\frac{dpH_s}{d\varphi_{eo}}\frac{d\varphi_{eo}}{dpH_b} - 1\right) \tag{10.21}$$

Rearranging Eq. (10.21) yields

$$\frac{d\varphi_{eo}}{dpH_b} = \frac{2.3\frac{kT}{q}}{2.3\frac{kT}{q}\frac{dpH_s}{d\varphi_{eo}} - 1} \tag{10.22}$$

Combining Eqs. (10.20) and (10.22), we obtain

$$\frac{d\varphi_{eo}}{dpH_b} = -2.3\frac{kT}{q}\frac{1}{2.3\frac{kT}{q^2}\frac{C_s}{\beta_s} + 1} = -2.3\alpha\frac{kT}{q} \tag{10.23}$$

where

$$\alpha = \frac{1}{2.3\frac{kT}{q^2}\frac{C_s}{\beta_s} + 1} \tag{10.24}$$

is a dimensionless sensitivity parameter which varies between 0 and 1 depending on the intrinsic buffer capacity and the differential capacitance.

The pH sensitivity of an ISFET with an inorganic oxide as the gate material can then be described through the parameter α, which approaches unity for large values of β_s/C_s. This means that surfaces with a large buffer capacity (large value of N_s and low value of ΔpK) and a low value of C_s (low electrolyte concentration) show at best a maximum response of 59.3 mV at 25°C.[17] This value is achieved by oxides with a large intrinsic buffer capacity, such as Ta_2O_5. On the other hand, an oxide such as SiO_2 exhibits a low pH sensitivity.

10.3 ENZYME FIELD-EFFECT TRANSISTOR (ENFET)

The enzyme field-effect transistor (ENFET) is a bioelectronic device which belongs to a class of chemical potentiometric sensors that take advantage of the high selectivity and sensitivity of biologically active materials (here enzymes). In these devices, the enzyme is immobilized on the insulator of the ISFET. In literature, a dual-gate ISFET is often proposed, where one of the FETs can act as a reference system and is assembled in the same way as the sample FET, but it contains a blank enzyme-free gel membrane (Fig. 10.7). This arrangement allows some automatic compensation of fluctuations in solution pH and temperature.

The enzyme-substrate system controls the specificity of the ENFET operation, and the reaction kinetics that take place in the biologically active materials (enzyme) determine how fast the substrate is converted into the product. To get an understanding of the ENFET operation, the reaction kinetics of the biological-recognition processes must be considered, together with the mass transport theory (Fig. 10.8).

A detailed discussion, however, is beyond the aim of this chapter; in addition, solutions to the transport equations for ENFET structures need a detailed knowledge of the diffusion coefficients for all species involved in the reactions and the geometry of the sensor structure. In addition, the partial differential equations usually require numerical solutions.

10.3.1 System Definition

Consider the system consisting of a sensor surface, lying at $x = 0$, coated with an immobilized enzyme layer of thickness L. Beyond this lies the transport boundary layer. Beyond the transport boundary layer, the concentration of analyte, the enzyme's substrate S, has a defined value $[S]$ and that of the product $[P]$ is taken to be zero (see Fig. 10.9). Notice that these bulk concentrations of substrate and product represent two basic boundary conditions of the system. Enzyme-catalyzed reaction takes place in the immobilized enzyme layer and, for the present description, will be considered to follow Michaelis-Menten kinetics (see Sec. 5.4.2). This reaction depletes the substrate and gen-

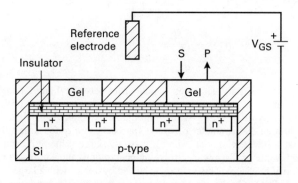

FIGURE 10.7 Schematic representation of a dual-gate ENFET chip. (*From S. D. Caras and J. Janata.*[19])

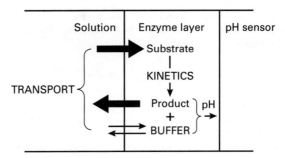

FIGURE 10.8 Mechanisms involved in the response of pH-based enzyme sensors. (*From B. H. van der Schoot and P. Bergveld.*[7])

erates the product in the enzyme layer, causing the concentration gradients which drive the mass transport process. The product is measured potentiometrically, without being consumed, at the sensor surface. A steady state is reached when the rate of reaction in the immobilized enzyme layer is balanced by mass transport of reactant and product to and from it.

10.3.2 Enzyme-Catalyzed Reaction of Substrate

For the sake of self-consistency and for readers' convenience, some concepts already introduced in Chap. 5, will be considered here again. Note that here, the enzyme is assumed to be immobilized in a region of space, and then mass transfer mechanisms (in particular diffusion) must be taken into account.

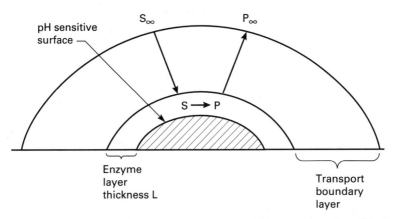

FIGURE 10.9 Schematic representation of the potentiometric biosensor using an immobilized enzyme layer.

Let us consider a single enzyme acting on a substrate molecule, where any other reactants are assumed to be in excess so as not to limit the reaction. The reaction then follows the Michaelis-Menten kinetics (see Sec. 5.4.2)

$$S + E \underset{k_{-1}}{\overset{k_1}{\rightleftharpoons}} ES \overset{k_2}{\longrightarrow} P + E \tag{10.25}$$

where E is the enzyme, S is the substrate, ES is the bound intermediate enzyme-substrate complex (indicated as C in Sec. 5.4.2), P is the product and k_1, k_{-1}, k_2 are the reaction-rate constants for the enzyme-substrate binding (forward) reaction, the unbinding (backward) of the enzyme-substrate complex, and the formation of the product (or complex decomposition), respectively.

The rate of product formation is given by

$$v_f = \frac{\partial [P]}{\partial t} = k_2 [ES] \tag{10.26}$$

where the rate v_f has units of mol/l-s and t is the time. To relate the rate of product formation with the substrate concentration, it is assumed that the reaction is in steady state; that is, by assuming that the rates of formation and breakdown of the complex are equal. This assumption, which is valid almost immediately after the reaction has begun, leads us to write the rate equation for the intermediate complex as

$$\frac{\partial [ES]}{\partial t} = k_1 [E][S] - k_{-1} [ES] - k_2 [ES] = 0 \tag{10.27}$$

Since the total concentration of enzyme $[E]_o$ present at all times will be the sum of concentrations in free and complexed forms $[E] + [ES]$, we substitute $[E] = [E]_o - [ES]$ into Eq. (10.27), solve it for $[ES]$, and substitute into Eq. (10.26), obtaining

$$v_f = \frac{k_2 [ES]_o [S]}{([S] + K_M)} = k_2 [ES] \tag{10.28}$$

where we have introduced the Michaelis constant defined as

$$K_M \equiv \frac{(k_{-1} + k_2)}{k_1} \tag{10.29}$$

The Michaelis constant K_M is not a true equilibrium constant but is the ratio of rate constants for complex formation and disappearance, the latter comprising both simple dissociation back to reactants plus decomposition into products.

Few enzymes follow the Michaelis-Menten kinetics over a wide range of experimental conditions. Most enzyme-substrate interactions have more complex kinetics, for example, the reverse reaction in which complex formation between the enzyme and product must be taken into account, multiple substrates with additional reaction steps, and inhibition. However, for these cases with complex kinetics, K_M is still a useful measure of the enzyme-substrate interaction.

When the enzyme is brought in contact with a substrate (glucose, for example), a complex concentration profile develops, which is a function of both diffusion and reaction processes. In the reaction region of the enzyme, the product concentration is achieved by mass transfer given by Fick's second law as well as by enzyme kinetics. The mass transport has been discussed in Sec. 5.1.

10.3.3 Enzyme-Modified Ion-Sensitive Field-Effect Transistor (ENFET)

For a gel membrane-immobilized system, the coupled mass transport and kinetic reaction for the depletion of substrate across the gel layer must be considered.

The enzyme reaction rate follows the Michaelis-Menten kinetics, Eq. (10.25), and the concentration of the substrate and the product in the enzyme membrane are described by the diffusion-reaction equation, which is just Fick's second law with a term added to account for the consumption or production of a species,[19] i.e.,

$$\frac{\partial C}{\partial t} = D\frac{\partial^2 C}{\partial x^2} \pm R(C) \qquad (10.30)$$

where D is the diffusion coefficient of the species, $R(C)$ is the reaction term, and the sign depends on whether species is produced (product) or consumed (substrate). Considering the steady-state response, the time derivative will be set to zero. Using the Michaelis-Menten kinetics, Eq. (10.28), as the reaction term, Eq. (10.30) becomes, for the *substrate*,

$$D_S\frac{\partial^2 [S]}{\partial x^2} - \frac{k_2[E]_o[S]}{K_M + [S]} = 0 \qquad (10.31)$$

and for the *product*,

$$D_P\frac{\partial^2 [P]}{\partial x^2} + \frac{k_2[E]_o[S]}{K_M + [S]} = 0 \qquad (10.32)$$

where D_S and D_P are the diffusion coefficients of the substrate in the gel layer and of the product, respectively.

An exact solution to Eqs. (10.31) and (10.32) needs numerical techniques. However, analytical solutions can be found when we consider the two limiting cases, $[S] \ll K_M$ and $[S] \gg K_M$. The first case represents enzyme kinetics that are much faster than the transport through the membrane, the substrate concentration being the limiting factor. The second case accounts for very high substrate concentration that saturates the enzyme.

In the first limiting case, $[S] \ll K_M$, Eq. (10.31) becomes

$$\frac{\partial^2 [S]}{\partial x^2} - \alpha[S] = 0 \qquad (10.33)$$

where

$$\alpha = \frac{k_2[E]_o}{K_M D_S} \qquad (10.34)$$

is the *enzyme loading factor* or *diffusion modulus*, which represents the effective concentration of the enzyme, including the concentration of enzyme in the membrane, the kinetics of the enzyme reaction, and the mass transport (diffusion) of the substrate molecules through the membrane. The boundary conditions we use to solve Eq. (10.33) are

$$\left.\frac{\partial [S]}{\partial x}\right|_{x=0} = 0 \qquad (10.35a)$$

$$[S]|_{x=L} = [S]_L \qquad (10.35b)$$

where $[S]_L$ is the substrate concentration at the outer surface of the immobilized enzyme layer at $x = L$.

The physical meaning of Eqs. (10.35) is that, at the transducer's surface ($x = 0$), there is no transport flux (the substrate is neither consumed nor generated there, so that with no reaction at the surface, there can be no reaction flux), and at the outer edge of the membrane ($x = L$), the substrate concentration is fixed at the value of the analyte solution at the membrane's outer surface. The solution for the substrate concentration is[20–22]

$$[S] = \left(\frac{\cosh(x\sqrt{\alpha})}{\cosh(L\sqrt{\alpha})}\right)[S]_L \quad (10.36)$$

where x defines the distance across the gel-membrane layer, which extends from the interface between the gel and the transducer's surface at $x = 0$, to the gel-solution interface at $x = L$. The term we are interested in is the product concentration at the transducer's surface ($x = 0$). To find a relationship of product concentration from the substrate concentration, we add Eqs. (10.31) and (10.32), i.e.,[20–22]

$$D_S \frac{\partial^2[S]}{\partial x^2} + D_P \frac{\partial^2[P]}{\partial x^2} = 0 \quad (10.37)$$

In this way the kinetic term due to the enzyme catalyzed reaction disappeared. Equation (10.37) can be integrated once to give

$$D_S \frac{\partial[S]}{\partial x} + D_P \frac{\partial[P]}{\partial x} = \text{constant} \quad (10.38)$$

Equation (10.38) represents the diffusion fluxes in the enzyme layer. At the membrane's outer surface ($x = L$), the product and substrate fluxes must balance in the steady state, since no material is being created or destroyed. Thus, the value of the constant must be zero throughout the membrane. Integrating Eq. (10.38) between $x = 0$ and $x = L$, results in the mass balance equation for the membrane:

$$D_S[S]_L + D_P[P]_L = D_S[S] + D_P[P] = \text{constant} \quad (10.39)$$

Combining Eq. (10.39) and Eq. (10.36), and solving for [P], we obtain[20–22]

$$[P] = \frac{D_S}{D_P}\left[1 - \frac{\cosh(x\sqrt{\alpha})}{\cosh(L\sqrt{\alpha})}\right][S]_L + [P]_L \quad (10.40)$$

Setting $x = 0$ in Eq. (10.40) gives the surface concentration of product, which is proportional to the transducer's output signal in an ENFET, in terms of the substrate concentration in the solution

$$[P]|_{x=0} = \frac{D_S}{D_P}\left(1 - \frac{1}{\cosh(L\sqrt{\alpha})}\right)[S]_L + [P]_L \quad (10.41)$$

Equation (10.41) indicates that, for the limiting case $K_M \gg [S]_L$, the surface concentration is directly proportional to the substrate concentration at the enzyme layer outer surface. This description of the system involves the parameter $L\sqrt{\alpha}$, which includes the important kinetic variables of the system.

Equation (10.41) shows that there is a term for the concentration of product just outside the membrane. This term can derive from the slow diffusion of the product into the bulk solution or from the presence of the product as part of the solution.

In the second limiting case, $[S] \gg K_M$, Eqs. (10.31) and (10.32) become

$$D_S \frac{\partial^2 [S]}{\partial x^2} - k_2[E]_o = 0 \qquad (10.42)$$

and

$$D_P \frac{\partial^2 [P]}{\partial x^2} + k_2[E]_o = 0 \qquad (10.43)$$

Equations (10.42) and (10.43) can be integrated, and both are subject to the boundary conditions of Eqs. (10.35). The solution for the substrate concentration is then

$$[S] = [S]_L + \frac{k_2[E]_o}{2D_S}(x^2 - L^2) \qquad (10.44)$$

and for the product concentration

$$[P] = [P]_L + \frac{k_2[E]_o}{2D_P}(L^2 - x^2) \qquad (10.45)$$

From Eq. (10.45), we can notice that the product concentration at the surface ($x = 0$) of the transducer is a constant that depends on the immobilized enzyme concentration, the reaction kinetics, the diffusion mass transport, and it is independent of substrate concentration. Therefore the output of the biosensor is constant. In this case the substrate concentration is so large that it has saturated the enzyme. Then there is no sufficient amount of enzyme in the membrane to detect changes in such a large substrate concentration.

The above analysis shows that, for a given amount of enzyme in the membrane, the response of the sensor will go from an approximately linear dependence on the analyte concentration to a saturation value for an increase of the analyte concentration. We remind the reader that the above analysis has been performed in absence of buffer. In practice, a buffer is always present and the introduction of feedback can take care of it.[7]

Analytical solutions of the differential equations written above, are described in detail in Ref. 22.

We now consider the response of an ISFET, modified by an immobilized enzyme layer E that catalyzes the reaction of the substrate S, leading to formation of the acid HA/($H^+ + A^-$), according to the reaction

$$S \xrightarrow{E} HA \rightleftharpoons H^+ + A^- \qquad (10.46)$$

In the steady state, a balance between the rates of mass transport of the substrate from bulk solution to the enzyme layer, production of the acid by the enzyme layer and transport of the product acid into the bulk solution will be achieved, leading to a stable local pH change in the region of the immobilized enzyme layer.

The mass transport of product acid from the surface of the immobilized enzyme layer into bulk solution is a complex problem because of proton association and dissociation reactions in solution. First, the proton dissociation and association reactions of the acid itself must be taken into account, i.e.,

$$H^+ + A^- \underset{k_{-a}}{\overset{k_a}{\rightleftharpoons}} HA \qquad (10.47)$$

Second, the reaction of any buffer system, B^-/BH, must be considered, i.e.,

$$H^+ + B^- \underset{k_{-b}}{\overset{k_b}{\rightleftharpoons}} HB \qquad (10.48)$$

Complete analytical solution of the set of differential equations describing mass transport coupled with the kinetics of acid protonation and deprotonation reactions is not possible, either in the presence or absence of buffer.

Concentration profiles for the involved species across the enzyme layer, and in particular the H$^+$ concentration at the pH-sensitive surface can be derived by adding to Eq. (10.31) the following relations[23]

$$D_{H+}\frac{\partial^2[H^+]}{\partial x^2} + \frac{k_2[E]_o[S]}{K_M + [S]} - k_b[B^-][H^+] + k_{-b}[BH] = 0 \tag{10.49}$$

$$D_{BH}\frac{\partial^2[BH]}{\partial x^2} + k_b[B^-][H^+] - k_{-b}[BH] = 0 \tag{10.50}$$

$$D_{B-}\frac{\partial^2[B^-]}{\partial x^2} - k_b[B^-][H^+] + k_{-b}[BH] = 0 \tag{10.51}$$

where D = diffusion coefficients of the various species in the immobilized enzyme layer
k_2 and K_M = constants for Michaelis-Menten enzyme kinetics as previously described
[E] = enzyme concentration in the immobilized layer
k_b and k_{-b} = forward and backward rate constants for the buffer protonation reaction

Solution of these equations with suitable boundary limits gives a model for the response of an enzyme-modified ISFET in buffer solution. For further details, the reader is referred to the work of Eddowes,[23] where an analytical solution for the steady-state response of an enzyme-modified pH-sensitive ion-selective device in the presence of pH buffer is derived.

10.3.4 Transient Responses of the ISFET

So far, we have described the pH response of ISFET devices in presence of a fixed electrolyte concentration. We now consider a different situation where a transient variation in such a concentration takes place. To fix ideas, let us assume that an ISFET is immersed in a KCl solution, which increases at $t = t_0$ from concentration C_1 to concentration C_2 during the time interval $\Delta t = t_1 - t_0$. The whole process includes two distinct situations, namely: an *equilibrium* situation for $t < t_0$ and for $t > t_1$, and a *nonequilibrium* situation in the interval Δt. In both situations, the specific value of the ion concentration will affect the surface electric potential φ_{eo}. Under equilibrium conditions, the surface potential can be calculated by solving the corresponding Poisson-Boltzmann equation, developed in Sec. 7.2.1. It can be shown[24] that *very large* changes in the electrolyte concentration, result into *very small* changes in the potential. This means that the static pH response of an ISFET having the appropriate insulator (e.g., Si_3N_4, Al_2O_3, or Ta_2O_5) should be not influenced by the value of the electrolyte concentration.

The transient situation can be quite different from the equilibrium one if the electrolyte concentration is increased in a fast way (i.e., with an "ion step"[24]). In this case, the integral diffuse capacitance $C_d = \sigma_d/\varphi_{eo}$ increases very fast because of the sudden increase in the electrolyte concentration. The charge density σ_d at the insulator surface [see Eq. (10.18)], being unaffected by the ion step, this will result in a decrease in the surface potential φ_{eo}, according to

$$\sigma_d = \varphi_{eo} C_d \tag{10.52}$$

We can then expect a decrease in the surface concentration of protons, H_s^+, as a consequence of the decrease in the potential. However, the ISFET insulator acts as a very good buffer for H_s^+ and it will keep the H_s^+ concentration constant by dissociating more A-OH groups, causing a change in σ_d, so that a new equilibrium is reached when σ_d/C_d reaches the same value as before the ion step. The speed of this process will depend on the diffusion of the H^+ ions and also on the buffer capacity of the system. Data reported in literature indicate that the transient response of a Ta_2O_5 ISFET to a 10- to 50-mM KCl step develops for about 2 s, with a peak value of about 35 mV.[24] This analysis can then be extended to study the ISFET transient response in the presence of biological material. For instance, the presence of layers of charged proteins entrapped in proximity to the ISFET insulator should modify the transient response of an ISFET to an ion step. This statement is based on the fact that the fixed protein charge should introduce a Donnan equilibrium potential (see Sec. 7.3.4) into the electrolyte-ISFET system. As a consequence, the ion-step method could be used to measure and characterize specific proteins (e.g., antibodies).

Up to now we considered ion steps as an input introduced by the experimenter. On the other hand, ion steps can be introduced spontaneously by living biological systems. As will be discussed in detail in Chap. 11, the electrophysiological activity of a neuron is precisely based on transient charge movements at the surface of its membrane. This means that an ISFET brought in proximity to an active neuron should be able to measure its activity, thus forming a *neuroelectronic junction*. This fact, first indicated in a pioneering paper by Bergveld[14] have been confirmed both experimentally[25] and via computer simulation.[26]

In view of its bioelectronic relevance, this topic will be further considered in Secs. 11.6 and 12.4.1.

10.3.5 Modes of Operation of the ISFET

From the analysis of the FET-based biosensor structures, it follows that the main characteristic of the device response is a change in the drain current due to the activity at the insulator gate region. Under the condition of strong inversion (see Sec. 9.1), a change in the activity of the protons in solution gives rise to a change in the concentration of mobile carriers in the channel and thus, for a given constant drain voltage V_{DS}, a change in the drain current I_{DS}.

We can consider two modes of operation:

- V_{GS} and V_{DS} constant; I_{DS} measured.
- I_{DS} and V_{DS} constant; V_{GS} measured.

Operation with Constant $\mathbf{V_{GS}}$. Under this mode of operation, the externally applied voltages (V_{DS} and V_{GS}) are kept constant and I_{DS} is measured by an operational amplifier in the current-to-voltage converter mode (Fig. 10.10a), with the gain control of the output signal given with the feedback resistor R_F, such that

$$V_{out} = -R_F I_{DS} \tag{10.53}$$

In practice, however, V_{out} does not directly reflect I_{DS}, but takes into account also the series resistance of the source and drain regions, causing deviations from linear behavior. The device must therefore be arranged in order to account for these deviations.

Operation with Constant $\mathbf{I_{DS}}$. The circuit of Fig. 10.10a can be modified (Fig. 10.10b), so that the output voltage V_{out} provides the control of V_{GS} to keep I_{DS} constant. The volt-

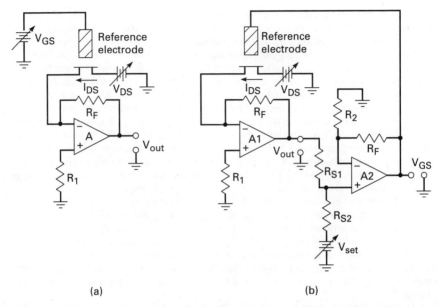

FIGURE 10.10 Modes of operation for MOSFET-based biosensors. Basic circuit for operation with (a) constant V_{GS}; (b) constant I_{DS}.

age V_{out} is fed through a voltage divider into a differential amplifier, where the inverting input is kept at 0 V, such that 0 V at the noninverting input must be achieved by

$$V_{out} + V_{set} = 0 \qquad (10.54)$$

and since

$$I_{DS} = -\frac{V_{out}}{R_F} \qquad (10.55)$$

then

$$I_{DS} = \frac{V_{set}}{R_F} \qquad (10.56)$$

and the required I_{DS} can be set by acting on V_{set}, so that the output from the differential amplifier will control V_{GS} to obtain the desired constant drain current.

This mode of operation has the advantage that changes in the interface potential can be measured directly. Any change in the value of the term $(kT/q) \ln [H^+]$ contributing to I_{DS} will be balanced by an adjustment of V_{GS}.

10.4 CELL-BASED BIOSENSORS AND SENSORS OF CELL METABOLISM

In the mid-1970s it was found that not only enzymes but also whole living microbial cells with their enzyme activity could be used to make biosensors. The first cell-based biosensor was the microbial electrode with the cells of *Acetobacter xylinum* immobilized in a cellulose membrane on the surface of an oxygen electrode. It was used to measure ethanol concentrations on the basis of the measurement of oxygen consumption during ethanol assimilation in bacterial cells. In the following years other microbial biosensors were developed, using the high activity of specific enzymes within the cells of various bacterial and yeast strains. Later on, other types of living cells, and even tissue slices from organs of animals and higher plants, were also used to make biosensors.

The reader can find a wide description of these devices in Ref. 27. In this section we introduce only the main properties of the cell-based biosensors, pointing out their differences with respect to the enzyme-based biosensors.

Cell-based biosensors are then analytical devices, where suitable cells (used as a receptor) are coupled with a transducer. The high enzyme activity within the cells is usually used; the enzyme catalyzes the conversion of a given substrate, and the transducer follows the resulting product with a change of its signal, which is registered by ad hoc electronic equipment. The function of most cell-based biosensors is thus analogous to that of biosensors with immobilized isolated enzymes. However, the use of whole cells has some advantages in comparison with enzyme biosensors. They can be summarized as follows[27]:

1. Cultivation of microbial cells, or preparation of tissue slices in the laboratory, is easy and cheap compared to the preparation of a pure enzyme.
2. Isolation, purification, and immobilization of enzymes are often very difficult; these problems can be eliminated by the use of the whole cells.
3. Some enzymes can lose their activity during isolation or immobilization process; this problem is also eliminated by the use of the whole cells.
4. Enzymes in the cell's natural environment are usually extremely stable.
5. Multistep enzyme reactions in intact cells can be used, making it possible to avoid the preparation of complicated artificial multienzyme systems.
6. Coenzymes and activators are often present in the cells, and thus it is not necessary to add them into the system; the cell itself usually regenerates them.

The use of whole cells for preparation of biosensors has, however, also some disadvantages that can be summarized as follows[27]:

1. The enzyme-catalyzed reaction can take place more slowly because the substrate must cross the cell membrane in order to reach the enzyme, as does the product in order to leave the cell and reach the transducer. Then substrates, especially macromolecular compounds, that are not able to cross the membrane cannot be used.
2. Other metabolic pathways in the cells can be the source of the side products that are also measured by a transducer; this leads to the decrease of the biosensor selectivity.

Cell-based biosensors are usually classified according to:

- *Type of cells* used: the most common are bacterial or yeast cells (*microbial biosensors*); multicellular organisms such as lichens, tissues of higher plants, and animals are also used (*tissue biosensors*).
- *Relation of the cells* to the transducer: *membrane biosensors,* which have the biocatalyzer in the form of a membrane containing the cell suspension or in the form of a tissue slice that is kept on the transducer surface, and *reactor biosensors,* which usually use a suspension of bacterial cells or of plant tissues in a vessel reactor.
- *Transducer type:* the transducer is usually represented by an ion-selective electrode that changes its potential according to the increase of the ion concentration due to the

enzyme, which catalyzes the conversion of the measured substrate. Another possibility is an amperometric detection of substances that can be oxidized on an anode or reduced on a cathode. A further possibility of electrochemical detection is represented by the use of the ISFET.

A different way of considering cell-based biosensors involves detecting changes in the physiological state of cultured living cells by monitoring the rate at which they excrete acidic products of metabolism. Metabolic rate can be defined in various ways: for example, one can monitor the rate of uptake of reactants such as glucose or O_2, the rate of production of products such as heat, or the acidic products of metabolism, lactic acid and CO_2. An alternative technique, proposed in Ref. 28, consists in measuring extracellular acidification rates: this choice has been made also because of the availability of MOSFET-based biosensors, a technology that is particularly well suited for such measurements. Thus, one measures the rate of pH change of the extracellular medium, which depends on the rate of proton production by cells (e.g., on the order of $10^8 H^+ s^{-1}$ per cell in mammalian cells) and on the buffer capacity of the medium.

Since cells excrete protons into buffered solutions and potentiometric methods detect pH rather than $[H^+]$ directly, it is important to establish the relationship between the number of protons n excreted into a volume V and the resulting pH change. If we assume that the medium is buffered by a weak acid HA present in total concentration $A_{tot} = [HA] + [A^-]$, with dissociation constant K, the buffer capacity β is given by (see Sec. 4.2.2)

$$\beta \equiv -\frac{d(n/V)}{d\text{pH}} = \ln(10) \, A_{tot} \frac{10^{(\text{p}K-\text{pH})}}{(1 + 10^{(\text{p}K-\text{pH})})^2} \qquad (10.57)$$

10.5 LIGHT-ADDRESSABLE POTENTIOMETRIC SENSOR (LAPS)

We have seen in the previous sections that biochemical reactions can be measured potentiometrically through changes in pH. An electronic device, based on the measure of an alternate photocurrent through an electrolyte-insulator-semiconductor (EIS) interface, has been proposed[29] to provide a highly sensitive means to measure such potential changes: such a structure is called *light-addressable potentiometric sensor* (LAPS).

In the following, we provide only the principles of operation mechanisms of LAPS, rather than a quantitative analysis of the device behavior.

10.5.1 LAPS Operation Mechanisms

The LAPS consists mainly of an insulator (usually Si_3N_4) which separates a silicon substrate from an electrolyte (assumed to have a resistance R_e), which is in contact with a controlling electrode to which a bias potential V_{bias} is applied with respect to the substrate of the semiconductor, as shown in Fig. 10.11a.

One or more infrared light-emitting diodes (LEDs) are placed just below the semiconductor, and when they are modulated by an alternating signal, a current of photogenerated hole-electron pairs is induced in the circuit. Discrete chemistries can be located on different regions of the insulating surface, producing variations in the local surface potential that can be determined by selective illumination with one or another of the light-emitting diodes (*light addressability*). The photocurrent I_m is the parameter being measured (usually as an rms value).

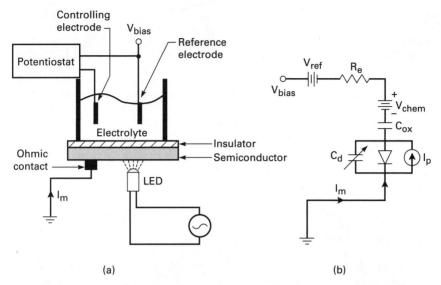

FIGURE 10.11 LAPS. (*a*) Schematic structure and (*b*) simplified equivalent electronic circuit.

In *absence of illumination,* the LAPS behaves like an MOS capacitor (see Chap. 8). For an *n*-type semiconductor, a positive bias potential V_{bias} applied to the structure of Fig. 10.11*a* will drive the semiconductor into accumulation (see Sec. 8.2). Majority carriers (electrons) will distribute uniformly in the semiconductor except at the semiconductor-insulator (SI) interface, where negative charges accumulate to balance the charge at the electrolyte-insulator (EI) interface. In accumulation, the structure is then analogous to a capacitor C_{ox} made of two parallel metal plates separated by the insulator, whose value is given by Eq. (8.2).

If now a negative bias potential V_{bias} is applied to the structure, electrons are driven away from the SI interface, causing a region depleted of majority carriers to appear. We say that the semiconductor is driven into depletion (see Sec. 8.3). The depletion region can be modeled as an insulator with a width that is a function of the bias potential, and it can be represented by a depletion region capacitance C_d. As V_{bias} becomes more negative, the depletion region width [see Eq. (8.17)] increases (and C_d decreases) until it reaches its maximum value.

In *presence of illumination* of the semiconductor of the LAPS with infrared light, there is a generation of hole-electron pairs in the semiconductor, where they can diffuse, recombine, or be separated in an electric field.

When the semiconductor is driven into depletion, hole-electron pairs that have diffused into the depletion region, or that are produced in this region, are separated in the electric field on a time scale short compared to that for recombination.

When the semiconductor is illuminated with a *constant-intensity light* source, charge separation of photogenerated hole-electron pairs in the depletion region gives rise to a transient current which acts to collapse the depletion region and charge the insulator. This current decays as the depletion region width reaches a new steady-state value. Turning off the illumination produces a transient current of opposite polarity as the insulator discharges and the depletion region returns to its original width.

If the *intensity* of the light source is *modulated* with a period short with respect to the decay time constants of the transient currents, an alternating current is produced because there is never enough flow of charges in either direction to significantly modulate the depletion region width.[30]

Figure 10.11b shows a simplified equivalent electronic circuit for the LAPS under photoexcitation produced by an intensity-modulated LED. Modulating the light intensity at a frequency ν results in alternating charging currents through the insulator and depletion region capacitances at the same frequency. The amplitude $|I_m|$ of the insulator charging current through the external circuit is the quantity to be measured. Under *strong depletion*, the measured alternating photocurrent $|I_m|$ is a function of the rate at which hole-electron pairs form in, or diffuse into, the depletion region and of the capacitances of the illuminated area, as given by[30]

$$|I_m| = \frac{C_{ox}}{C_{ox} + C_d}|I_p| \qquad (10.58)$$

where $|I_p|$ is the alternating component of I_p, which represents the photogeneration of hole-electron pairs.

When the semiconductor is biased into *accumulation*, although charge separation takes place, there is a small current in the external circuit due to the large value of C_d, and the low resistance to charge recombination across C_d (represented by the diode in Fig. 10.11b).

Figure 10.12 shows typical sigmoidal shape curves of $|I_m|$ versus V_{bias}. At the most positive bias potential, the semiconductor is in accumulation and no photocurrent is measured. As the semiconductor is driven into depletion, $|I_m|$ increases as C_d decreases. When the maximum depletion width is reached, $|I_m|$ reaches a maximum value.

The LAPS can be used to measure chemically sensitive surface potentials (V_{chem} in Fig. 10.11b) on the insulator surface. For example, the potential of a surface with proton binding capacity, will increase on addition of H$^+$; therefore, on dropping the pH, the bias potential must be decreased to keep the same electrical field in the semiconductor and hence the same photocurrent I_m. As a consequence, there is a shift of the $|I_m|$ versus V_{bias} curves along the V_{bias} axis due to changing the pH, as shown in Fig. 10.12.

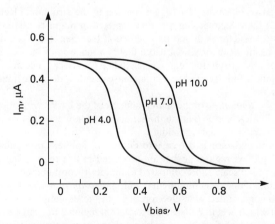

FIGURE 10.12 Alternating photocurrent I_m as a function of bias potential V_{bias} for different values of pH. The curves shown are for n-type silicon; for p-type silicon the shape of the curve is reversed, left to right.

10.6 CONTRIBUTIONS OF MICROFABRICATION TECHNOLOGIES TO THE FIELD OF BIOSENSORS

Two issues will be shortly addressed in this section, both related to silicon microfabrication techniques, namely:

1. The use of photolithographic techniques and surface chemistry to transform a solid surface (e.g., silicon or gold) into a pattern of micrometer-sized regions with different sensing properties.
2. The exploitation of the mechanical properties of silicon microcantilevers to obtain sensors able to transduce energy changes into bending.

10.6.1 Surface Patterning

One of the key issues in the development of biosensors and arrays of biosensors is the *patterning of surfaces* with complex organic functional groups, including distinct attachment points for proteins, peptides, and carbohydrates.

Surface patterning can be achieved by making use of *photolithography*, a technique routinely used in microelectronics for the fabrication of integrated circuits. In the photolithographic process, a surface is exposed to UV light through a mask in such a way that predetermined, micrometer-sized regions of the surface are alternately exposed to or protected from UV light. In this way, molecules anchored to specific regions of the surface are appropriately modified by the UV light, while other molecules belonging to other regions are not, according to the pattern of the mask. A sequence of masks can be used to generate, as a final result, an array of micrometer-sized regions, each of them with a specific chemistry on it. This methodology is known as *combinatorial chemistry*, and it combines solid-phase synthesis, photolabile protecting groups, and photolithography. Some features of this methodology are illustrated in Fig. 10.13.[31]

Of special interest for the field of biosensors is the use of *self-assembled monolayers* (SAMS) to functionalize the patterned regions. The system of self-assembled monolayers of *alkanethiols*[32] on gold is probably the best chemical system to be utilized in the development of arrays of biosensors. The structure of these monolayers is as follows: the chemisorption of the *thiolsulfur* atoms onto the gold substrate drives the self-assembly of the alkanethiols. The density of the sulfur head groups brings the alkyl backbones of these molecules into close contact, causing the chains to order into a densely packed monolayer with very few defects. Monolayers of alkanethiols on gold are stable for a period of several months in air, or in contact with water or ethanol. They are sufficiently stable for many applications dealing with biosurfaces and have been used for studies of protein adsorption and cell adhesion in aqueous media. Both aspects are of great relevance for the field of biosensors. The adsorption of specific proteins is achieved via the insertion of appropriate functional groups on the portion of the thiols exposed to the solution (see Fig. 10.14). Such functional groups include alkyl, perfluoroalkyl, amide, ester, alcohol, nitrile, carboxylic acid, phosphoric acid, boric acid, amine, and heterocycle groups.

10.6.2 Silicon Microcantilevers

The scanning force microscope (SFM), also known as the atomic force microscope (AFM), was first introduced in 1986.[33] Very briefly, it consists of a Si_3N_4 microcantilever, typically

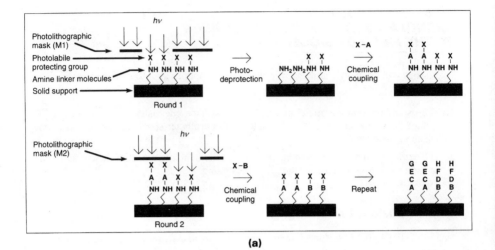

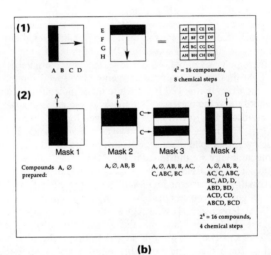

FIGURE 10.13 (a) Light-directed parallel chemical synthesis. A surface is derivatized with amine linkers that are blocked by a photochemically cleavable protecting group. The surface is selectively irradiated with light to liberate free amines, which can be coupled to photochemically protected building blocks. The process is repeated with different regions of the synthesis surface being exposed to light, until a desired array of compounds is prepared. The patterns of photolysis and the order of addition of building blocks define the products and their locations. (b) Synthesis strategies: (1) Orthogonal-stripe method. A layer of monomers is formed by photolyzing stripes for each building block. Dimers are formed by photolyzing stripes orthogonal to the first set, preparing n^2 compounds in $2n$ chemical steps. (2) Binary synthesis. Half of the synthesis surface is photolyzed during each coupling step, with subsequent photochemical steps overlapping one-half of the previous synthesis space. With this strategy, 2^n compounds are made in n chemical steps. (*From J. W. Jacobs and S. P. A. Fodor.*[31])

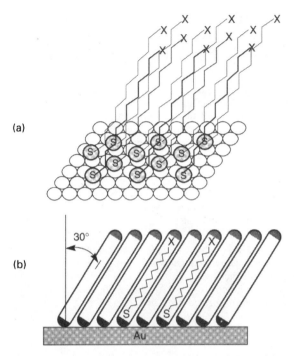

FIGURE 10.14 Representation of a self-assembled monolayer (SAM) of alkanethiolates on the surface of gold. (*a*) The sulfur atoms (S) of the alkanethiolates coordinate to the hollow three-fold sites of the gold (1,1,1) surface; the gold atoms (open circles) are arranged in a hexagonal relationship. The alkyl chains are close packed and tilted approximately 30° from the normal to the surface. (*b*) The properties of the SAM are controlled by changing the length of the alkyl chain and the terminal functional group X of the precursor alkanethiol. (*From M. Mrksich and G. M. Whitesides.*[32])

100 to 300 μm long, 10 to 40 μm wide, and with a thickness in the order of 0.5 to 1 μm. Microelectronic techniques such as etching are utilized in order to form a silicon nitride *tip* at the end of the microcantilever, with radius of the tip (approximated as a half-sphere) in the order of 50 nm. The cantilever-tip system is then driven (with a precision better than 0.1 nm) by piezoelectric motors to scan a surface, and the tip-surface interaction forces, typically in the order of a few nanonewtons, are optically detected by exploiting the fact that the force-induced bending of the microcantilever is amplified by deflecting a light beam generated by a laser impinging on the back of the cantilever. For the purposes of this section, from now on we will focus on the silicon microcantilever *bending* as a consequence of its interactions with the surroundings, without discussing any more the operating principles and the remarkable achievements of SFM, which are described, for example, in Ref. 34. A microcantilever can be viewed as a highly sensitive sensor which can be used to detect the presence of very small amounts of materials both in the gaseous and liquid phase. Two interesting examples of application of this concept are the following.

Detection of Temperature Changes. Aluminum (Al) and silicon nitride (Si_3N_4) present different thermal expansion coefficients. Because of this difference, an Si_3N_4 microcantilever with an Al-coated surface will bend when heat is produced in its proximity. For uniform heating, the bending z is proportional to the absorbed heating power and is expressed by

$$z = \frac{5}{4}(\alpha_1 - \alpha_2)\frac{t_1 + t_2}{Kt_2^2}\frac{l^3}{W(\lambda_1 t_1 + \lambda_2 t_2)}P \qquad (10.59)$$

where[35]
K = device parameter
α = expansion coefficient
t and λ = thickness and thermal conductivities of the two layers (i.e., Al and Si_3N_4), respectively
l and W = length and width of the microcantilever
P = total power impinging on the bi-material cantilever

Such a cantilever can then be used as an *ultrasensitive calorimeter* for investigating surface catalytic reactions. Further applications include the detection of the heat involved in phase transitions of picoliter volumes of solid materials, with a resolution in the order of femtojoules under ambient conditions.[35]

Detection of Surface Stress. Measuring the bending of a plate to determine the surface stress σ in thin films is a common technique. Given a microcantilever with two different opposite faces (e.g., one face Si_3N_4, the other one coated with gold), experiencing two different surface stress variations $\Delta\sigma_1$ and $\Delta\sigma_2$, the radius of curvature R of the cantilever is related to the differences in surface stress by

$$\frac{1}{R} = 6\frac{1-\nu}{Et^2}(\Delta\sigma_1 - \Delta\sigma_2) \qquad (10.60)$$

where E = Young elasticity modulus of the basic cantilever material (i.e., Si_3N_4)
ν = Poisson ratio
t = thickness of the cantilever

Equation (10.60) implies that a microcantilever with appropriately prepared opposite faces will *bend* as a result of changes of stress on the two surfaces. Suppose an Si_3N_4 cantilever is coated with gold on the upper surface and then is immersed in a liquid solution with a given pH. As already discussed in Chaps. 7 and 10, an Si_3N_4 surface presents a charge density which is a function of the H^+ ions present in solution. We can further assume negligible pH-dependent charges on the Au surface. As a whole, we can then predict changes in cantilever deflection and surface stress when varying the pH in an aqueous solution. The cantilever thus can become an unconventional, micrometer-sized pH sensor.[36] Moreover, the gold surface can be coated with a self-assembled monolayer of alkanethiol molecules appropriately functionalized with functional groups (see previous section). The interaction with specific molecules present in a liquid solution can then be revealed by a differential stress-induced cantilever bending. As a subsequent step, by evaporating gold also on the free Si_3N_4 surface, monolayers of thiols with different biosensing properties can be assembled on the two surfaces of the cantilever, thus increasing the biological selectivity of the cantilever-based biosensor.[36–38] Arrays of microcantilevers with different coating and thus different sensing properties could result into a micrometer-sized integrated multiple-sensor system.

PROBLEMS

10.1 Consider a Ta_2O_5 ISFET exposed to a 1:1 electrolyte. Assuming $pK_a = 4$, $pK_b = -2$, and a number of surface sites $N_s = 10 \times 10^{14}/cm^2$, find, at $T = 300$ K, *a.* The proton concentration $[H^+]_s$ at the surface in the case of equilibrium.

    b. The value of the intrinsic buffer capacity of the surface β_s.

10.2 For the Ta_2O_5 ISFET of Prob. 10.1, assuming a bulk solution $pH_b = 2$, a surface capacitance $C_s = 20\ \mu F/cm^2$, and an electrolyte concentration in the bulk $c_0 = 10$ mM, find, at $T = 300$ K,

    a. The value of the surface potential φ_{eo}.

    b. The charge density σ_d in the diffuse layer, when the electrolyte permittivity $\varepsilon_e = 5\varepsilon_0$.

    c. The value of the sensitivity parameter α.

10.3 Derive Eq. (10.28), and show that the Michaelis constant K_M is given by Eq. (10.29). The constant K_M is the ratio of the rate of formation and the rate of consumption of the enzyme-substrate complex [ES]. Plot the rate of product formation v_f versus the substrate concentration [S] using the asymptotes.

10.4 Find an expression for the photogeneration of hole-electron pair current density J_p in a LAPS, assuming that (1) the structure operates at low-level injection; (2) the depletion region width depends only on $V_{GB} = V_{bias}$; (3) the field in the neutral region is negligible; (4) the illumination is perpendicular to the surface of the insulator, which is considered to be transparent to radiation; (5) the photogenerated hole-electron pairs are generated in the depletion region only. Also assume a photogeneration rate $G(x, t) = \alpha\phi_1 \exp(-\alpha x) \exp(j\omega t)$, where α is the absorption coefficient and ϕ_1 is the flux density of photons. (*Hint:* $J_p = J_{\text{diff}} + J_{\text{phgen}}$).

REFERENCES

1. P. Bergveld, "The impact of MOSFET-based sensors," *Sensors and Actuators*, 8: 109–127, 1985.
2. M. J. Madou and S. R. Morrison, *Chemical sensing with solid state devices*, New York: Academic Press, 1989.
3. S. M. Sze, (ed.), *Semiconductor sensors*, New York: John Wiley & Sons, 1994.
4. A. P. F. Turner, I. Karube, and G. S. Wilson (eds.), *Biosensors: Fundamentals and applications*, New York: Oxford University Press, 1987.
5. E. A. H. Hall, *Biosensors*, Buckingham: U.K.: Open University Press, 1990.
6. L. C. Clark Jr. and C. Lyons, "Electrode systems for continuous monitoring in cardiovascular surgery," *Ann. N.Y. Acad. Sci.*, 102: 29–45, 1962.
7. B. H. van der Schoot and P. Bergveld, "ISFET based enzyme sensors," *Biosensors*, 3: 161–186, 1988.
8. F. Sternberg, C. Meyerhoff, F. J. Mennel, F. Bischof, and E. F. Pfeiffer, "Subcutaneous glucose concentration in humans: Real estimation and continuous monitoring," *Diabetes Care*, 18: 1266–1269, 1995.
9. A. K. Covington, F. Valdes-Perezgasga, P. A. Weeks, and A. Hedley Brown, "pH ISFETs for intramyocardial pH monitoring in man," *Analusis*, 21(2): M43–M46, 1993.
10. J. H. Yun, L.-M. Lee, J. A. Wahr, B. Fu, M.E. Meyerhoff, and V. C. Yang, "Clinical application of disposable heparin sensors. Blood heparin measurements during open heart surgery," *Asaio J.*, M661–M664, 1995.
11. F. Cespedes et al., "Fermentation monitoring using a glucose biosensor based on an electrocatalytically bulk-modified epoxy-graphite biocomposite integrated in a flow system," *Analyst*, 120(8): 2255–2258, 1995.
12. J. L. Besombes et al., "A biosensor as warning device for the detection of cyanide, chlorophenols, atrazine and carbamate pesticides," *Analytica Chimica Acta*, 311(3): 255–263, 1995.

13. G. A. Zhylyac et al., "Application of urease conductometric biosensor for heavy-metal ion determination," *Sensors and Actuators B*, 24(1–3): 145–148, 1995.
14. P. Bergveld, "Development, operation and application of the ion-sensitive field-effect transistor as a tool for electrophysiology," *IEEE Trans. Biomed. Eng.*, BME-19: 342–351, 1972.
15. L. Bousse, "Single electrode potentials related to flat-band voltage measurements on EOS and MOS structures," *J. Chem. Phys.*, 76: 5128–5133, 1982.
16. D. E. Yates, S. Levine, and T. W. Healy, Site-binding model of the electrical double layer at the oxide/water interface, *J. Chem. Soc. Faraday Trans.*, (70): 1807–1819, 1974.
17. P. Bergveld, R. E. G. van Hal, and J. C. T. Eijkel, "The remarkable similarity between the acid-base properties of ISFETs and proteins and the consequences for the design of ISFET biosensors," *Biosensors & Bioelectronics*, (10): 405–414, 1995.
18. L. Bousse, N. F. DeRooij, and P. Bergveld, "Operation of chemically sensitive field-effect sensors as a function of the insulator-electrolyte interface," *IEEE Trans. on Electron Devices*, ED-30(10): 1263–1270, 1983.
19. S. D. Caras and J. Janata, "pH-based enzyme potentiometric sensors," *Anal. Chem.*, 57: 1917–1920, 1985.
20. M. J. Eddowes, "Response of an enzyme-modified pH-sensitive ion-selective device; consideration of the influence of the buffering capacity of the analyte solution," *Sensors and Actuators*, 7: 97–115, 1985.
21. A. S. Dewa and W. H. Ko, "Biosensors," in S. M. Sze (ed.), *Semiconductor sensors*, New York: John Wiley & Sons, 415–472, 1994.
22. M. J. Eddowes, "Theoretical methods for analyzing biosensors performance," in A. E. G. Cass (ed.), *Biosensors: A practical approach*, 211–263, New York: Oxford University Press, 1990.
23. M. J. Eddowes, "Response of an enzyme-modified pH-sensitive ion-selective device; analytical solution for the response in the presence of pH buffer," *Sensors and Actuators*, 11: 265–274, 1987.
24. J. C. van Kerkhof, J. C. T. Eijkel, and P. Bergveld, "ISFET responses on a stepwise change in electrolyte concentration at constant pH," *Sensors and Actuators B*, 18/19: 56–59, 1994.
25. P. Fromherz, A. Offenhausser, T. Vetter, and J. Weiss, "A neuron-silicon junction: a Retzius cell of the leech on an insulated gate-field-effect transistor," *Science*, 252: 1290–1293, 1991.
26. A. Cambiaso, M. Grattarola, G. Arnaldi, S. Martinoia, and G. Massobrio, "Detection of cell activity via ISFET devices: modelling and computer simulations," *Sensors and Actuators*, B1: 373–379, 1990.
27. J. Racek, *Cell-based biosensors*, Lancaster, PA: Technomic Publishing, 1995.
28. J. C. Owicki and J. W. Parce, "Biosensors based on the energy metabolism of living cells: The physical chemistry and cell biology of extracellular acidification," *Biosensors & Bioelectronics*, 7: 255–272, 1992.
29. D. G. Hafeman, J. W. Parce, and H. M. McConnell, "Light-addressable potentiometric sensor for biochemical systems," *Science*, 240: 1182–1185, 1988.
30. G. B. Sigal, D. G. Hafeman, J. W. Parce, and H. M. McConnell, "Electrical properties of phospholipid bilayer membranes measured with a light-addressable potentiometric sensor," in R. W. Murray, R. E. Dessy, W. R. Heineman, J. Janata, and W. R. Seitz (eds.), *Chemical sensors and microinstrumentation*, 47–63, from a symposium sponsored by the Division of Analytical Chemistry at the 196th National Meeting of the American Chemical Society, Los Angeles, Sept. 25/30, 1988.
31. J. W. Jacobs and S. P. A. Fodor, "Combinatorial chemistry. Applications of light-directed chemical synthesis," *TIBTECH*, 12: 19–26, 1994.
32. M. Mrksich and G. M. Whitesides, "Patterning self-assembled monolayers using microcontact printing: A new technology for biosensors?" *TIBTECH*, 13: 228–235, 1995.
33. G. Binning, C. F. Quate, and C. Gerber, "Atomic force microscope," *Phys. Rev. Letters*, 56: 930–933, 1986.

34. D. Sarid, *Scanning force microscopy*, New York: Oxford University Press, 1991.
35. J. R. Barnes, R. J. Stephenson, M. E. Welland, Ch. Gerber, and J. K. Gimzewski, "Photothermal spectroscopy with femtojoule sensitivity using a micromechanical device," *Nature*, 372: 79–81, 1994.
36. H. J. Butt, "A sensitive method to measure changes in the surface stress of solids," *J. Colloid and Interface Science*, 179: 113–120, 1996.
37. R. Raiteri and H. J. Butt, "Measuring electrochemically induced surface stress with an atomic force microscope," *J. Phys. Chem.*, 99: 15728–15732, 1996.
37. R. Raiteri, M. Grattarola, and H. J. Butt "Measuring electrostatic double-layer forces at high surface potentials with an atomic force microscope," *J. Phys. Chem.*, 100: 16700–16705, 1996.

CHAPTER 11
NEURONS AND NEURONAL NETWORKS

A series of papers published in 1952 by A. C. Hodgkin and A. F. Huxley opened the door to the detailed comprehension of how *electrophysiological signals are transmitted* inside the nervous system. The model described in these papers has become a milestone for all the subsequent electrophysiological studies. In more recent years, great knowledge has been accumulated on the rules which govern *signal generation* at the molecular level. Then, starting from the 1980s, an increasing merging between neurobiology and formal studies in areas of physics, mathematics, and informatics has taken place, originating a very successful field known as *neural* (or *neuronal*) *networks*. Nowadays, a topic such as "neurons and neuronal networks" is a very broad and complex one, which, by itself, could deserve a whole book.

Among the many possible choices in developing this chapter, we tried to follow the sequence just summarized above. After a short overview of neurobiological notions, the model proposed by Hodgkin-Huxley (H-H) for the generation and propagation of action potentials is described and then the focus is shifted to simplified neuronal models, synaptic connections, and neuron-to-neuron coupling. The chapter ends with two relevant examples of the bioelectronic approach, i.e., the neuron-transducer junction and the silicon neuron.

11.1 SHORT OVERVIEW OF THE BIOLOGY OF THE NEURON

The human body is made of about 10^{13} cells; roughly 10^{11} of them are neurons. They, together with *glia cells*, are the main cellular components of the brain, where each neuron, on the average, makes 10^3 to 10^4 contacts with other neurons. We can classify the neurons forming the nervous system of any organism into three broad categories:

1. Neurons which receive stimuli from the external environment, i.e, *sensory neurons*.
2. Neurons which control the actions of the organism mainly via contact with muscle cells, i.e., *motor neurons*.
3. Neurons which bring in contact other neurons, i.e., *interneurons*.

Qualitatively, we can say that the number of interneurons increases as the complexity of the organism increases. The brain is mostly made of interneurons.

In its essence, any neuron is made of three main components: a *cell body,* named *soma,* which contains the nucleus and the machinery for synthethizing proteins (see Chap.

4), one long (up to several dm) *arborization,* named *axon;* and several *short arborizations,* named *dendrites.* This scheme identifies a *multipolar* neuron, typical of the central nervous system of complex organisms, such as mammalians. It is sketched in Fig. 11.1.

In most of these neurons, a layer of lipids (not indicated in Fig. 11.1), the *myelin sheet,* insulates the axon from its environment, with periodic interruptions known as *nodes of Ranvier.* Neurons of invertebrates can have simpler structures, such as *monopolar* neurons, where the soma does not present dendrites and a single arborization branches into an axon and several dendrites.[1] In *bipolar* neurons two distinct processes originate from the soma: one axon and a branched dendrite. Most bipolar neurons are sensitive ones, as the *bipolar cells* of the *retina,* or those of the *olfactive epithelium.*

It should be appreciated that a neuron inside the nervous tissue is not an isolated entity, but it rather is densely packed together with other neurons and glia cells (which are 10 to 50 times more numerous than neurons and whose role is still not fully understood[2]). This packing results in the above-mentioned 10^3 to 10^4 contacts (up to 1.5×10^5 for the *cerebellum Purkinje* neurons) from the axons of other neurons, mostly made on the dendritic arborizations and on the soma, through specialized structures named *synapses.*

Multipolar neurons display a remarkable variety of shapes, and the reader is invited to consult Refs. 1 and 2 to fully appreciate it. In spite of this variability, we can state that, from an operational point of view, all the neurons share the same rule: dendrites and soma receive and elaborate electrochemical signals (input) while the axon transmits them (output) through synaptic connections to other neurons, muscle cells, or gland cells.

Among sensory neurons, olfactory neurons are periodically replaced during the life span of the organism to which they belong[2] and the presence of multipotent stem cells (i.e., duplicating cells, see Sec. 4.8) has been suggested in the adult vertebrate central nervous system.[3] On the other hand, most neurons survive as long as the organism to which they belong and can retain stable electrical properties for much of the organism lifetime. This implies that the ion channels of the neuron membrane (see Chap. 4) are

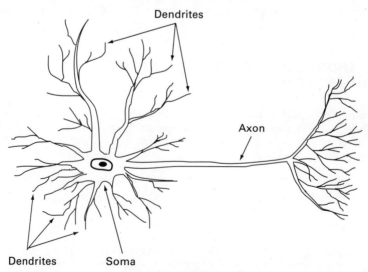

FIGURE 11.1 Morphology of a (multipolar) neuron. Soma (with the cell nucleus), dendrites, and axon are indicated. (*Kindly provided by Brunella Tedesco.*)

replaced by protein turnover and that the neuron may even change size or shape in time. So, the study of the behavior of a neuron, on a time scale longer than minutes, should take into account the presence of several metabolic pathways active in the various cell compartments. On the other hand, if we are interested in the generation and propagation of electrochemical signals in the subminute time scale, then we can simplify our approach by focusing on membrane properties only. In this context, the description of the electrical potential of the neuronal membrane will be the main issue of the following section.

11.2 BIOPHYSICAL DESCRIPTION OF THE ACTION POTENTIAL

As a one-statement summary of the previous section, we can say that in a neuron, dendrites and soma receive and *elaborate* information, while the axon safely *propagates* it. The propagation tool, for which the axon is specialized, is the *action potential* (or *spike*), a nonlinear change in the axon membrane which temporarily (millisecond scale) sustains a 100-mV positive potential variation. This quite dramatic phenomenon was explained, at the beginning of the 1950s, as the result of the research of several scientists including Cole, Katz, Hodgkin, and Huxley. Hodgkin and Huxley produced in 1952 a seminal paper[4] modeling the generation and propagation of the action potential in the squid giant axon. In the subsequent 40 years, the H-H model was successfully utilized, with small modifications, for many other neurobiological membranes. In describing some of its basic features, let us start by considering a 1- by 1-μm patch of axon membrane.

As a first approximation, the membrane of a nerve cell is permeable to sodium, potassium, and chloride. Table 11.1 shows the distribution of these ions inside a patch of the membrane of the squid axon and outside it. Values for a typical mammalian cell are also given for comparison.

At physiological pH (i.e., around pH = 7.3), the inside of a neuron contains fixed negative charges (see Chap. 4) and one could think that the ion distribution indicated in Table 11.1 is the result of a Donnan equilibrium only (see Chap. 7). This is incorrect. A living cell is not in equilibrium and metabolic energy is used by the cell to maintain itself in an *out-of-equilibrium steady state*. If the ion permeabilities are known, then an estimate of

TABLE 11.1 Intracellular and Extracellular Concentration of Na^+, K^+ and Cl^-; Ion Concentrations in Seawater Are Also Given

Medium	Concentration, mM		
	K^+	Na^+	Cl^-
Squid axon			
Intracellular	400	50	40–60
Extracellular	20	440	560
Seawater	10	460	540
Typical mamalian cell			
Intracellular	140	5–15	5–15
Extracellular	5	145	110

the membrane *resting* (i.e., no stimuli) potential V_m can be obtained by using the *Goldmann-Hodgkin-Katz* (GHK) *equation* (see Chap. 7), i.e.,

$$V_m = \frac{kT}{q} \ln \frac{[K^+]_o + r[Na^+]_o}{[K^+]_i + r[Na^+]_i} \tag{11.1}$$

where the chloride ion was neglected because it was assumed to be approximately in equilibrium and r is the Na$^+$ (sodium) permeability divided by the K$^+$ (potassium) permeability. By assuming $r = 0.04$ (Ref. 2) and by using the ion concentrations given in Table 11.1 for the inside of the axon and seawater, respectively, we obtain

$$V_m = -67 \text{ mV} \tag{11.2}$$

This value is not exact. Indeed, (besides not considering Cl$^-$) in our scheme, we did not consider that an *active mechanism* should be present that allows Na$^+$ and K$^+$ to maintain their given concentrations (of course the inside of the neuron is not an infinite compartment). This mechanism is known as the *sodium-potassium active transport pump*. By taking into account also the sodium-potassium pump, a slightly different value for V_m is obtained, in the order of -73 mV (Ref. 2). By following the schemes developed in Sec. 7.3, the above considerations can be translated into a useful electrical circuit model, where equilibrium potentials and conductances are utilized instead of concentrations and permeabilities. Such a translation is depicted in Fig. 11.2, where $P_{Na,K}$ represents the active transport pump and the capacitor C_m is introduced to take into account the charging capability of the phospholipid insulating layers (see Chaps. 4 and 7).

As stated, the equivalent-circuit approach is very useful. Nevertheless, it must be always remembered that, in contrast to standard electrical circuits, here we are dealing with

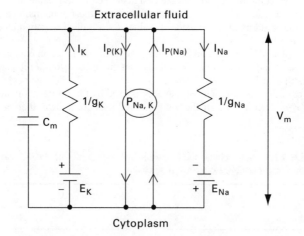

FIGURE 11.2 Equivalent-circuit representation of a patch of neuron membrane. E_K and E_{Na} represent the equilibrium (Nernst) potentials of potassium and sodium ions, respectively; g_K and g_{Na} are the potassium and sodium conductances. I_K and I_{Na} are the potassium and sodium currents driven by the respective electrochemical gradients. $I_{p(K)}$ and $I_{p(Na)}$ are the potassium and sodium currents driven by the sodium/potassium active pump $P_{Na,K}$. The capacitor C_m takes into account the membrane charging capability.

ionic currents. Moreover, different ionic species are supposed to flow through physically separated branches. It can be worth reading again Sec. 7.3.3.

The general equation governing the total current I_{tot} crossing the patch of membrane is

$$I_{tot} = C_m \frac{dV}{dt} + g_K(V - E_K) + g_{Na}(V - E_{Na}) + I_p \quad (11.3)$$

where I_p takes into account the Na/K active pump; it can be further split into two separate currents $I_p(K)$, $I_p(Na)$. Under steady conditions, with the membrane potential fixed at its resting value V_m and no currents injected from the outside, the total current I_{tot} should be zero and no capacitive current should be present. Moreover, in consideration of its small contribution, we can also neglect the current generated by the active pump. Strictly speaking, this approximation is somehow not satisfactory: each of the two pump-driven currents should be exactly opposite to the corresponding electrochemical current. Nevertheless, the argument becomes of little relevance in the presence of externally injected currents, as it will be shown in the following. We can then write

$$0 = g_K(V - E_K) + g_{Na}(V - E_{Na}) \quad (11.4)$$

and rearranging gives the *resting membrane potential*

$$V_m = \frac{E_K + NKE_{Na}}{1 + NK} \quad (11.5)$$

where NK is the sodium/potassium conductance ratio. Equation (11.5) is equivalent to Eq. (11.1).

Suppose now that we inject an ionic current I_{ext} inside the cell through the membrane. If we still neglect I_p, Eq. (11.3) becomes

$$I_{ext} = C_m \frac{dV}{dt} + g_K(V - E_K) + g_{Na}(V - E_{Na}) \quad (11.6)$$

Rearranging gives

$$C_m \frac{dV}{dt} = -g_K(V - E_K) - g_{Na}(V - E_{Na}) + I_{ext} \quad (11.7)$$

Then a constant current I_{ext} applied for a short time Δt will shift the membrane potential to the new steady value:

$$V_m^* = \frac{E_K + NKE_{Na}}{1 + NK} + \frac{I_{ext}}{g_K + g_{Na}} \quad (11.8)$$

Conversely, after turning off the current, the membrane potential will decay back to its original resting value V_m. This behavior is illustrated in Fig. 11.3.

Experimental evidence accumulated in the 1940s clearly indicated that, under appropriate stimulus conditions, the response of an axon membrane (typically from the squid) was much more dramatic than the one shown in Fig. 11.3. This response, known as *action potential* or *spike,* is schematically depicted in Fig. 11.4.

The key idea to interpret this experimental evidence was to assume that the *sodium and potassium conductances are a function of the membrane potential.* In order to appreciate this statement, let us consider in more detail the conductances g_K and g_{Na}. As suggested by Hodgkin and Huxley,[4] at a microscopic level these conductances can be consid-

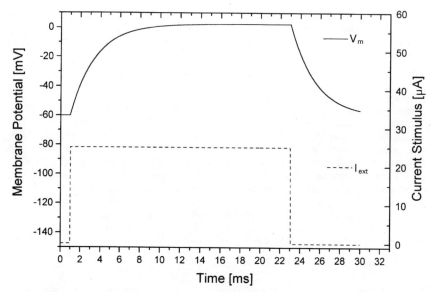

FIGURE 11.3 Passive response of a patch of membrane to the injection of the current I_{ext}.

ered as the result of the presence of many protein channels, which can be assumed to be selective to the potassium and sodium ions, respectively (see Secs. 4.7 and 7.3). We can further hypothesize that these channels can be either *open* or *closed* and that in any specific physiological situation a given fraction of them is open. Formally,

$$g_{Na} = \bar{g}_{Na} a \tag{11.9}$$

and
$$g_K = \bar{g}_K b \tag{11.10}$$

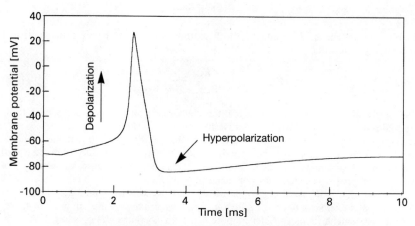

FIGURE 11.4 Sketch of an action potential.

where g_{Na} and g_K are maximum conductances, proportional to the maximum number of sodium and potassium channels, respectively, and a, b are the corresponding fractions of open channels. This fraction can then be described with the kinetic scheme with α and β as rate constants (see Chap. 5),

$$\text{Fraction of open channels} \underset{\beta}{\overset{\alpha}{\rightleftarrows}} \text{Fraction of closed channels} \qquad (11.11)$$

which gives for the two kinds of channels

$$a \underset{\beta_a}{\overset{\alpha_a}{\rightleftarrows}} 1 - a \qquad (11.12)$$

and

$$b \underset{\beta_b}{\overset{\alpha_b}{\rightleftarrows}} 1 - b \qquad (11.13)$$

As discussed in Chap. 5, Eqs. (11.12) and (11.13) can be transformed into two differential equations,

$$\frac{da}{dt} = -\alpha_a a + \beta_a (1 - a) \qquad (11.14)$$

and

$$\frac{db}{dt} = -\alpha_b b + \beta_b (1 - b) \qquad (11.15)$$

The respective equilibrium fractions of channels will then be

$$a = \frac{\beta_a}{\alpha_a + \beta_a} = \frac{1}{\frac{\alpha_a}{\beta_a} + 1} \qquad (11.16)$$

and

$$b = \frac{\beta_b}{\alpha_b + \beta_b} = \frac{1}{\frac{\alpha_b}{\beta_b} + 1} \qquad (11.17)$$

For the sake of briefness, let us now focus on the fraction of sodium channels only. We note that the ratio α_a/β_a is precisely the *equilibrium constant* of Eq. (11.12), therefore it depends exponentially, via a Boltzmann distribution, on the energy associated to the transition.

Let us now further assume that, besides other energy factors, it specifically depends on the actual value of the membrane potential (more precisely, on the potential energy associated with it) in the form

$$a = \frac{1}{e^{-(V - V_{th})/M} + 1} \qquad (11.18)$$

where M has the dimension of a voltage (it is actually equal to kT divided by a charge), V is the actual value of the membrane potential and V_{th} is a *threshold* potential. The behavior of the fraction a as a function of V is given in Fig. 11.5 for two possible values of M (corresponding to two hypothetical charge values).

For appropriate values of M, Fig. 11.5 suggests that most of the sodium channels are closed for $V_m < V < V_{th}$ and most of them open as soon as V overcomes V_{th}. A large increase in open sodium channels means a large increase in the inward flux of sodium ions. This current will unbalance Eq. (11.4), by further shifting the membrane potential toward

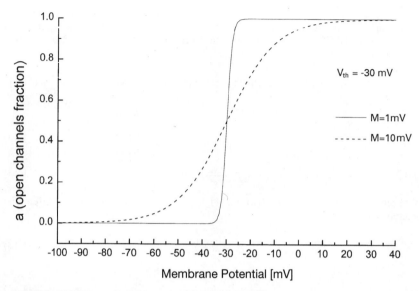

FIGURE 11.5 Sigmoidal behavior of the variable a (fraction of open channels) as a function of membrane potential. For appropriate values of the parameter M, a displays a sudden transition from 0 to 1 in correspondence to a narrow range of membrane potential values, centered around V_{th}.

the sodium equilibrium potential and so on. Let us hypothesize that, at the end of this process, the membrane reaches a new steady state characterized by

$$\bar{g}_{Na}a \cong \bar{g}_{Na} \gg \bar{g}_K b \tag{11.19}$$

Then the new resting potential can be easily calculated by reducing Eq. (11.4) to

$$0 \cong \bar{g}_{Na}(V - E_{Na}) \tag{11.20}$$

That is (by Nernst potential with values given in Table 11.1),

$$V_m \cong E_{Na} \cong 60 \text{ mV} \tag{11.21}$$

This prediction is consistent with the experimental evidence that, during the *raising* phase of the action potential, the membrane *depolarizes* toward the sodium equilibrium potential E_{Na}. In a few milliseconds it then goes back to its original resting potential, with even a transient *hyperpolarization* below it. During this hyperpolarization period, the axon is unable to generate another action potential (*refractory period*). Of course this behavior cannot be predicted by the very crude model just described. It is presented in detail by the H-H model. The *complete model equation* reads

$$I_{ext} = C_m \frac{dV}{dt} + \bar{g}_{Na}m^3h(V - E_{Na}) + \bar{g}_K m^4(V - E_K) + \bar{g}_l(V - E_l) \tag{11.22}$$

where m, h, and n can be interpreted as channel fractions and E_l, g_l refer to a "leakage" ion, typically interpreted as Cl$^-$. In this detailed model, the role of the "didactic variable" a is taken by the product m^3h, where m^3 takes care of the Na channels' opening process, and h of the delayed process of their closing. The power of 3 suggests the presence of three *independent gating mechanisms,* which control the opening of the sodium channel. The quantity n^4 is introduced to describe the (minor) hyperpolarizing effect due to the late opening of potassium channels. The consequent high potassium conductance, together with the residual inactivation of the sodium conductance (quantity h), allows a refractory period during the falling part of the action potential and during a few subsequent milliseconds.[2] The three channel fractions m, h, n (which can also be interpreted as probabilities) satisfy three kinetic schemes governed by the kinetic constants α_m, β_m, α_h, β_h, α_n and β_n, resulting in the following differential equations:

$$\frac{dm}{dt} = \alpha_m(1-m) - \beta_m m \tag{11.23a}$$

$$\frac{dh}{dt} = \alpha_h(1-h) - \beta_h h \tag{11.23b}$$

$$\frac{dn}{dt} = \alpha_n(1-n) - \beta_n n \tag{11.23c}$$

where

$$\alpha_m = \frac{0.1(V+25)}{e^{(V+25)/10} - 1} \tag{11.24a}$$

$$\beta_m = 4e^{V/18} \tag{11.24b}$$

$$\alpha_n = \frac{0.01(V+10)}{e^{(V+10)/10} - 1} \tag{11.24c}$$

$$\beta_n = 0.125 e^{V/80} \tag{11.24d}$$

$$\alpha_h = 0.07 e^{V/20} \tag{11.24e}$$

$$\beta_h = \frac{1}{e^{(V+30)/10} - 1} \tag{11.24f}$$

The numerical values for the kinetic constants were explicitly chosen by Hodgkin and Huxley to fit the experimental data concerning the squid axon, and they do not have any special meaning for the purposes of this section. The interested reader can find detailed analysis of Eq. (11.22) and of the related kinetic equations (11.23) in Refs. 4 to 6.

The action potential predicted by Eq. (11.22), together with the predicted time evolution of the conductances g_{Na} and g_K [Eqs. (11.23) and (11.24)] are shown in Fig. 11.6.

The structure of Eq. (11.22) has been used, with slight modifications, to successfully reproduce the electrophysiological activity of neurons (including mammalian ones), muscle cells, and pancreatic β-cells. It is then evident that Eq. (11.22) is a quite appropriate description of an action potential generated in *any* equipotential patch of an excitable membrane. The *propagation* of the action potential along the axonal membrane can then be modeled by transforming Eq. (11.22) into[4,6]

$$\frac{r_a}{2Rv^2}\frac{d^2V}{dt^2} = C_m \frac{dV}{dt} + \bar{g}_K n^4(V - V_K) + \bar{g}_{Na} m^3 h(V - V_{Na}) + \bar{g}_l(V - V_l) \tag{11.25}$$

where r_a is the radius of the axon, assumed to be a cylinder, R is the specific resistance of the inside of the axon (i.e., the axoplasm), and v is the velocity of propagation of the spike. Equation (11.25) is based on the assumption that, during steady propagation, the shape of

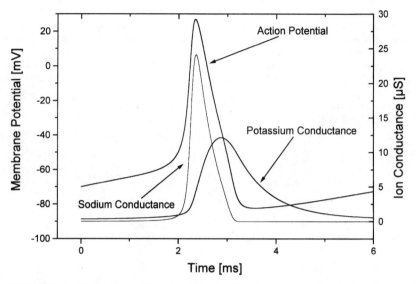

FIGURE 11.6 Time evolution of an action potential and corresponding variations in the sodium and potassium conductances, as predicted by the H-H equations.

potential V against time at any position along the axon is similar to that of V against distance at any time, i.e.,

$$\frac{\partial^2 V}{\partial x^2} = \frac{1}{v^2} \frac{\partial^2 V}{\partial t^2} \qquad (11.26)$$

It is worth mentioning that there is experimental evidence that, in some neurons, there is also an active backpropagation into the dendrites.

We notice that the H-H equations do not include active mechanisms to reequilibrate ion concentrations after each spike. A similar observation was already discussed in regard to the Goldmann equation for calculating the resting potential [Eq. (11.5)], but here one could argue that a 100-mV change (from resting potential up to about 40 mV) can have dramatic effects from the point of view of unbalancing ion concentrations inside the axon. This is not the case, as is shown by the following example.

Example 11.1 Calculate the number of Na$^+$ ions involved in the generation of an action potential inside a small patch of an axon membrane.

Answer Let us assume a capacitance per unit area of 1 μF/cm^2 for a typical neuronal membrane. By using the relationship

$$C = \frac{\Delta Q}{\Delta V} \qquad (E11.1)$$

we can estimate that a change in membrane potential of 100 mV corresponds to a charge movement of 0.001 pC through 1 μm^2 of membrane, corresponding to about 6000 monovalent ions. If we consider the raising phase of the action potential, these 6000 ions can be identified as Na$^+$ ions, entering the axon. Inside a 1-μm^3 volume of cytoplasm there are roughly 3×10^7 Na$^+$ ions, and therefore the spike-induced variation is 2×10^{-2} percent of the sodium present. We can conclude that this variation has no practical effect on the sodium concentration.

Let us overview the neuronal properties considered so far. A "typical" neuron has a soma, several dendrites, and one axon. As a rule, action potentials (or spikes) are generated in the segment of the axon near the cell body and propagate in the axon. The soma of several neurons has the capability of generating an action potential, but the threshold voltage to be reached is much greater than that of the initial segment of the axon, where there is a high density of Na$^+$ channels whose proteic structure makes them voltage-sensitive.[1,5]

Dendrites are usually unable to generate an action potential, even though exceptions have been reported.[1]

In motor neurons and interneurons *input signals* arrive in the form of ionic currents injected into dendrites and soma through synapses made by the axon terminals of other neurons. Sensory neurons receive physical or chemical *input signals* from the environment, which again result in ionic currents. As several patches of the membrane of dendrites and soma receive input currents, an *integration process* takes place and, as a result, the membrane potential of the whole neuron, including its axon, changes with time. If the change is appropriate both in sign and intensity, then it allows the axon membrane potential to overcome the threshold for the *generation* of an action potential, which is then *propagated* through the axonal membrane. Finally, the signal is *transmitted* through *secretion* from the axon terminals to a target cell.

In conclusion, four communication phenomena [input, integration, generation and conduction (action potential), and secretion] happen in four regions of the neuron membrane, as summarized in Fig. 11.7.[1]

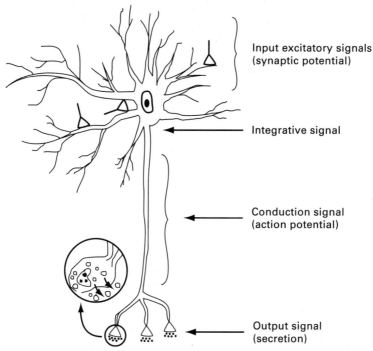

FIGURE 11.7 Four functional elements in a neuron are characterized by four different signals: an input signal (defined as synaptic potential in interneurons and motoneurons and receptor potential in the sensitive neurons), an integrative signal, a conduction signal, and an output signal. In some interneurons the conduction signal can be absent.

Equations (11.22), (11.23), and (11.24) describe in detail the generation of an action potential, i.e., the *firing* of a neuron; Eq. (11.25) describes its propagation. Input and integration aspects are lumped into the current term I_{ext}. Secretion has not yet been taken into account.

11.3 THE NEURON AS A THRESHOLD DEVICE

In the previous section we discussed how to model the outbreak of an action potential. We can conclude that a *single* action potential is a kind of *stereotypical response of any neuron* to any stimulus which is strong enough to allow the membrane potential to overcome a critical threshold. In other words, an action potential is a very efficient way of spreading around, along the nervous tissue, the information that a stimulus or (better) an integration of stimuli has forced the membrane potential of a neuron to cross a threshold. Then, the *coding* of this information is in the rules which dictate how this threshold is crossed and crossed again in time, rather than in the stereotypical shape of the action potential itself. This statement shifts the focus of our study from the detailed characterization of a single action potential to the study of the mechanisms which govern the *sequence* of firings of a network of connected neurons. Toward this goal we first consider simplified versions of the mathematical structure of the H-H equations.

From a mathematical view point, the Hodgkin-Huxley equations are quite complicated to handle. The detailed structure of Eq. (11.22), together with its associated kinetic reactions [Eqs. (11.23) and (11.24)], are necessary for quantitative data fitting, but, on the other hand, some mathematical details are not crucial in order to study the basic properties of the action potential which is, in its essence, a nonlinear oscillator. As already noted, Eq. (11.22) is so constituted that a small increase in V brings about an influx of sodium current that, raising V further, *autocatalyzes* (see Chap. 5) an accelerating increase of V.

Fitzhugh (1961), and Nagumo, Arimoto, and Yoshizawa (1962), derived a simplified version of the H-H model which maintains its essence. The model, known as the *Fitzhugh-Nagumo model*,[7] is based on the following system of nonlinear ordinary differential equations

$$I = f(V) + w \quad (11.27a)$$

$$\frac{dw}{dt} = \gamma V \quad (\gamma > 0) \quad (11.27b)$$

$$f(V) = V(\delta - V)(1 - V) \quad (0 \le \delta \le \tfrac{1}{2}) \quad (11.27c)$$

In Eqs. (11.27) there is only one internal variable, w, replacing the three variables m, h, and n of the H-H model. The permeability function $f(V)$ has the autocatalytic property which assures the above-described autocatalytic mechanism.

More recently, Morris and Lecar[8] provided another simplified version of the H-H model. Even though the model was originally conceived for the barnacle giant muscle fiber, nevertheless the temporal evolution of membrane potential fits quite well many other excitable systems. The model is described by the following equations (note that the Na^+ ion is no longer considered and that, instead, the Ca^{++} ion now plays a fundamental role):

$$C_m \frac{dV}{dt} = \bar{g}_{Cl}(E_{Cl} - V) + \bar{g}_{Ca}m(E_{Ca} - V) + \bar{g}_K n(E_K - V) + I_{ext} \quad (11.28)$$

$$\frac{dm}{dt} = \lambda_m(V)[M_\infty(V) - m] \tag{11.29}$$

$$\frac{dn}{dt} = \lambda_n(V)[N_\infty(V) - n] \tag{11.30}$$

where E_{Cl}, E_{Ca}, E_K are the Nernst equilibrium potentials for Cl^-, Ca^{2+}, and K^+, respectively. The parameters λ_m, λ_n, M_∞, and N_∞ are given by the following relations:

$$\lambda_m = \cosh\left(\frac{V-V_1}{2V_2}\right) \tag{11.31a}$$

$$\lambda_n = \frac{1}{15}\cosh\left(\frac{V-V_3}{2V_4}\right) \tag{11.31b}$$

$$M_\infty = \frac{1}{2}\left[1 + \tanh\left(\frac{V-V_1}{V_2}\right)\right] \tag{11.32a}$$

$$N_\infty = \frac{1}{2}\left[1 + \tanh\left(\frac{V-V_3}{V_4}\right)\right] \tag{11.32b}$$

where $V_1 = -1$ mV; $V_2 = 15$ mV; $V_3 = 10$ mV; $V_4 = 14.5$ mV.

An example of periodic oscillatory behavior of the membrane potential V, caused by a stimulating current is shown in Fig. 11.8.

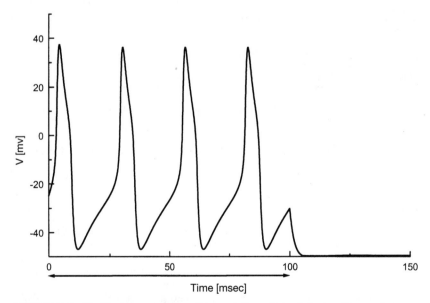

FIGURE 11.8 The Morris-Lecar model predicts a periodic oscillatory behavior of the membrane potential of a neuron stimulated by an appropriate current density I_{ext} (13 μA/cm²). (*Kindly provided by Michele Giugliano.*)

This model can be further simplified by assuming "instantaneous" dynamics for the m variable. Under this condition m can be replaced by its steady value M_∞ and Eq. (11.28) reduces to

$$C_m \frac{dV}{dt} = \overline{g}_{Cl}(E_{Cl} - V) + \overline{g}_{Ca} M_\infty (E_{Ca} - V) + \overline{g}_K n(E_K - V) + I_{ext} \quad (11.33)$$

It is worth summarizing the main features of this model: the *integrative properties* of the membrane are described by Eq. (11.33), which takes into account the membrane capacitive term. On the other hand, *refractoriness* and *action potential generation* are taken into account by Eq. (11.30). No further simplification is possible if we wish to maintain all of these properties (i.e., integration, refractoriness, and action potential shape). On the other hand, if we decide to give up with details about the shape of the action potential, then we can further reduce the model, by preserving the structure of Eq. (11.33) and by setting the membrane potential equal to its resting value every time it overcomes a given threshold. This choice is summarized in the following equations:

$$C \frac{dV}{dt} = g(V_m - V) + I_{ext} \qquad \text{for } V < V_{th} \quad (11.34)$$

$$V = V_m \qquad \text{for } V \geq V_{th} \quad (11.35)$$

Equations (11.34) and (11.35) describe a class of formal neurons known as *integrate-and-fire neurons*, which are a powerful tool used in the field of *formal neural networks*. They are threshold (i.e., nonlinear) devices which reset instantaneously to their resting potential V_m every time I_{ext} reaches an appropriate value. In order to further study them, it is worth considering in more detail the *generation* of I_{ext} and its *effects* on the membrane potential. These two topics will be covered in the next two sections. A rigorous, more detailed treatment of the mathematical operations which transform the H-H model into the integrate-and-fire model is given in Ref. 9.

11.4 SYNAPSES

This section address separately two interlinked questions:

- How chemical synapses work and can be modeled
- How the membrane potential of a neuron surrounded by other cells is affected by its environment

11.4.1 Chemical Synapses

Neuronal signals are transmitted, in an indirect way, from a *presynaptic* cell to a *postsynaptic* cell, in specialized sites of contact known as *synapses*. The two cells are usually separated by a narrow (20 to 30 nm) region of extracellular fluid, known as *synaptic cleft*. When an action potential propagates along the axon of the presynaptic cell, Ca^{2+} channels allow calcium ions to flow from the extracellular medium into the terminal of the presynaptic cell from the synaptic cleft. These ions activate the intracellular machinery that results in the release of signaling molecules, known as *neurotransmitters*, previously stored in *synaptic vesicles*. Neurotransmitters diffuse across the synaptic cleft and bind to ion channels present on the membrane of the postsynaptic cell. The process goes on for a short time (about 1 ms), after which the neurotransmitter is rapidly removed, thus localizing the synaptic event both in time and space. Postsynaptic regions usually are a patch of the membrane of a dendrite or of the soma of a neuron, or of a muscle cell. These mechanisms are illustrated in Fig. 11.9.

As a result of the binding, ion channels are transiently opened, local permeability changes are produced, and hence changes of membrane potentials, whose intensity is re-

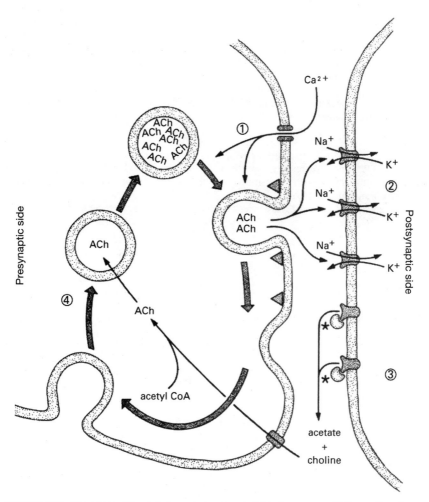

FIGURE 11.9 The process of synaptic transmission at an acetylcholine synapse. (1) Transmitter release. Upon membrane depolarization, Ca^{2+} enters the terminal and facilitates the binding of synaptic vesicles to the presynaptic membrane. Fusion of vesicles with the presynaptic membrane results in the release of transmitter (ACh) into the synaptic cleft. (2) Transmitter action. Once released, ACh diffuses to the postsynaptic membrane and binds to membrane channels that open and allow Na^+ and K^+ to cross the membrane. (3) Transmitter removal. Acetylcholinesterase (labeled with asterisks) breaks down ACh to acetate and choline. (4) Transmitter resynthesis and repackaging. Choline is transported back into the terminal and joined with acetyl coenzyme A (CoA, an activated form of acetate) to make ACh. ACh is concentrated in vesicles that reform by infolding of the terminal membrane. (*From J. E. Dowling.*[12] *Used by permission.*)

lated to how much neurotransmitter is released and how long it persists. If the change is a depolarizing one, an action potential can then be triggered from the synaptic site if a sufficient number of voltage-gated Na$^+$ channels are opened in the axon of the postsynaptic cell as a result of this change.

Of course, the changes in the membrane potential can be either depolarizations or hyperpolarizations, depending on the sign of the ionic current elicited. Synapses resulting in a depolarization (eventually leading to an action potential in the axon of the postsynaptic cell), are said to be *excitatory*. Synapses causing a hyperpolarization are said to be *inhibitory*. Some neurotransmitters, such as *acetylcholine,* have receptors (i.e., channels) on the membrane of the postsynaptic cell that have little selectivity among cations, and the relative contributions of the different cations to the current through the channel depends mostly on their concentration and on their electrochemical gradient. For example, under physiological conditions, the acetylcholine receptors at the neuromuscular junction cause the flow of Na$^+$ ions (up to 30,000 ions per channel each second), with a consequent membrane depolarization of the muscle cell. The behavior of a synapse can be modeled by the kinetic scheme[10]

$$R + T \underset{\beta}{\overset{\alpha}{\rightleftharpoons}} TR^* \tag{11.36}$$

where R, TR*, and T represent, respectively, the unbound, the bound form of the postsynaptic receptor, and the neurotransmitter; α and β are the forward and backward rate constants for transmitter binding.

We can assume that the total concentration of the postsynaptic receptor is fixed, i.e.,

$$[R] + [TR^*] = [A] \tag{11.37}$$

and we can define the fraction r of bound receptors (or, equivalently, of open channels) as

$$r = \frac{[TR^*]}{[A]} \tag{11.38}$$

The time evolution of r is then given by

$$\frac{dr}{dt} = \alpha[T](1-r) - \beta r \tag{11.39}$$

where [T] is the concentration of the transmitter. If we assume that [T] occurs as a rectangular pulse, then we can write

$$[T](t) = [T]_{max}[u(t-t_0) - u(t-t_0-t_1)] \tag{11.40}$$

where $u(t)$ is the step function and $\Delta t = t_1 - t_0$ the duration of neurotransmitter release.

The analytical solution of Eq. (11.39) is

$$r(t) = \begin{cases} [r(t_0) - r_\infty]e^{-(t-t_0)/\tau_r} + r_\infty & t_0 < t < t_1 \\ r(t_1)e^{-\beta(t-t_1)} & t > t_1 \end{cases} \tag{11.41}$$

where

$$r_\infty = \frac{\alpha[T]_{max}}{\alpha[T]_{max} + \beta} \tag{11.42}$$

and

$$\tau_r = \frac{1}{\alpha[T]_{max} + \beta} \tag{11.43}$$

If the binding of transmitter to a postsynaptic receptor directly results in the opening of an ion channel, then the total conductance $G(t)$ through all channels is given by

$$G(t) = \bar{g}_{syn} r(t) \tag{11.44}$$

where $\bar{g}_{syn}$ is the maximum conductance of the synapse.

The synaptic current resulting from the process is

$$I_{syn} = \bar{g}_{syn} r(t)(E_{syn} - V) \tag{11.45}$$

where V is the membrane potential of the postsynaptic cell and E_{syn} is the synaptic *reversal potential*, whose meaning will be better clarified in the next subsection. The process of synaptic transmission, as modeled by the previous equations, is simulated in Fig. 11.10. In Fig. 11.10a, two subsequent action potentials (V_{pre}), generated by the presynaptic neuron, are superimposed on the related fraction $r(t)$ of receptors, present on the membrane of the postsynaptic neuron, which are bound by the neurotransmitter as a result of the presynaptic action potentials. In Fig. 11.10b, the two presynaptic action potentials (V_{pre}) are overimposed on the two resulting changes in postsynaptic potential (V_{post}).

It can be shown[11] that $G(t)$ defined by Eq. (11.44) satisfies the equation

$$\frac{1}{\beta} \frac{dG}{dt} = -G + (G_{sat} - G) \frac{1}{\beta} \alpha Q \delta(t - t_0) \tag{11.46}$$

where α and β have the same meaning as in Eq. (11.36), G_{sat} is the saturation value of the conductance G, $\delta(t - t_0)$ is the Dirac delta function (see Sec. 5.1), and $Q\delta(t - t_0)$ is the neurotransmitter released at time t.

If we further assume that the rise time of the conductance response to an input spike is short enough, then first-order linear dynamics for $G(t)$ can be used, namely

$$\tau_G \frac{dG}{dt} = -G(t) + G_{sat} W E(t) \tag{11.47}$$

where

$$E(t) = \sum_k \delta(t - t_k) \tag{11.48}$$

The quantity $E(t)$ represents the input spike train that drives $G(t)$. The parameter W (with dimension of time) is introduced to model the *efficacy* of the synapse in transforming arriving spikes into variations in conductance. The quantity τ_G is the time constant of the exponential relaxation of $G(t)$ in the absence of external inputs. In the case of a single presynaptic spike, Eq. (11.47) becomes

$$\tau_G \frac{dG}{dt} = -G(t) + G_{sat} W \delta(t - t_0) \tag{11.49}$$

By further considering instantaneous the time dynamics of $G(t)$, we reduce Eq. (11.49) to

$$G(t) = G_{sat} W \delta(t - t_0) \tag{11.50}$$

Equation (11.50) can be introduced into an integrate-and-fire postsynaptic neuron,

$$C_m \frac{dV}{dt} = -G_m V + V_{rev} G_{sat} W \delta(t - t_0) \tag{11.51}$$

where C_m = membrane capacitance

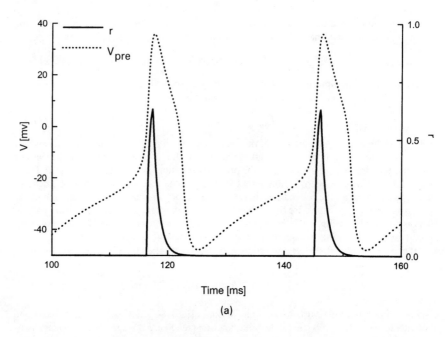

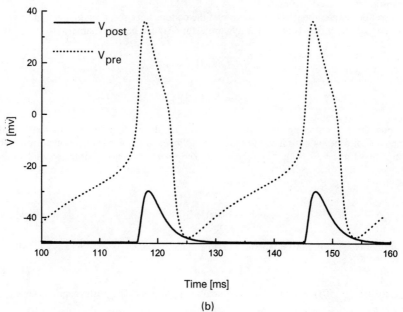

FIGURE 11.10 (a) Two presynaptic action potentials (V_{pre}) and fraction $r(t)$ of neurotransmitter bound receptors on the postsynaptic membrane. (b) The same presynaptic action potentials are compared to the two resulting changes in the postsynaptic membrane potential V_{post}. (*Kindly provided by Michele Giugliano.*)

G_m = conductance
V_{rev} = constant which drastically simplifies $(E_{syn} - V)$

Finally, if the membrane potential $V(t)$ is reduced to a dimensionless variable $\bar{v}(t)$, i.e.,

$$\bar{v}(t) = \frac{V(t)}{V_{th}} \qquad (11.52)$$

we then obtain from Eq. (11.51)

$$\frac{d\bar{v}}{dt} = -\frac{G_m}{C_m}\bar{v} + W^s \delta(t - t_0) \qquad (11.53)$$

where

$$W^s = \frac{V_{rev}}{V(t)} \frac{W}{C_m} G_{sat} \qquad (11.54)$$

Equation (11.53) is a classical model for a neuron of a formal neural network, where the reduced membrane potential $\bar{v}(t)$ is directly driven by input spikes. The efficacy of the transduction operated by the synapse is modeled by the dimensionless coefficient W^s.

After starting with a chemical kinetic description of a synapse [(Eq. (11.36)], we arrived, at the end of a reduction process, at the schemes utilized in the field of formal neural networks [Eq. (11.53)].

Summarizing, the alternative choices of utilizing *ion channel kinetics,* or *synaptic conductance dynamics,* or *scalar synaptic coefficients* are all available to readers, depending on their interest in biological details versus mathematical tractability. Interested readers can find further discussion in Refs. 10 and 11.

Before closing this subsection, it is worth mentioning that another kind of synapse does exist, namely, *electrical synapse*. In this case, there is a direct connection from one neuron to another by way of a channel or *gap junction*. This junction has two effects: it reduces the resistance across the two terminal membranes and it eliminates the electrical shunting by the extracellular fluid. Direct connection implies better transfer of electrical events but less plasticity of the system. Gap junctions are very important in cell synchronization (e.g., the *syncitia* of heart cells) and also for signal processing in the retina.

11.4.2 Excitation, Inhibition, Shunting Inhibition, and Integration

As a result of the synaptic mechanisms described in the previous section, ions move across the membrane of the postsynaptic cell with a consequent variation in its membrane potential [Eq. (11.45)]. *Excitatory postsynaptic potentials* (EPSPs) depolarize the membrane potential. Experimental evidence[12] indicates that EPSPs typically depolarize the membrane potential up to about −10 mV, after which EPSP becomes *hyperpolarizing*. The membrane potential corresponding to the reversing of the effect (i.e., from depolarizing to hyperpolarizing) is known as the *reversal potential* (which has been already utilized in the previous section as E_{syn}). Because none of the ions commonly found inside and outside neurons has an equilibrium potential at −10 mV, it is reasonable to assume that both Na^+ and K^+ contribute to the response. When membrane potential is more negative than −10 mV (i.e., closer to the K^+ equilibrium potential), more Na^+ ions enter the cell than K^+ ions leave it and depolarization results. When membrane potential is more positive than −10 mV (i.e. closer to the Na^+ equilibrium potential), more K^+ ions leave the cell than Na^+ ions enter it, and hyperpolarization results. When membrane voltage is at −10 mV, equal amounts of Na^+ and K^+ ions cross the membrane, and no net potential change occurs. This is the *reversal potential*.

It is possible for membrane channels to allow one ion to flow more easily across the membrane than other ions, and so reversal potentials for EPSPs could be in principle at any value between resting potential (−80 to −70 mV) and Na⁺ equilibrium potential (40 to 55 mV). However, EPSPs generally have reversal potentials between 0 and −20 mV, which indicates that these channels become about equally permeable to Na⁺ and K⁺ ions when they are activated.

Inhibitory postsynaptic potentials (IPSPs) have reverse membrane potentials in the order of −70 mV or even more negative values, suggesting the role of Cl⁻ anions and K⁺ cations.

Occasionally, synaptic terminals make a synapse (T_1) onto another synaptic terminal (T_2). These interactions are often inhibitory and terminal T_2 decreases the EPSP induced by terminal T_1. Evidence suggests that usually terminal T_2 releases a neurotransmitter that opens Cl⁻ channels in the membrane of terminal T_1. As a result, membrane resistance of terminal T_1 decreases because of the membrane's increased permeability, even if the membrane potential is closed to the Cl⁻ equilibrium potential. When a spike subsequently reaches T_1, the voltage across the membrane is reduced because of the lowered membrane resistance. As a result, the current flowing into the terminal as a consequence of the action potential is not modified, but the voltage will be smaller because membrane resistance is lower. Since the amount of neurotransmitter released depends on membrane voltage, the resulting EPSP induced in the neuron postsynaptic to terminal T_1 is reduced. Other situations deal with reduced neuronal responsiveness originated by decreased membrane resistance. This general phenomenon is known as *shunting inhibition*.[12]

All the neurons of the central nervous system are continuously stimulated by thousands of synaptic signals coming from other neurons, some of them excitatory, some others inhibitory. Generally speaking, no presynaptic neuron is able to send a single signal so strong as to allow the axon of the postsynaptic neuron to reach the firing threshold. On the contrary, EPSPs and IPSPs affecting over time a neuron in different regions, with different intensities, are *integrated* by the neuron. Usually, the integration takes place in the axonal segment proximal to the cell body, where the density of voltage-dependent Na⁺ channels is maximal.[1] The result of this integration is called the *grand postsynaptic potential* (grand PSP). If EPSPs predominate, the grand PSP will be a depolarization; if IPSPs predominate, the grand PSP will be a hyperpolarization. The grand PSP is then translated, or *encoded*, into the *frequency* of firing of action potentials by the postsynaptic neuron.

By linking these neurobiological observations to the formal modeling of Sec. 11.4.1, it is possible to write the synaptic current entering a neuron connected to N neurons as[9]

$$I = \frac{1}{2} \sum_{j=1}^{N} W_j (S_j + 1) \qquad (11.55)$$

where W_j are synaptic weights and S_j is a binary variable with possible values +1 and −1, corresponding to an action potential and to the resting potential, respectively.[13]

11.4.3 Membrane Potential Fluctuations

Fluctuations can be further divided into *environmental fluctuations* and *intrinsic fluctuations*. As we already know, a generic term for the sum of currents governing the membrane state can be written as

$$I_i = g_i(V - E_i) \qquad (11.56)$$

Any mechanism which results in a fluctuation either in E_i or in g_i will then result in a transient current with inhibitory/excitatory effects.

Let us first consider the equilibrium potential E_i, which is related to the concentration of the ion i via the well-known equation (Nernst equilibrium potential):

$$E_i = \frac{RT}{z_i \mathscr{F}} \ln \frac{[c_i]_\text{out}}{[c_i]_\text{in}} \tag{11.57}$$

where $[c_i]_\text{out}$ and $[c_i]_\text{in}$ represent the extracellular and intracellular concentrations of ion i, respectively, and z_i is the ion valence. Equation (11.57) implies that any variation in these concentrations will modify the corresponding equilibrium potential, thus generating a current. It is worth observing that the local environment surrounding a neuron (both *in vivo* and *in vitro*) can be densely occupied by other neurons and/or glia cells, which can produce ion fluxes. These fluxes can then result in local values of extracellular ion concentrations which may be transiently different from bulk values, thus causing transient currents in the neuron and therefore spontaneous activity. For example, local increases in $[K^+]_\text{out}$ have been taken into account to model the spontaneous synchronous activity in developing retinas. In this model, K^+ is extruded from the soma of a ganglion cell (i.e., a type of neuron of the retina) during action potential activity. This extrusion increases $[K^+]_\text{out}$, leading to the depolarization of neighbor cells. As a result, excitation is assumed to propagate from cell to cell in a wavelike manner. Other cells then remove K^+ from the extracellular space and reset the environment. More details can be found in Ref. 14.

Let us now consider the conductance g_i as a source of fluctuations. This conductance is a macroscopic quantity resulting from the integration of a large number of microscopic entities (*channels*) which are inherently *stochastic devices*.

This statement should sound familiar to the reader: It may be worth recalling that the concentration of electrons and holes in any semiconductor is the result of generation-recombination phenomena which, at a microscopic level, satisfy a statistical law and do fluctuate over time (the so-called *generation-recombination noise*, see Chap. 2). Generally speaking, any chemical reaction can be interpreted in terms of statistical processes (see Chap. 5). Let us apply these considerations to the sodium and potassium conductances, as described by the Hodgkin-Huxley formalism.

We recall that, according to this formalism, the axon membrane potential satisfies the equation

$$C_m \frac{dV}{dt} = -[\overline{g}_\text{Na} m^3 h(V - E_\text{Na}) + \overline{g}_K n^4(V - E_K) + \overline{g}_l(V - E_l)] + I_\text{ext} \tag{11.58}$$

As already described, in the resting condition most of the sodium channels are closed and their opening is the fundamental event characterizing the rising phase of the action potential. In Eq. (11.58) it is the variable m that controls this rising phase. For this reason and for the sake of simplicity, we will limit our discussion to the stochastic interpretation of the variable m only. Under this approximation, we can assume that the Na channel has three identical m gates, which could be open or shut. This picture gives rise to the four states m_0, m_1, m_2, m_3. The state m_0 is characterized by having all three gates closed, and there is only one way of obtaining it; the state m_1, which corresponds to two closed gates, has three identical substates. The gates being identical, the transition from m_0 to m_1 can be then performed in three ways. This is easily visualized if we indicate with a 0 any closed gate and with a 1 any open one. Then m_0 is identified by (0, 0, 0) and the three equivalent substates of m_1 are given by (1, 0, 0), (0, 1, 0), and (0, 0, 1). State m_0 can be reached from any of these substates in exactly one way. We can translate the previous considerations into the following kinetic scheme which summarizes the transitions between m_0 and m_1:

$$m_0 \underset{\beta_m}{\overset{3\alpha_m}{\rightleftarrows}} m_1 \tag{11.59}$$

where α and β are the rate constants of the Hodgkin-Huxley equations, already described in Sec. 11.2. The reader can verify that the complete chain of transitions is given by[5]

$$m_0 \underset{\beta_m}{\overset{3\alpha_m}{\rightleftarrows}} m_1 \underset{2\beta_m}{\overset{2\alpha_m}{\rightleftarrows}} m_2 \underset{3\beta_m}{\overset{\alpha_m}{\rightleftarrows}} m_3 \qquad (11.60)$$

The m_0, m_1, m_2, m_3 variables can now be interpreted as stochastic ones. In particular, the stochastic variable m_3, which describes the open state, corresponds to the deterministic variable m^3 of the H-H equation. Let us indicate with m_∞^3 the equilibrium value of the deterministic variable m^3. Then, we can assume

$$\langle m_3 \rangle = m_\infty^3 \qquad (11.61)$$

That is, the deterministic variable m_∞^3 corresponds to the ensemble average $\langle m_3 \rangle$ of the stochastic variable m_3. Following this reasoning, and further characterizing the statistical properties of m_3, the deterministic Hodgkin-Huxley equation can be transformed into a stochastic one, where the sodium conductance is allowed to *fluctuate over time* with known statistics. It is worth observing that Eq. (11.58) now resembles a Langevin-type equation, already introduced in describing the Brownian motion (see Sec. 5.2).

The neurobiological implications of these formal considerations are that, given a number of Na channels "not too elevated,"[15] then spontaneous activity can result in a neuron in the absence of any externally stimulating current, just because of "natural" fluctuations in the fraction of open channels. Nature may well use fluctuations to encode information, through stochastic resonance processes.

11.5 NETWORKS

So far we have analyzed the operation of single neurons: how they carry information with action potentials and how they receive/transmit stimuli at synapses. The subsequent step in the analysis is to study, model, and reproduce the way networks of nerve cells produce meaningful patterns of neural activity leading, as the final result, to *behavior*. This is a formidable goal, presently pursued by *integrative neuroscience*. In this section we will just scratch the surface of this topic. Whenever appropriate, indications of books for further reading will be given.

In this section the word *networks* is intended to mean *biological networks*, that is, populations of neurons which, as a part of a nervous system, present *integrative activity*, leading to *behavior*.

Biological networks are, as expected, simpler in invertebrate nervous systems than in the vertebrate ones. In invertebrates, brains consist of much fewer nerve cells than are found in vertebrates. Moreover, invertebrate neurons are larger and typically organized into discrete aggregates, called *ganglia*, which are distributed throughout the animal. Much activity in simple invertebrates is *rhythmic*. The neural circuitry that underlies a rhythmic activity is termed a *central pattern generator* (CPG). For example, the central pattern generator controlling swimming in the sea slug *Tritonia* is based on 12 interneurons, which can be classified into three types of cells connected by inhibitory and excitatory synapses. The leech (*Hirudo medicinalis*) presents another example of widely studied CPGs.[12] The leech nervous system has on the order of 10^4 neurons, distributed among 21 ganglia. The heartbeat of the leech is controlled by a well-analyzed CPG. In this animal, blood is pumped through the circulatory system by two contractile lateral vessels, or heart tubes, one on each side of the animal. The periodic contraction of the muscles on the walls of the heart tubes is the heartbeat that provides the pumping action that circulates the blood. Motor neurons of ganglia (from the third to eighteenth) innervate the heart tubes. All the interneuronal synapses in the circuit are inhibitory.

Model networks connecting three or four neurons via excitatory/inhibitory synapses have been proved to display rhythmic activity. An example is given in Fig. 11.11a, where a network of three neurons, N_1, N_2, and N_3, is considered. The electrophysiological activity of the three neurons can be described by H-H equations, resulting in the patterns of spikes shown in Fig. 11.11b. The synaptic connections of the network (Fig. 11.11a) are as follows: Neurons N_1 and N_2 are linked by reciprocally excitatory synaptic connections, and neuron N_3, in turn, is linked by inhibitory synaptic connections to N_1 and N_2. Neuron

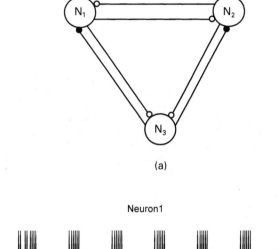

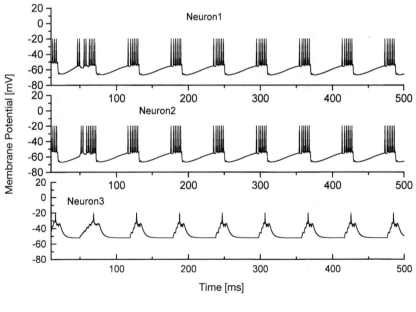

FIGURE 11.11 (a) A three-neuron network. (b) The resulting behavior. (○ = inhibitory synapsis; ● = excitatory synapsis.) (*Kindly provided by Marco Bove*).

N_3 has a high threshold for action potential generation, which is reached only when N_3 receives a high level of excitatory input due to high-frequency spike activity in N_1 and N_2. The resulting circuit generates a rhythm of concurrent action potential bursts in N_1 and N_2 (see Fig. 11.11b), provided by the capability of N_3 to hyperpolarize N_1 and N_2. Detailed discussions of these kinds of networks can be found in Refs. 16 and 17, with specific reference to the leech and to the sea slug, respectively.

More complex functions, such as *memory* and *learning,* have been studied in invertebrates. For example, the phenomenon of *abituation* has been extensively studied in *Aplysia californica.* Further analysis of these topics can be found in Refs. 2 and 12.

The vertebrate central nervous system is far more complex than the invertebrate one. It consists of the spinal cord and the brain[12] (see Fig. 11.12).

Higher neural processing is carried out within the brain; the spinal cord contains the circuitry for rhythmic motor activity, mediates many reflexes, and serves to conduct the information flow from the brain to the periphery of the body and vice versa. Most of the sensory information entering the spinal cord is somatosensory; other sensory information enters the brain directly via the cranial nerves (from the eye and ear, for example). The cranial nerves also carry motor information that controls eye, face, and tongue movements and visceral function. Motor information controlling the rest of the body passes down the spinal cord and out to the periphery via motor nerves.

Dorsal roots allow sensory information to enter the spinal cord. Ventral roots let motor information exit the spinal cord.

The brain can be conveniently divided into three regions: the hindbrain, midbrain, and forebrain. The forebrain is far more developed in mammals and especially in primates than it is in nonmammalian species. The forebrain in mammals is the site of more complex sensory processing, sensory-motor integration, and motor initiation, whereas in nonmammalian species these activities are centered in the midbrain. The hindbrain consists of the medulla, pons, and cerebellum. The medulla contains numerous axon tracts that carry information between higher brain centers and the spinal cord. The pons contains neurons that provide information from the cortex to the cerebellum, and the cerebellum coordinates and integrates motor activity. It receives information from all sensory systems, as well as input from the cortex concerning the body's movement.

The hypothalamus is part of the forebrain. It is a regulatory center. The hypothalamus is concerned with basic controls and acts, such as eating, drinking, sexual behavior, body temperature, and so forth. Certain hypothalamic neurons release small peptides into the circulation that act as hormones, either by regulating the release of other hormones from the pituitary gland or by acting directly on target tissues. The thalamus, like the hypothalamus, is part of the forebrain. It provides mainly sensory information to the cerebral cortex, and also information about ongoing motor activity.

The cerebral cortex, in the forebrain, is the seat of sensation, consciousness, memory, and intelligence in humans. Skilled movement also originates in the cortex. The cortex is divided into four lobes, each concerned with specific functions. The frontal lobe deals with movement and smell, the parietal lobe with somatic sensation, the occipital lobe with vision, and the temporal lobe with audition and memory. Models of the brain are out of the scope of the book and will not be discussed.

11.6 NEUROBIOENGINEERING: NEUROELECTRONIC JUNCTIONS

Cultures of dissociated neurons constitute a promising method for characterizing the autoorganization properties of populations of neurons under controlled physicochemical

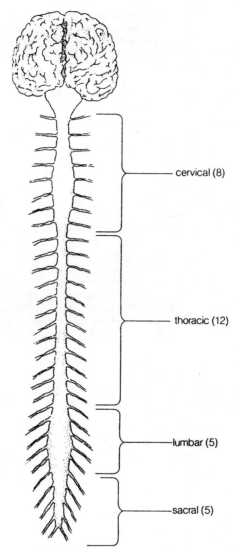

FIGURE 11.12 Schematic diagram of the central nervous system. (*From J. E. Dowling.*[12] *Used by permission.*)

conditions. Neurons can survive for weeks in culture, where they reorganize into two-dimensional networks. Especially in the case of populations obtained from vertebrate embryos, these networks cannot be regarded as faithful reproductions of *in vivo* situations, but rather as new rudimentary neurobiological systems whose activity can change over time spontaneously or as a consequence of chemical/physical stimuli. A technique appropriate for recording the electrical activity of networks of cultured neurons lies in using substrate transducers, i.e., arrays of planar microtransducers forming the adhesion surface for the reorganizing networks. This nonconventional electrophysiological method has several advantages, over standard intracellular recording, that are related to the possibility of monitoring/stimulating noninvasively the electrochemical activities of several cells, independently and simultaneously for a long time.[18]

This technique is a valuable proof of cross-fertilization between microelectronic technologies and neuroscience and it represents a nice example of the bioelectronic approach. Let us briefly review its key features in comparison with other more traditional electrophysiological methods. Intracellular and patch-clamp recordings are single-cell electrophysiological techniques requiring that a thin glass capillary be brought near a cell membrane. Intracellular recording involves a localized rupture of the cell membrane. Patch-clamp methods can imply the rupture and (possible) isolation of a small membrane patch or, in the case of the so-called "whole-cell-loose-patch configuration," the sealing between the microelectrode tip and the membrane surface.[2,12]

The use of planar substrate microtransducers presents some similarities to the latter technique, in the sense that it involves the sealing between the neuron membrane and the underlying planar microtransducer surface the neuron is growing on.

Peculiar to this technique are at least two features: (1) Several (i.e., tens to hundreds) neurons are brought into simultaneous contact with several underlying microtransducers, with a neuron-to-microtransducer correspondence which can be supported by mechanical and/or chemical means. Simultaneous multisite recording from units at well-localized positions is a unique feature of this technique, though an exact one-to-one coupling between neurons and electrodes is not always feasible. (2) Recording/stimulation can be protracted for days. During this period, which is very long compared to the typical time intervals allowed by intracellular techniques, the neuronal population in culture is continuously developing and synaptic contacts change in the presence of different physiological conditions, producing changes in the network functions and dynamics.[18]

Two possible experimental situations are illustrated in Fig. 11.13. Figure 11.13a shows a network of 6 neurons guided to perfectly contact 6 underlying planar microelectrodes, and Fig. 11.13b shows a larger network of neurons randomly growing on the top of the microelectrodes.

The neuron-to-transducer junction can be appropriately characterized by using an equivalent-circuit approach, as previously done for other, more traditional, electrophysiological methods. Such a characterization is especially effective, in consideration of the above-mentioned feature 2, which implies that, during the very long period (days) of an experiment, neurons are alive (i.e., possibly showing continuous changes in shape, adhesion, and arborizations) on top of the recording transducer; therefore, the sealing might change over time, with possible variations in the recorded signals. Most of the data reported in the literature refer to noble metal microelectrodes, i.e., *passive transducers*. An interesting exception lies in the use of field-effect transistors deprived of the gate metalization, that is, the H^+-FET sensors already considered in Chap. 10. These two distinct categories of microtransducers will therefore be considered in the following.

The coupling strength between a neuron and a planar metallic substrate microelectrode is a very critical parameter in determining the "quality" (shape and amplitude) of a recorded signal.[19] To systematically address this issue, a circuit model can be developed to simulate a nerve cell membrane patch coupled to a noble metal planar microelectrode. Such a circuit model is shown in Fig. 11.14.

The meanings of the symbols are the following: R_i denotes the cytoplasmic resistance connecting two adjacent compartments; R_{Cl} and V_{Cl} are the chloride resistance and the chloride equilibrium potential, respectively; I_{act} represents the sum of the sodium, potassium, and calcium currents; C_{me} is the cell membrane-electrolyte capacitance; R_{seal} and

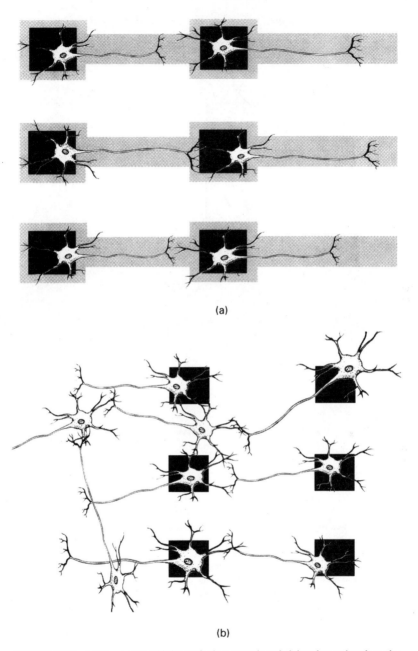

FIGURE 11.13 (a) Six neurons guided to perfectly contact six underlying planar microelectrodes and to contact each other (via microchannels) in a predetermined way. (b) A population of neurons developing a spontaneous network randomly contacting six underlying planar microelectrodes. (*Kindly provided by Monica Fortin.*)

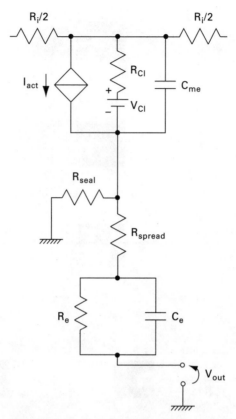

FIGURE 11.14 Equivalent circuit of the junction between a neuron membrane patch and a planar noble metal microtransducer.

R_{spread} denote the sealing resistance between a cell and a microelectrode and the spreading resistance, respectively; R_e and C_e are the equivalent elements of the electrode-electrolyte interface. The sealing impedance is represented by the RC coupling circuit consisting of C_{me}, R_{seal}, and R_{spread}.

The capacitor C_{me} models the capacitive component of the membrane-electrolyte interface and it can be represented as a series of at least three equivalent capacitances (see Fig. 11.15): the membrane capacitance C_m, the Helmholtz capacitance C_h (i.e., the capacitance modeling the inner Helmholtz plane (IHP) and the outer Helmholtz plane (OHP) layers just outside the cell membrane) and the diffuse layer capacitance C_d, (i.e., the capacitance modeling the diffuse layer of ions in the solution), as seen in Chap. 7. The resistance R_{seal} models the resistive component of the thin layer of solution between the cell membrane and the microelectrode. Given the model, computer simulations of the signal transduction operated by the microelectrode can be produced (see Chap. 12).

Figure 11.16a shows a simulation result for $C_{me} = 0.7$ pF and $R_{seal} = 0.2$ MΩ, which corresponds to a weak coupling between a membrane patch of the neuron and a microelectrode. For this coupling, we can hypothesize that the small C_d dominates, and that

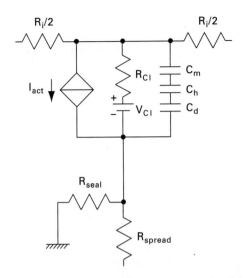

FIGURE 11.15 A detailed circuital representation of the neuron-electrode junction of Fig. 11.14.

the total capacitance is lower than the capacitance of the membrane itself. The amplitudes and durations of signals increase as the values of the parameters C_{me} and R_{seal} increase, too. Simulation results corresponding to increasingly stronger contact conditions between a neuronal membrane patch and a microelectrode for $C_{me} = 1$ pF and $R_{seal} = 1$ MΩ and for $C_{me} = 1.4$ pF and $R_{seal} = 4$ MΩ are shown in Fig. 11.16b and c, respectively.

Whenever an "insulated" FET is utilized as a transducer, the sealing region is bounded by the patch of membrane on one side and by the surface of an insulator (typically Si_3N_4 or SiO_2) on the other side. The electrophysiological use of insulated (or ion-sensitive) FETs was proposed by Bergveld[20] in a pioneering paper, and again later by Fromhertz et al.,[21] and it was simulated by Grattarola et al.[22]

To characterize this neuroelectronic interface, let us introduce the following parameters:

1. *Thickness d.* The thickness d is the (average) patch-to-insulator distance. It affects the sealing resistance R_{seal} through the relation

$$R_{seal} = \frac{\rho_{seal}}{d} \frac{l}{w} \qquad (11.62)$$

where ρ_{seal} is the sealing resistivity and l and w are the length and width of the portion of the insulated FET coupled to the patch of neuronal membrane.

2. *Resistivity of the sealing region ρ_{seal}.* For a weak sealing, we can consider a ρ_{seal} value typical of an electrolyte solution (i.e., 0.7 Ω-m). For a tight sealing, we can assume that a considerable portion of space is occupied by the cell glycocalix and its associated fixed charge; therefore a larger value of ρ_{seal} (e.g., 1 to 5 Ω-m) can be utilized.

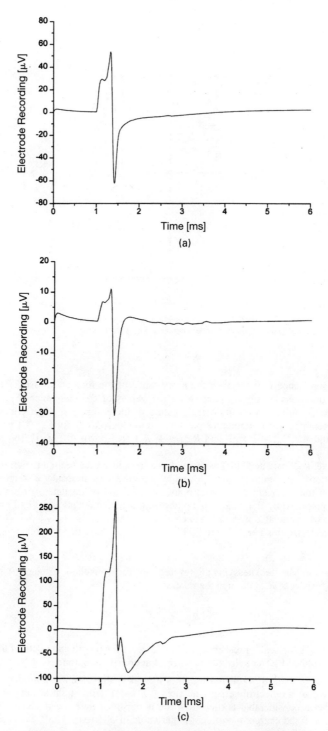

FIGURE 11.16 Simulation results for different values of the sealing impedance: (*a*) $C_{me} = 0.7$ pF and $R_{seal} = 0.2$ MΩ; (*b*) $C_{me} = 1$ pF and $R_{seal} = 1$ MΩ; (*c*) $C_{me} = 1.4$ pF and $R_{seal} = 4$ MΩ.

3. *Water permittivity* ε. Water permittivity affects the inner Helmholtz plane, the outer Helmholtz plane, and diffuse layer capacitances, which model the polarization of the electrolyte solution in front of the membrane patch and in front of the insulator through the relation,

$$C_h = \frac{\varepsilon_{IHP}\varepsilon_{OHP}}{\varepsilon_{OHP}d_{IHP} + \varepsilon_{IHP}d_{OHP}} \qquad (11.63)$$

where ε_{IHP} and ε_{OHP} are the dielectric permittivities of the inner and the outer Helmholtz planes, respectively (the values 6 and 32 are assumed,[23] respectively, for the relative permittivities); $d_{IHP} = 0.2$ nm, and $d_{OHP} = 0.7$ nm are the insulator–nonhydrated ion and the insulator-hydrated ion distances, respectively. A value $\varepsilon_w = 78.5$ is assumed for the relative permittivity of the bulk solution.

Figure 11.17 shows the equivalent circuit of the patch-to-FET sealing; the meanings of the symbols are the same as those in Fig. 11.14, and V_{GS} is the bias voltage of the insulated FET. The neuron membrane can be described according to the Hodgkin-Huxley model (see Sec. 11.2).

Figure 11.18 gives the results of simulations of the transduction of action potentials obtained by considering different sealing conditions between the neuron and the transistor.

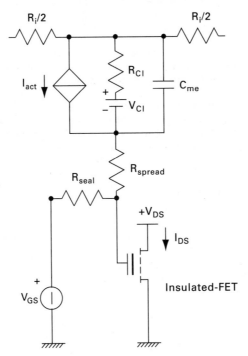

FIGURE 11.17 Equivalent circuit of the junction between a patch of neuronal membrane (described by the compartment model) and an insulated FET microtransducer.

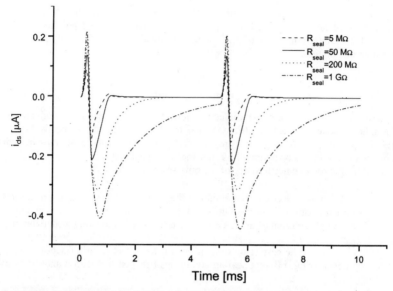

FIGURE 11.18 Simulation of action potentials transduced by the insulated FET for different sealing conditions (R_{seal} = 5 MΩ, 50 MΩ, 200 MΩ, and 1 GΩ).

Finally, Fig. 11.19 shows the effects induced on the recorded signals by variations in the densities of the ionic channels of the membrane patch coupled to the transistor. The simulations were based on the following alternative assumptions:

1. A very tight sealing does preclude channel opening.
2. Under looser sealing conditions, channel migration phenomena are induced by the presence of a component of the local electric field parallel to the membrane patch, thus causing a local increase in the density of the ionic channels.[24]

11.7 SILICON NEURONS

As pointed out by C. Mead,[25] there are intriguing similarities between the physical mechanisms inherent in the behavior of a (subthreshold-mode) MOSFET (see Chap. 9) and those underlying the behavior of a neuron. In a broad sense, both devices are characterized by having channel conductances controlled by a voltage. More specifically, as discussed in Chap. 12, the voltage-dependent conductances of neuronal membranes have a conductance-voltage relation shape which is similar to the current-voltage relation generated by MOSFETs arranged to form a *differential pair* (see Chaps. 9 and 12).

Starting from this statement, simple CMOS circuits can be designed which faithfully reproduce the sodium and potassium currents occurring in a neuron membrane in correspondence to an action potential. In other words, unlike the hardware of "formal" neural networks, the basic circuits of the silicon neuron are intended to emulate biological behavior, rather than simulate biology by means of conventional design principles. This approach has several implications. A complete silicon neuron can occupy less than 0.1 mm^2

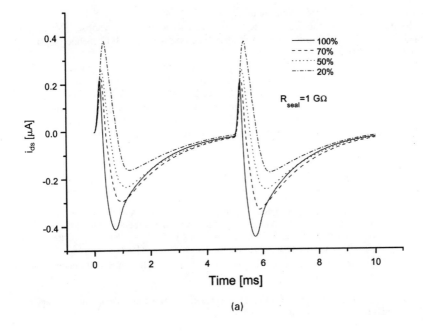

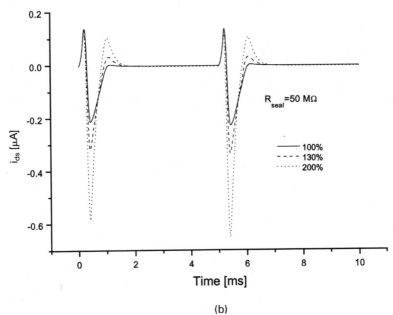

FIGURE 11.19 Simulation of action potentials transduced by the insulated FET (*a*) for a constant tight sealing ($R_{seal} = 1$ GΩ) and decreasing densities of the ionic channels. (*b*) For a constant loose sealing ($R_{seal} = 50$ MΩ) and increasing densities of the ionic channel.

and a linear array of 100 silicon neurons could be fabricated on a 1 × 1-cm die, with the second dimension of the die left free for dendritic structures and synapses.[26,27] In principle, a silicon neuron could operate with the speed typical of CMOS technology, i.e., 10^6 times faster than a biological neuron.

As a result, it can be foreseen that in the future, chips containing thousands of neurons could result into tools for neurobiologists, as they allow the user to simulate on a single chip, in a quite short time, the effects induced on the network by various biologically meaningful synaptic contacts, quickly generated by on-chip reprogramming the network organization. Moreover, from the neural network viewpoint, the use of silicon neurons could result in general-purpose, biologically inspired, hardware architectures.

PROBLEMS

11.1 *a.* Consider a patch of membrane under resting conditions. Assume that the ions to be considered are Na and K. On the basis of the following numerical values: $V_m = -68.8$ mV; $g_{Na} = 0.5 \times 10^{-6}$ S; $E_{Na} = 55$ mV; $g_K = 10 \times 10^{-6}$ S; $E_K = -75$ mV, calculate the sodium current I_{Na} and the potassium current I_K (approximate to the picoampere).

b. On the basis of the previous results, discuss the need of introducing two pump-driven currents, $I_{p,Na}$ and $I_{p,K}$.

11.2 With reference to the data of Fig. 11.5, calculate (for $M = 1$ mV) the fraction a of open channels for $V = -35$ mV and $V = -25$ mV.

11.3 By making use of Eq. (E11.1), and by assuming $C_m = 1$ μF/cm², calculate the number of charges separated by a spherical cell with diameter of 20 μm and membrane potential of -60 mV.

11.4 Calculate the membrane potential $V = V(t)$ by solving the differential equation

$$C_m \frac{dV}{dt} = -g(V - E) + I$$

where C_m, g, E, I are given constants and the initial condition is $V = V_m$.

11.5 Calculate the average membrane potential $\langle V \rangle$ and its variance $\langle V^2 \rangle - \langle V \rangle^2$ by solving the Langevin-type equation

$$\frac{dV}{dt} = -\beta V + A$$

where A is a white noise and the initial condition is $V = V_m$. (*Hint:* Read again Sec. 5.2.)

REFERENCES

1. E. R. Kandel, J. H. Schwartz, and T. M. Jessell, *Principles of neural science*, 3d ed., New York: Elsevier Science, 1991.
2. J. C. Nicholls, A. R. Martin, B. G. Wallace, *From neuron to brain*, 3d ed., Sunderland, Mass.: Sinauer Associates, 1992.
3. R. McKay, "Stem cells in the central nervous system," *Science*, 276(4): 66–71, 1997.
4. A. L. Hodgkin and A. F. Huxley, "A quantitative description of membrane current and its application to conduction and excitation in nerve," *J. Physiol.*, 117: 500–544, 1952.

5. B. Hille, *Ionic channels of excitable membranes,* Sunderland, MA: Sinauer Associates, 1992.
6. D. Johnston and S. Miao-Sin Wu, *Foundations of cellular neurophysiology,* Cambridge, Mass.: MIT Press, 1995.
7. L. A. Segel (ed.), *Mathematical models in molecular and cellular biology,* New York: Cambridge University Press, 1980.
8. C. Morris and H. Lecar, "Voltage oscillations in the barnacle giant muscle fiber," *Biophys. J.,* 35: 193–213, 1981.
9. L. F. Abbott and T. B. Kepler, "Model neurons: from Hodgkin-Huxley to Hopfield," in *Statistical mechanics of neural networks,* L. Garrido (ed.), pp. 5–18, Barcelona: Springer Verlag, 1990.
10. A. Destexhe, Z. F. Mainen, and T. J. Sejnowski, "An efficient method for computing synaptic conductances based on a kinetic model of receptor binding," *Neural Comp.,* 6: 14–18, 1994.
11. F. Chapeau-Blondeau and N. Chambet, "Synapse models for neural networks: from ion channel kinetics to multiplicative coefficient w_{ij}," *Neural Comp.,* 7: 713–734, 1995.
12. J. E. Dowling, *Neurons and networks,* Cambridge, Mass.: The Belknap Press of Harvard University Press, 1992.
13. J. J. Hopfield, "Neural networks and physical systems with emergent collective computational abilities," *Proc. Natl. Acad. Sci. USA,* 81: 3088–3092, 1982.
14. P. Y. Burgi and N. M. Grzywacz, "Model based on extracellular potassium for spontaneous synchronous activity in developing retinas," *Neural Comp.,* 6: 983–1004, 1994.
15. C. C. Chown and J. A. White, "Spontaneous action potentials due to channel fluctuations," *Biophys. J.,* 71: 3013–3021, 1996.
16. G. S. Stent and W. B. Kristan, Jr., "Neural circuits generating rhythmic movements," in *Neurobiology of the leech,* K. J. Muller, J. G. Nicholls, G. S. Stent (eds.), New York: Cold Spring Harbor Laboratory, 1981.
17. P. A. Getting, "Reconstruction of small neural networks," in *Methods in neuronal modeling: from synapses to networks,* C. Koch and I. Segev (eds.), Cambridge, Mass.: MIT Press, 1989.
18. G. W. Gross, J. M. Kowalski, "Experimental and theoretical analysis of random nerve cell network dynamics," in *Neural Networks,* Vol. 4, P. Antognetti and V. Milutinovic (eds.), Englewood Cliffs, N.J.: Prentice Hall, 1991.
19. M. Grattarola and S. Martinoia, "Modeling the neuron-microtransducer junction: from extracellular to patch recording," *IEEE Trans. Biomed. Eng.,* BME-40: 35–41, 1993.
20. P. Bergveld, "Development, operation and application of the ion-sensitive field-effect transistor as a tool for electrophysiology," *IEEE Trans. Biomed. Eng.,* BME-19: 342–351, 1972.
21. P. Fromhertz, A. Offenhauser, T. Vetter, and J. Weis, "A neuron-silicon junction: a Retzius cell of the leech on an insulated-gate-field effect transistor," *Science,* 252: 1290–1293, 1991.
22. M. Grattarola, S. Martinoia, G. Massobrio, A. Cambiaso, R. Rosichini, and M. Tetti, "Computer simulations of the responses of passive and active integrated microbiosensors to cell activity," *Sensors and Actuators,* B4: 261–265, 1991.
23. O. M. Bockris and A. K. N. Reddy, *Modern electrochemistry,* Vol. 2, 3d ed., New York: Plenum Press, 1977.
24. P. Fromhertz, "Self-organization of the fluid mosaic of charged channel proteins in membranes," *Proc. Natl. Acad. Sci. USA,* 85: 6353–6357, 1988.
25. C. Mead, *Analog VLSI and neural systems,* Reading, Mass.: Addison-Wesley, 1989.
26. M. Mahowald and R. Douglas, "A silicon neuron," *Nature,* 354: 515–518, 1991.
27. M. Grattarola, M. Bove, S. Martinoia, and G. Massobrio, "Silicon neuron simulation with SPICE: tool for neurobiology and neural networks," *Med. Biol. Eng. & Comput.,* 33: 533–536, 1995.

CHAPTER 12
MODELS OF BIOELECTRONIC DEVICES AND COMPUTER SIMULATIONS

For almost all bioelectronic devices, silicon technology is one of the most promising for their development. Silicon technology offers the advantage of integration, i.e., the opportunity to integrate several different devices and the related circuits for signal amplification and processing on the same chip. Moreover, powerful simulation tools are available which, originally introduced for designing electronic circuits, can be adapted to design silicon-based bioelectronic devices. These tools will be considered in the following, with reference to the modifications made in the program SPICE in order to simulate chemically sensitive field-effect transistors (CHEMFETs), biosensors (BIOFETs), and neurons.

12.1 SPICE SIMULATOR

SPICE (an acronym which stands for *Simulation Program with Integrated Circuit Emphasis*)[1,2] is a general-purpose electronic circuit simulation program. SPICE can perform various analyses of electronic circuits, e.g., the operating point of transistors, time-domain response, small-signal frequency response, temperature and noise analysis. SPICE contains models for the common circuit elements, active as well as passive, and it is able to simulate most electronic circuits. The input syntax for SPICE is a free-format type that does not need data to be entered in fixed column locations. SPICE assumes reasonable default values for the device model parameters that are not specified in the .MODEL statement. In addition, it provides error checking to ensure that the circuit has been entered correctly.

The circuit to be analyzed is described to SPICE by a set of element statements which define the circuit topology and element values, and a set of control statements which define the model parameters and the run controls. Each element in the circuit is specified by an element statement that contains the element name, the circuit nodes to which the element is connected, and the values of the parameters that determine the electrical characteristics of the element. The first letter of the element name specifies the element type.

SPICE is a versatile program and is widely used both in industries and universities. Originally, SPICE was available only on mainframe computers; now it runs also on PCs.

In this chapter we describe how SPICE can be used to simulate the behavior of bioelectronic devices and biological systems such as biosensors and neurons and to design small networks of such neurons. Since the considered bioelectronic devices are mainly based on the MOSFET structure, we first present a summary of the MOSFET models implemented in SPICE. A wide and complete treatment of the semiconductor device models implemented in SPICE, as well as in other SPICE-based commercially available versions of the program, can be found in Ref. 3.

12.2 MOSFET MODELS IN SPICE

SPICE model of an n-channel MOSFET is shown in Fig. 12.1. The static (or dc) model and the small-signal model obtained from Fig. 12.1 and implemented in SPICE, are shown in Figs. 12.2 and 12.3, respectively.

SPICE model equations of the MOSFET are described in Ref. 3, which the reader can refer to.

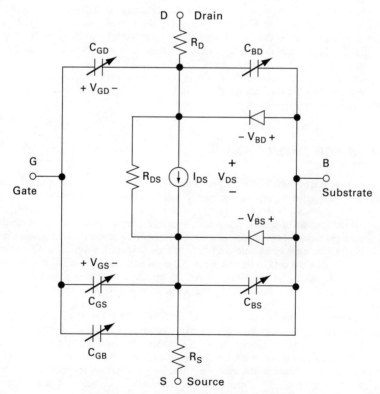

FIGURE 12.1 SPICE n-channel MOSFET model.

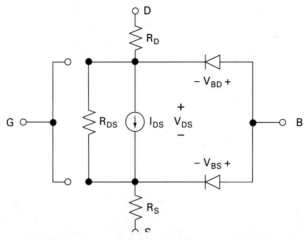

FIGURE 12.2 Static *n*-channel MOSFET model in SPICE.

The model statement of an *n*-channel MOSFET has the form

.MODEL   MNAME   NMOS   (P1=V1   P2=V2   P3=V3   ...   PN=VN)

and the statement of a *p*-channel MOSFET has the form

.MODEL   MNAME   PMOS   (P1=V1   P2=V2   P3=V3   ...   PN=VN)

where MNAME is the model name; NMOS and PMOS are the keys of *n*-channel and *p*-channel type MOSFETs, respectively; P1, P2, ... and V1, V2, ... are the model parameters and their values, respectively.

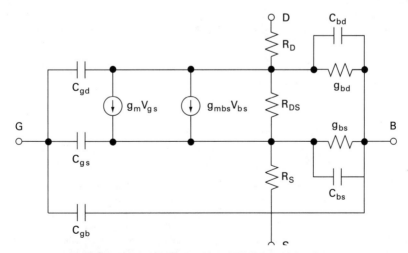

FIGURE 12.3 Small-signal *n*-channel MOSFET model in SPICE.

The key for the MOSFET is M. Thus, the name of a MOSFET must begin with M followed by an alphanumeric one to eight character string, and takes the general form of

M(name) ND NG NS NB MNAME
+ [L=(value)]   [W=(value)]
+ [AD=(value)]   [AS=(value)]
+ [PD=(value)]   [PS=(value)]
+ [NRD=(value)]   [NRS=(value)]
+ [NRG=(value)]   [NRB=(value)]
+ [IC=VDS, VGS, VBS]

where ND, NG, NS, and NB are the drain, gate, source, and substrate (or bulk) nodes, respectively. MNAME is the model name. Positive current flows into a terminal; i.e., the current flows from the drain node through the device to the source node for an n-channel MOSFET.

The model parameters of the MOSFET and their default values assigned by SPICE (when they are not specified in the .MODEL statement) are given in Table 12.1.

With reference to Table 12.1 and the MOSFET name statement, L and W are the channel length and width of the MOSFET, respectively, and AD and AS are the drain and source diffusion areas. The length L is decreased by twice LD to get the effective channel length. The width W is decreased by twice WD to get the effective channel width. The values of L and W can be specified on the device, the model, or on the .OPTIONS statement. The value on the device supersedes the value on the model, which supersedes the value on the .OPTIONS statement.[4] The parameters PD and PS are the drain and source diffusion perimeters. The drain-substrate and source-substrate saturation currents can be specified either by JS, which is mulitplied by AD and AS, or by IS, which is an absolute value. The zero-bias depletion capacitances can be specified by CJ, which is multiplied by AD and AS, and by CJSW, which is multiplied by PD and PS. Alternatively, these capacitances can be set by CBD and CBS, which are absolute values.

The MOSFET is modeled as an intrinsic MOSFET with ohmic resistances in series with the drain, source, gate, and substrate. In some versions, there is also a shunt resistance RDS in parallel with the drain-source channel. The parameters NRD, NRS, NRG, and NRB are the relative resistivities of the drain, source, gate, and substrate in squares. These parasitic (ohmic) resistances can be specified either by RSH, which is multiplied by NRD, NRS, NRG, and NRB, respectively; alternatively, the absolute values of RD, RS, RG, and RB can be specified directly. The parameters NRB and RB are present only in early versions of the code. The parameters PD and PS default to 0; NRD and NRS default to 1; NRG and NRB default to 0. Default values for L, W, AD, and AS can be set in the .OPTIONS statement. If AD or AS default values are not set, they also default to 0. If L or W default values are not set, they default to 100 μm.

The dc characteristics of MOSFETs are defined by the parameters VTO, KP, LAMBDA, PHI, and GAMMA, which can be computed by SPICE by using the fabrication-process parameters NSUB, TOX, NSS, NFS, TPG, UO, etc.[3] The values of VTO, KP, LAMBDA, PHI, and GAMMA, which are specified on the .MODEL statement, supersede the values calculated by SPICE from fabrication-process parameters. The parameter VTO is positive for n-channel enhancement-mode MOSFETs and for p-channel depletion-mode MOSFETs; VTO is negative for p-channel enhancement-mode MOSFETs and for n-channel depletion-mode MOSFETs.

SPICE provides three MOSFET basic models.[3] The LEVEL parameter selects among different models for the MOSFET. If LEVEL = 1, the Schichman-Hodges model is used. If LEVEL = 2, an advanced version of the Schichman-Hodges model, which is a geometry-based analytical model and contains extensive second-order effects, is used. If LEVEL

TABLE 12.1 Model Parameters of the MOSFET in SPICE (pn = pn Junction)

Name	Model parameters	Units	Default	Example
LEVEL	Model type (1, 2, or 3)		1	
L	Channel length	m	DEFL	
W	Channel width	m	DEFW	
LD	Lateral diffusion length	m	0	
WD	Lateral diffusion width	m	0	
VTO	Zero-bias threshold voltage	V	0	0.1
KP	Transconductance	A/V^2	2×10^{-5}	2.5×10^{-5}
GAMMA	Bulk threshold parameter	V$^{1/2}$	0	0.35
PHI	Surface potential	V	0.6	0.65
LAMBDA	Channel-length modulation (LEVEL = 1 or 2)	V^{-1}	0	0.02
RD	Drain ohmic resistance	Ω	0	10
RS	Source ohmic resistance	Ω	0	10
RG	Gate ohmic resistance	Ω	0	1
RB	Bulk ohmic resistance	Ω	0	1
RDS	Drain-source shunt resistance	Ω	∞	
RSH	Drain and source diffusion sheet resistance	Ω/square	0	20
IS	Bulk pn saturation current	A	1×10^{-14}	1×10^{-15}
JS	Bulk pn saturation current/area	A/m^2	0	1×10^{-8}
PB	Bulk pn potential	V	0.8	0.75
CBD	Bulk-drain zero-bias pn capacitance	F	0	5×10^{-12}
CBS	Bulk-source zero-bias pn capacitance	F	0	2×10^{-12}
CJ	Bulk pn zero-bias bottom capacitance/area	F/m^2	0	
CJSW	Bulk pn zero-bias perimeter capacitance/length	F/m	0	
MJ	Bulk pn bottom grading coefficient		0.5	
MJSW	Bulk pn sidewall grading coefficient		0.33	
FC	Bulk pn forward-bias capacitance coefficient		0.5	
CGSO	Gate-source overlap capacitance/channel width	F/m	0	
CGDO	Gate-drain overlap capacitance/channel width	F/m	0	
CGBO	Gate-bulk overlap capacitance/channel length	F/m	0	
NSUB	Bulk doping density	cm^{-3}	0	
NSS	Surface state density	cm^{-2}	0	
NFS	Fast surface state density	cm^{-2}	0	
TOX	Oxide thickness	m	1×10^{-7}	
TPG	Gate material type: +1 = opposite of substrate −1 = same as substrate 0 = aluminum		+1	
XJ	Metallurgical junction depth	m	0	
UO	Surface mobility	cm^2/V-s	600	
UCRIT	Mobility degradation critical field (LEVEL =2)	V/cm	1×10^4	
UEXP	Mobility degradation exponent (LEVEL =2)		0	
VMAX	Maximum drift velocity of carriers	m/s	0	
NEFF	Channel charge coefficient (LEVEL =2)		1	
XQC	Fraction of channel charge attributed to drain		1	
DELTA	Width effect on threshold voltage (LEVEL = 2 or = 3)		0	
THETA	Mobility modulation (LEVEL=3)	V^{-1}	0	
ETA	Static feedback (LEVEL=3)		0	
KAPPA	Saturation field factor (LEVEL=3)		0.2	
KF	Flicker noise coefficient		0	1×10^{-26}
AF	Flicker noise exponent		1	1.2

= 3, a modified version of the Schichman-Hodges model, which is a semiempirical short-channel model, is used. These and other models are described in Ref. 3.

The LEVEL 1 model, which uses fewer fitting parameters, provides approximate results. However, it is useful for a quick and rough estimate of the circuit performance and it is usually adequate for the analysis of basic electronic circuits. The LEVEL 2 model requires a great amount of CPU time for the calculations and could cause convergence problems. The LEVEL 3 model introduces a smaller error than the LEVEL 2 model and the CPU time is also approximately 25 percent less. The LEVEL 3 model is developed for MOSFETs with short and narrow channels.

12.3 USE OF SPICE FOR MODELING SILICON-BASED CHEMICAL SENSORS*

In this section we will present the development and the implementation in SPICE of models for CHEMFET and BIOFET sensors (see Chap. 10).[5,6]

The simplest CHEMFET is an H^+-sensitive FET, in which the sensitive surface is an insulator layer (e.g., Si_3N_4 or Al_2O_3). The response of these devices to pH is explained by considering H^+-specific binding sites at the surface of the insulator exposed to the electrolyte. The model has been designed to take into account also nonideal effects such as drift, hysteresis, and the limiting case of partial insensitivity to pH, i.e., pseudoreference field-effect transistor (REFET) structures.

The CHEMFET model will be then expanded by incorporating a biological sensing element (a membrane incorporating an enzyme) close to the insulator. The use of a biological element will allow us to develop a model which simulates biosensors (BIOFETs).

12.3.1 CHEMFET Model Theory

Static Model. To develop the static model for the CHEMFET, we first derive a set of equations to characterize the electrolyte-insulator-semiconductor (EIS) structure, which is the heart of the chemical sensor (CHEMFET) shown in Fig. 12.4. To this purpose, we consider a *p*-type semiconductor and a Si_3N_4 insulator (with two kinds of binding sites) exposed to an electrolyte (1:1 salt solution). The choice of 1:1 salt solution is based on the fact that for typical biological problems, electrolyte solutions can be well approximated by an NaCl (or KCl) solution. For more complex electrolyte solutions the model should be expanded in order to numerically solve the Poisson-Boltzmann equation.

The meaning of the potentials used in the model is self-explanatory in Fig. 12.4. The goal of this model is to obtain a relationship of the form pH = $f(\varphi_{eo})$, where $\varphi_{eo} = (\varphi_o - \varphi_g)$ is the potential of the electrolyte-insulator interface. For this purpose, the condition of *charge neutrality* of the system in Fig. 12.4 must be considered in conjunction with the *site-binding theory* and the *electrical double-layer theory*[7] (see Sec. 7.2).

The condition of *charge neutrality* for the system in Fig. 12.4 gives

$$\sigma_d + \sigma_o + \sigma_{mos} + \sigma_b = 0 \qquad (12.1)$$

where σ_d is the charge density in the diffuse layer (*Gouy-Chapman-Stern* theory), σ_o is

---

*Sections 12.3.1 to 12.3.8 from G. Massobrio, S. Martinoia, and M. Grattarola,[6] used by permission of Myu K. K., Tokyo. Sections 12.3.9 to 12.3.10 from G. Massobrio, S. Martinoia, and M. Grattarola,[12] used by permission of Elsevier Science S. A., Lausanne, Switzerland.

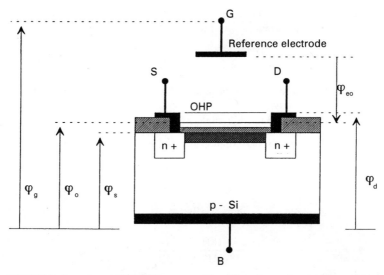

FIGURE 12.4 *n*-channel CHEMFET structure. The voltage drops relevant to the model are indicated. OHP indicates the outer Helmholtz plane.

the charge density at the electrolyte-insulator interface, σ_{mos} is the charge density in the semiconductor, and σ_b is the charge density of a buried layer. A buried layer (modeling drift effects) consists of a small fraction of binding sites (about 10 percent) positioned about 3 nm inside the insulator surface, so that OH$^-$ and H$^+$ groups can react with them at the end of a diffusion process. Each component of Eq. (12.1) is considered below.

For the *Gouy-Chapman layer*, the *charge density* σ_d is given by

$$\sigma_d = \sqrt{8\varepsilon_w kTc_o} \sinh\left[\frac{q(\varphi_g - \varphi_d)}{2kT}\right] \qquad (12.2)$$

where ε_w is the permittivity of the electrolyte, and c_o is the ion concentration in the electrolyte.

For the *Helmholtz layer*, the *charge density* σ_d is given by

$$\sigma_d = C_h(\varphi_d - \varphi_o) \qquad (12.3)$$

where C_h is the capacitance of the Helmholtz layer per unit area.

By combining Eqs. (12.2) and (12.3), we obtain

$$\varphi_g = \varphi_d + \frac{2kT}{q} \sinh^{-1}\left[\frac{C_h(\varphi_d - \varphi_o)}{\sqrt{8\varepsilon_w kTc_o}}\right] \qquad (12.4)$$

The *charge density* σ_o at the *electrolyte-insulator interface* can be written as

$$\sigma_o = C_h(\varphi_o - \varphi_d) - \sigma_{mos} \qquad (12.5)$$

where σ_{mos} is defined later [see Eq. (12.14)].

The charge density σ_o can also be obtained from the site-binding theory.[7] If we assume only silanol sites and basic primary amine sites to be present on the insulator surface after oxidation in electrolyte solution,[8] we can rewrite Eq. (12.5) in the form

$$\frac{\sigma_o}{qN_s} = \left(\frac{[H^+]_s^2 - K_+K_-}{[H^+]_s^2 + K_+[H^+]_s + K_+K_-}\right)\frac{N_{sil}}{N_s} + \left(\frac{[H^+]_s}{[H^+]_s + K_{N+}}\right)\frac{N_{nit}}{N_s} \quad (12.6)$$

where N_{sil} and N_{nit} = surface densities of silanol sites and primary amine sites, respectively
N_s = total number of binding sites
K_+, K_-, K_{N+} = dissociation constants
$[H^+]_s$ = proton concentration at the electrolyte-insulator interface

We evaluate the surface dissociation constants by considering the chemical reactions that occur at the surface, i.e.,

$$SiOH + H^+ \underset{k_{b1}}{\overset{k_{f1}}{\rightleftarrows}} SiOH_2^+ \quad (12.7)$$

with

$$\frac{[SiOH][H^+]_s}{[SiOH_2^+]} = \frac{k_{f1}}{k_{b1}} = K_+ \quad (12.8)$$

and

$$SiO^- + H^+ \underset{k_{b2}}{\overset{k_{f2}}{\rightleftarrows}} SiOH \quad (12.9)$$

with

$$\frac{[SiO^-][H^+]_s}{[SiOH]} = \frac{k_{f2}}{k_{b2}} = K_- \quad (12.10)$$

and

$$SiNH_2 + H^+ \underset{k_{b3}}{\overset{k_{f3}}{\rightleftarrows}} SiNH_3^+ \quad (12.11)$$

with

$$\frac{[SiNH_2][H^+]_s}{[SiNH_3^+]} = \frac{k_{f3}}{k_{b3}} = K_{N+} \quad (12.12)$$

In Eqs. (12.7) through (12.12), k_{fi} and k_{bi} are the individual reaction rates for the generation and recombination of the charged surface groups. They play the same role as the generation-recombination rates encountered in Chap. 3 in dealing with semiconductors.

In addition, the relationship between the concentration $[H^+]_b$ of protons in the bulk electrolyte and the concentration $[H^+]_s$ of protons at the surface of the insulator follows the Boltzmann law

$$[H^+]_s = [H^+]_b e^{-q\varphi_{eo}/kT} \quad (12.13)$$

The *charge density* σ_{mos} in the *semiconductor* is given by Eqs. (8.56) and (8.58), i.e.,

$$\sigma_{mos} = \pm\sqrt{2\varepsilon_s kT p_{p0}}\left[\left(\frac{q\varphi_s}{kT} - 1 + e^{-q\varphi_s/kT}\right) + \frac{n_{p0}}{p_{p0}}\left(-\frac{q\varphi_s}{kT} - 1 + e^{q\varphi_s/kT}\right)\right]^{1/2} \quad (12.14)$$

where n_{p0} and p_{p0} are the equilibrium concentrations of electrons and holes, respectively.

The charge density σ_{mos} within the semiconductor is positive for $\varphi_s < 0$ and negative for $\varphi_s > 0$. The charge density, σ_{mos} can also be written as

$$\sigma_{mos} = C_{ox}(\varphi_s - \varphi_o) \quad (12.15)$$

where C_{ox} is the insulator capacitance per unit area.

The *charge density* σ_b of the *buried layer* describes nonideal effects observed in H$^+$-sensitive FETs (CHEMFETs), such as hysteresis and drift. The charge density σ_b can be written as[9]

$$\frac{\sigma_b}{qN_s} = \left(\frac{[H^+]_{bur}^2 - K_+K_-}{[H^+]_{bur}^2 + K_+[H^+]_{bur} + K_+K_-}\right)\eta \frac{N_{sil}}{N_s} + \left(\frac{[H^+]_{bur}}{[H^+]_{bur} + K_{N+}}\right)\eta \frac{N_{nit}}{N_s} \quad (12.16)$$

where η is the fraction of buried sites and $[H^+]_{bur}$ is the proton concentration in the buried layer, given by the relationship

$$[H^+]_{bur} = [H^+]_s \left[1 + \text{erf}\left(\frac{x}{2\sqrt{D_{eff}t}}\right)\right] \quad (12.17)$$

In Eq. (12.17), t is the diffusion time, x the diffusion direction, and D_{eff} the effective diffusion constant, which can be expressed as

$$D_{eff} = \frac{D}{K_{bur}(N_{sil} + N_{nit})\eta} \quad (12.18)$$

where K_{bur} is the reaction equilibrium constant and D is the diffusion constant of H$^+$ and OH$^-$ ions.

In all the above equations, N_s is given by

$$N_s = (1 + \eta)(N_{sil} + N_{nit}) \quad (12.19)$$

When $\eta = 0$ (e.g., no hysteresis present), the ideal pH CHEMFET results.

The equations obtained by substituting Eq. (12.14) into (12.15) and Eq. (12.5) into (12.6) form, together with Eqs. (12.4), (12.13), (12.16), and (12.17), a system that describes a CHEMFET structure with two kinds of binding sites (i.e., Si$_3$N$_4$) and nonideal effects. The system of equations is solved for the five variables φ_d, φ_o, φ_s, $[H^+]_s$, and $[H^+]_{bur}$, while φ_g is an externally defined parameter. This set of equations, from now on called the *EIS system*, will be used to calculate the electrolyte-insulator potential φ_{eo}.

The above model can also be used to simulate REFET structures. In fact, partially pH-insensitive materials such as Teflon or parylene can be modeled by considering the chemical equilibrium equation between the surface sites —COOH and —COH of the organic membrane and protons. This equation can be obtained in a straightforward manner by modifying Eq. (12.6), as

$$\frac{\sigma_o}{qN_s} = \left(\frac{K(COOH)}{[H^+]_s + K(COOH)}\right)\frac{N(COOH)}{N_s} + \left(\frac{K(COH)}{[H^+]_s + K(COH)}\right)\frac{N(COH)}{N_s} \quad (12.20)$$

where $K(COOH)$ and $K(COH)$ are the dissociation constants for the chemical reactions, and $N(COOH)$ and $N(COH)$ are the numbers of —COOH and —COH sites per unit area, respectively.

Finally, we combine the above-described EIS structure with the physics of the MOSFET. By taking into account the potential differences among the elements of the system in Fig. 12.4, this results in a modification of the MOSFET threshold voltage, which becomes

$$V_{TH}^*(\text{CHEMFET}) = (E_{ref} + \varphi_{ij}) - (\varphi_{eo} - \chi_e) - \left[\frac{Q_{ss} + Q_{sc}}{C_{ox}} - 2\varphi_F + \frac{\phi_{sc}}{q}\right]$$

$$= V_{TH}(\text{MOSFET}) + E_{ref} + \varphi_{lj} + \chi_e - \varphi_{eo} - \frac{\phi_m}{q} \quad (12.21)$$

where φ_F = Fermi potential of the semiconductor (*p*-type for the *n*-channel MOSFET considered here)
Q_{ss} = fixed surface-state charge per unit area at the insulator-semiconductor interface
Q_{sc} = semiconductor surface depletion region charge per unit area
E_{ref} = potential of the reference electrode (Ag/AgCl is considered here)
φ_{lj} = liquid-junction potential difference between the reference solution and the electrolyte
φ_{eo} = potential of the electrolyte-insulator interface
χ_e = surface dipole potential
ϕ_{sc} = semiconductor work function
ϕ_m = work function of the metal gate (reference electrode) relative to vacuum

The CHEMFET *static model* is then determined by solving the EIS system for ϕ_{eo} and by replacing the MOSFET threshold voltage V_{TH} with V^*_{TH} in the I_{DS}-V_{DS} equations of the MOSFET, here rewritten for readers' convenience,

$$I_{DS} = \frac{K}{2}(V_{GS} - V^*_{TH})^2 \quad \text{(saturation region)} \quad (12.22)$$

$$I_{DS} = K\left[(V_{GS} - V^*_{TH}) - \frac{V_{DS}}{2}\right]V_{DS} \quad \text{(nonsaturation region)} \quad (12.23)$$

where $K = \mu C_{ox} Z/L$, μ, Z and L being the electron mobility, channel width, and channel length, respectively.

Large-Signal Model. The insulator capacitance of the CHEMFET is sensitive to the ionic activity of the electrolyte, and consists essentially of two elements connected in series: the MOSFET insulator capacitance C_{ox} and the Helmholtz layer capacitance C_h. The CHEMFET large-signal model is then determined by replacing the MOSFET insulator capacitance C_{ox} with the equivalent oxide capacitance

$$C^* = \frac{C_{ox} C_h}{C_{ox} + C_h} \quad (12.24)$$

Temperature Model. The similarity between the MOSFET and CHEMFET can be exploited in the analysis of the temperature dependence of the CHEMFET. Because both drain current equations (12.22) and (12.23) depend only on the quantities K, V^*_{TH}, and applied voltages, any temperature dependence must be through the quantities K and V^*_{TH}. Neglecting the temperature dependence of the parameters of the CHEMFET common to the MOSFET, because its theory is well known and its implementation in SPICE already exists, attention is focused on the most significant parameters of the EIS structure. Thus, we consider the following parameters.[10]

- Potential of the Ag/AgCl reference electrode relative to the hydrogen electrode: The potential of the reference electrode is temperature-dependent by means of its component of potential E_{rel} relative to the standard hydrogen electrode, i.e.,

$$E_{rel}(T) = E_{rel}(T_{bio}) + 1.4 \times 10^{-4}(T - T_{bio}) \quad (12.25)$$

where $T_{bio} = (25 + 273.15)$ K. In Eq. (12.25), the temperature coefficient applies when the hydrogen electrode is kept at 25°C.

DEVICES AND CAD

- Liquid junction potential for the Ag/AgCl electrode: The temperature-dependence of the potential difference between reference solution and electrolyte is

$$\varphi_{lj}(T) = \varphi_{lj}(T_{bio}) + 10^{-5}(T - T_{bio}) \qquad (12.26)$$

where the temperature coefficient value holds for a liquid junction potential of 3 mV.

- Electrolyte surface dipole potential: The surface dipole potential of the electrolyte is assumed to depend on temperature through the relation

$$\chi_e(T) = \chi_e(T_{bio})[1 - e^{0.86\log(I)}] \times \left[1 - \frac{0.4 \times 10^{-3}}{\chi_e(T_{bio})}(T - T_{bio})\right] \qquad (12.27)$$

where I is the ionic strength of the solution and the temperature coefficient applies to an aqueous solution.

- Dissociation constants: The dissociation constants of the chemical reactions at the insulator-electrolyte interface are temperature-dependent according to the relation

$$K_X(T) = [K_X(T_{bio})]^{T_{bio}/T} \qquad (12.28)$$

where X must be considered as A or B for the dissociation constants of the amphoteric binding sites, and N for the dissociation constant of the amine binding sites.

- Dielectric constants: The dielectric constants of the insulator and semiconductor are assumed to be temperature-independent. This does not hold for the electrolyte solution, which can be modeled by three regions each with a temperature-dependent dielectric constant depending on the orientation of the water dipole (ε_{EL}, ε_{IHP}, ε_{OHP}). The temperature dependence for the three constant dielectric components is given by the empirical relation

$$\varepsilon_X(T) = \varepsilon_X(T_{bio})\varepsilon_0[1 - 4.6 \times 10^{-3}(T - T_{bio}) + 8.8 \times 10^{-6}(T - T_{bio})^2] \qquad (12.29)$$

where ε_0 is the permittivity of free space and X must be considered as EL for the bulk of the electrolyte, IHP for the electrolyte region in the inner Helmholtz plane, and OHP for the electrolyte region in the outer Helmholtz plane.

12.3.2 CHEMFET Model in SPICE

Static Model. The equations describing the MOSFET static model[3] in SPICE have been used for the CHEMFET static model by replacing the expression for V_{TH} of the MOSFET with Eq. (12.21) and using the φ_{eo} value resulting from the solution of the EIS system.

To this end, we have modified some subroutines within SPICE (such as SPICE, MOSFET, FIND, MODCHK, ADDELT, READIN, ERRCHK), and introduced new subroutines to solve the EIS system. These new subroutines, called by the MOSFET subroutine, are ZPHISF (it returns the value of φ_{eo}); Z0ISF (it computes the solution of the nonlinear equation resulting from the EIS system); ZERFC [it computes the value of the erfc (x) function to solve the diffusion equation (12.17)].

The nonlinear EIS system is solved numerically by using a first-guess value for φ_s and then by verifying the consistency of the applied voltage φ_g.

Thus, in addition to the MOSFET static model parameters, we have introduced 12 static parameters that the user of SPICE can specify in the .MODEL statement to characterize the CHEMFET static behavior. These parameters are c_o (zscon), K_{N+} (znkap), K_+ (zakap), K_- (zbkap), N_{sil} (znox), N_{nit} (znnit), pH (zph), E_{rel} (zerel), χ_e (zchieo), φ_{lj} (zphilj), η (zepsil), D (zdeff).

Moreover, in order to simulate devices with more than one insulator layer (SPICE has only built-in parameters for an SiO_2-type insulator), we have introduced two new parameters for the .MODEL statement: the added insulator thickness t_{oxl} (ztoxn) and its dielectric constant ε_{oxl} (zepsox). The parentheses contain the .MODEL keywords for the new parameters for the CHEMFET static model.

Large-Signal Model. The equation describing the MOSFET large-signal model[3] in SPICE has been used for the CHEMFET large-signal model by replacing the insulator capacitance C_{ox} of the MOSFET with the series of C_{ox} and C_h in the MOSFET subroutine. The MOSFET subroutine shares (through a "common" statement) Eq. (12.24) with the large-signal model subroutines[3] of SPICE according to the LEVEL parameter. To characterize the CHEMFET large-signal model behavior, the user of SPICE need not specify new parameters. In fact, C_h is computed internally to SPICE by

$$C_h = \frac{\varepsilon_{IHP}\varepsilon_{OHP}}{\varepsilon_{OHP}d_{IHP} + \varepsilon_{IHP}d_{OHP}} \quad (12.30)$$

where ε_{IHP} and ε_{OHP} are the inner and the outer Helmholtz plane dielectric constants, respectively, and d_{IPH} and d_{OPH} are the insulator-nonhydrated ion and insulator-hydrated ion distances, respectively. The parameters of Eq. (12.30) are set in SPICE to their typical values.

Temperature Model. The CHEMFET parameters common to the MOSFET maintain the same temperature dependence. In addition, we have introduced Eqs. (12.25) to (12.29) in the TMPUPD subroutine of SPICE to model the temperature dependence of these parameters.

In Eqs. (12.25) through (12.29), T_{bio} indicates that the values of each involved parameter are assumed to have been measured (input in the .MODEL statement) at 25°C. Since all the other parameter values specified in the .MODEL statement are usually assumed to have been measured at T = TNOM = 27°C (default value for TNOM), we have initialized and updated at TNOM the "bio" parameter value in the MODCHK subroutine before the temperature analysis loop is executed.

12.3.3 CHEMFET Simulation with SPICE

The input and output characteristics of "ideal" SiO_2-, Al_2O_3-, and Si_3N_4-gate CHEMFETs are simulated by introducing appropriate values for the equilibrium constant K's (see Table 12.2) into the EIS system via the .MODEL statement of SPICE, and by utilizing the other physicochemical parameters listed in Table 12.3. This table also shows that the user of SPICE must set the LEVEL parameter to 4 in order to select the CHEMFET structure.

Figure 12.5 shows the resulting I_{DS}-V_{GS} curves (with source and substrate grounded) for different values of pH and for a given V_{DS} (i.e., 2 V) in the case of Al_2O_3 as the insulator. The simulated curves are compared with the experimental ones.

An estimate of the pH average sensitivity is also obtained, according to Eq. (12.21), by calculating the shifts in V^*_{TH} from the SPICE outputs for different pH values. In the case of Al_2O_3, the value obtained is 54 mV/pH. Analogous simulations give, for SiO_2 and Si_3N_4, pH sensitivities of 35 mV/pH and 53 mV/pH, respectively.

If the CHEMFET operating point is changed (e.g. from V_{GS} = 1 V to V_{GS} = 5 V), a decrease in pH sensitivity is obtained. In the case of Al_2O_3, the pH sensitivity decreases by about 3.0 percent, and for SiO_2 and Si_3N_4 the decreases are equal to about 4.3 (worst case) and 3.4 percent, respectively.

TABLE 12.2 Insulator Electrochemical Parameters (25°C)

Insulator	Surface ionization constants*			Surface site density[†]		pH_{pzc}[‡]
	K_-	K_+	K_{N+}	$N_{sil/al}$	N_{nit}	
SiO_2	15.8	63.1×10^{-9}		5.0×10^{14}		3.0
Si_3N_4	15.8	63.1×10^{-9}	1×10^{-10}	3.0×10^{14}	2.0×10^{14}	6.8
Al_2O_3	12.6×10^{-9}	79.9×10^{-10}		8.0×10^{14}		8.5

*Moles per liter.
[†]Number per square centimeter.
[‡]pH_{pzc} = pH value at the point of zero charge (σ_o).

TABLE 12.3 .MODEL Statement for an *n*-channel Al_2O_3-Gate CHEMFET*

```
.MODEL isf NMOS LEVEL=4 TOX=100N LD=0.8U XJ=2U NSUB=2e15 TPG=1
+        UO=520 RD=85 RS=85 LAMBDA=0.02 NSS=3e10 XQC=1
+        IS=1F CGSO=40P CGDO=40P CGBO=200P CJ=200U
+        MJ=0.5 CJSW=80N MJSW=0.33
+           ztoxn=90N zepsox=28.7 zepsil=0.1
+           zph=7.0 zdeff=1E-18 zerel=0.205 zchieo=3M
+           zphilj=1M zscon=100M znkap=0.0 zakap=12.6N
+           zbkap=0.79N znox=8E14 znnit=0.0
MISPD 22 1 0 0 isf L=20U W=400U AD=100N AS=100N PS=900U PD=900U
```

*The lowercase parameters concern the CHEMFET model; the uppercase parameters are common to MOSFET and CHEMFET.

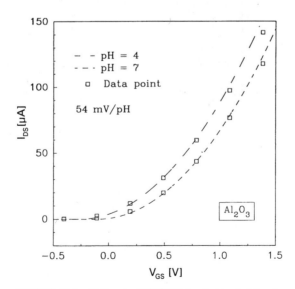

FIGURE 12.5 Al_2O_3-gate CHEMFET input characteristics for different pH values and for $V_{DS} = 2$ V. Hollow squares indicate experimental data.

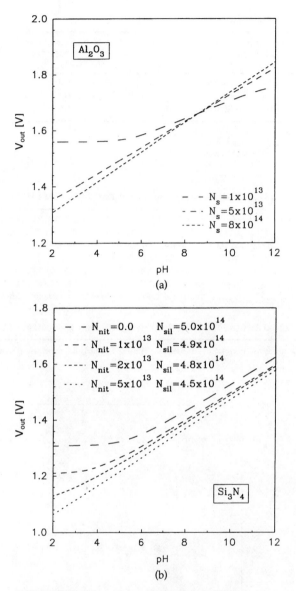

FIGURE 12.6 Effects of changing the number of binding sites on the output voltage. (*a*) Decrease in the total number of binding sites for an Al_2O_3-gate CHEMFET. (*b*) Decrease in the number of amine sites (with a fixed total number of binding sites) in the case of an Si_3N_4-gate CHEMFET.

In accordance with data reported in the literature, Si_3N_4 and Al_2O_3 pH sensitivities are "virtually" nernstian. This is not the case for SiO_2. These behaviors can be imputed to (besides other parameters) the density and type of the respective binding sites.

Figures 12.6a and b show the output voltage of the measuring circuit of Fig. 10.10b as a function of pH for Si_3N_4 and Al_2O_3, respectively. The outputs are straight lines when the site densities given in Table 12.2 are considered. In this case, the slopes of the curves directly give the pH sensitivities (54 and 53 mV/pH for Si_3N_4 and Al_2O_3, respectively).

The effect of decreasing the total number of binding sites N_s is quite evident for the Al_2O_3-gate CHEMFET. A similar effect is obtained by changing the ratio between silanol and amine sites for the Si_3N_4-gate CHEMFET. In the latter case, the limiting condition of zero amine sites describes the SiO_2 behavior. The reduction in the number of sites induces a departure from linearity, as expected.

We have simulated the input and output characteristics of Al_2O_3- and Si_3N_4-gate CHEMFETs over the pH range and a wide temperature range, using the values for the device parameters listed in the .MODEL statement of Table 12.4.

Figure 12.7 shows the simulation results of the I_{DS}-V_{DS} output curves (with source and substrate grounded) for the Si_3N_4-gate CHEMFET of Table 12.4 for the indicated pH and temperature values, at $V_{GS} = 1.5$ V and $V_{DS} = 1$ V.

12.3.4 REFET Simulation with SPICE

The EIS system can also be adapted to simulate surface groups other than silanol or amine, e.g., a REFET structure. The results obtained with this modified structure are shown in Fig. 12.8, which gives the output voltage of the measuring circuit of Fig. 10.10b for CHEMFET devices as a function of pH for various —COOH site densities (Fig. 12.8a), and for —COH site densities (Fig. 12.8b). A large region of pH insensitivity, as expected, is predicted by the CHEMFET model we have developed.

12.3.5 BIOFET Model Theory

An electrolyte–biological membrane–electrolyte–insulator–semiconductor structure (from now on indicated as EBEIS) can be considered as an extension of the EIS (electrolyte-insulator-semiconductor) structure. Like the EIS for the CHEMFET, the EBEIS can be considered the heart of any FET-based biosensor (BIOFET). In the following the membrane will contain an enzyme, thus modeling an ENFET.

TABLE 12.4 .MODEL Statement for an *n*-Channel Si_3N_4- Gate CHEMFET

```
.MODEL isfNaCl7 LEVEL=4 TOX=30n LD=2.85U XJ=3U NSUB=3.E16
+    TPG=0 UO=527 UCRIT=1.9e4 UEXP=0.0502 RSH=30
+    LAMBDA=0.02 XQC=1 PB=0.763 VTO=1.0 NFS=1.26E12 NEFF = 0.54
+    VMAX=5.4e4 DELTA=6.25 JS=7U CGSO=1.18n CGDO=1.18n CGBO=.276n
+    CJ=223U MJ=0.445 CJSW=1.52n MJSW=0.371
+    ztoxn=100n zepsox=66.4p zepsil=0.0 znkap=0.1n zph=7.02e0 zdeff=1
+    zakap=15.8 zbkap=63.1n zscon=0.1 znox=4.5e14 zdelet=10n znnit=0.5e14
+    zerel=0.205 zphilj=3e-3 zchieo=50e-3
```

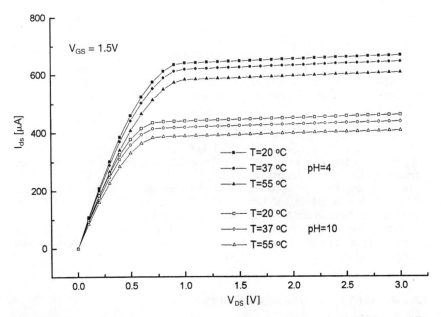

FIGURE 12.7 Simulation results of the I_{DS}-V_{DS} output curves for Si_3N_4-gate CHEMFET for two different pH values at various temperatures.

The EBEIS structure is shown in Fig. 12.9. It consists of a reference electrode, a "substrate" electrolyte solution (> 1 μl), a biological membrane a few nanometers thick, a thin (1 to 10 nm) layer of "local" electrolyte solution (0.004 pl to 0.04 pl), a layer (100 nm thickness, 20 μm length, 200 μm width) of insulator (e.g., Al_2O_3 or Si_3N_4) and a semiconductor (e.g., p-type silicon). In the following we shall consider only those structures where the thickness of the biological component (membrane) and its distance from the insulator are both in the nanometer range.

Then we assume that, when a chemical signal S (substrate) reaches the membrane from the bulk electrolyte, an enzyme reaction that can be described by a Michaelis-Menten mechanism takes place (see Sec. 10.3.2), resulting in the production of protons in the local buffered electrolyte compartment (between the membrane and the insulator). This transduction (into an electrical signal) mechanism (see Fig. 12.10) is described by the well-known relationship.

$$E + S \underset{K_{-1}}{\overset{K_1}{\rightleftarrows}} ES \xrightarrow{K_2} E + P \quad (12.31)$$

where S is the substrate, E is the enzyme in the free form, ES is the intermediate enzyme-substrate complex, and P is the product. K_1 and K_{-1} are the forward and backward rate constants for complex formation, and K_2 is the rate constant for complex decomposition into the product.

Assuming steady-state kinetics, Eq. (12.31) can be written in the form [Eq. (10.27)]

$$\frac{d[ES]}{dt} = K_1[E][S] - K_{-1}[ES] - K_2[ES] = 0 \quad (12.32)$$

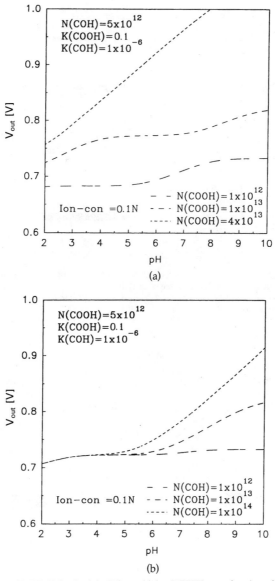

FIGURE 12.8 Partial pH insensitivity (REFET) as a function of (a) —COOH and (b) —COH site densities.

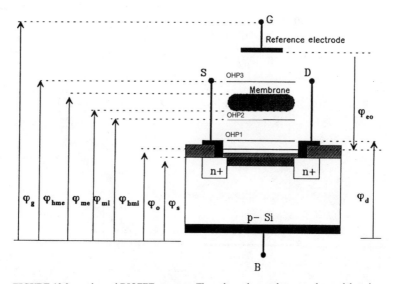

FIGURE 12.9 *n*-channel BIOFET structure. The voltage drops relevant to the model are indicated. OHP1, OHP2, and OHP3 are the outer Helmholtz planes of the sensor surface and of the membrane, respectively. three planes: (1) sensor surface, (2) inner membrane surface, (3) outer membrane surface.

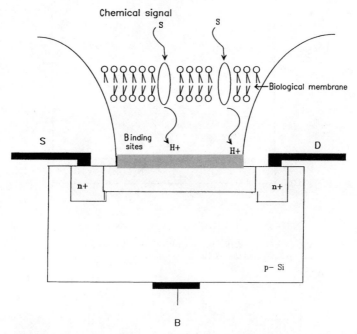

FIGURE 12.10 Chain of transduction mechanisms. (*Kindly provided by Paolo Massobrio.*)

or
$$\frac{d[ES]}{dt} = K_1[E_0][S] - (K_1[S] + K_{-1} + K_2)[ES] = 0 \quad (12.33)$$

where
$$[E_0] = [ES] + [E] \quad (12.34)$$

is the total concentration of the enzyme (defined as the sum of the concentrations in free and complexed forms). The solution of Eq. (12.33) gives (see Sec. 5.4.2)

$$[ES] = \frac{[E_0][S]}{K_M + [S]} \quad (12.35)$$

$$K_M = \frac{K_{-1} + K_2}{K_1} \quad (12.36)$$

where K_M is the Michaelis constant. Considering the flux equations describing the substrate mass transport and product mass transport, we can write

$$\frac{d_m K_2 [E][S]}{K_M + [S]} = K_d[P] \quad (12.37)$$

where d_m is the membrane thickness, K_d is a time-independent rate constant, and [P] is the product concentration.

We also assume that the "local" electrolyte compartment constitutes a microenvironment bound by the membrane on one side and the insulator on the other side. Protons interact with the insulator according to site-binding theory.[7]

Following the steps that led to the description of the EIS structure, we can also write the condition of *charge neutrality* for the EBEIS system as

$$\sigma_o + \sigma_{mos} + \sigma_d + \sigma_b + 2\sigma_m = 0 \quad (12.38)$$

where $2\sigma_m$ is the charge density (assumed to be negative and constant) on both sides of the membrane. The charge density σ_d is then given, from Eq. (12.3), by

$$\sigma_d = C_h(\varphi_{me} - \varphi_{hme}) \quad (12.39)$$

where the meaning of the voltage drops is the same as in Fig. 12.9.

The expression for φ_g [see Eq. (12.4)] becomes

$$\varphi_g = \varphi_{hme} + \frac{2kT}{q} \sinh^{-1}\left[\frac{C_h(\varphi_{hme} - \varphi_{me})}{\sqrt{8\varepsilon_w kTc_o}}\right] \quad (12.40)$$

As a simplifying condition, we assume a linear drop in potential to occur inside the thin layer of the local electrolyte compartment. As for the EIS, the proton concentration at the insulator surface, $[H^+]_s$, is related through a Boltzmann distribution to its concentration $[H^+]_m$ just outside the membrane, i.e.,

$$[H^+]_s = [H^+]_m e^{-q(\varphi_o - \varphi_{mi})/kT} \quad (12.41)$$

By applying Gauss' law to the structure, we obtain the voltage drops φ_{mi} and φ_{me} as

$$\varphi_{mi} = \varphi_o + 2\sigma_m\left(\frac{2}{C_h} + \frac{1}{C_i}\right) + (\varphi_{hme} - \varphi_{me})\left(2 + \frac{C_h}{C_i}\right) \quad (12.42)$$

$$\varphi_{\text{me}} = \frac{C_i(C_m + C_h)}{3C_iC_m + C_iC_h + C_iC_h} \left\{ \left[\varphi_o + 2\sigma_m \left(\frac{2}{C_h} + \frac{1}{C_i} \right) + \varphi_{\text{hme}} \left(2 + \frac{C_h}{C_i} \right) \right] \right.$$

$$\left. \cdot \frac{C_m}{C_m + C_h} + \frac{\sigma_m}{C_m + C_h} + \frac{C_h}{C_m + C_h} \varphi_{\text{hme}} \right\} \quad (12.43)$$

where C_m is the membrane capacitance per unit area and C_i is the capacitance per unit area of the local electrolyte compartment between the two OHP planes of the membrane and the insulator. All the other relationships obtained for the EIS structure also hold for the EBEIS system.

Another assumption implies that at any interface, the relative dielectric permittivity of water, ε_w, is modeled with a maximum of three compartments (piecewise linear shape): a first compartment of fully oriented dipoles (thickness = 0.2 nm; $\varepsilon_w = 6$); two layers of quasi-oriented dipoles (thickness = 0.4 nm; $\varepsilon_w = 32$), and a substrate compartment ($\varepsilon_w = 78.5$).

As for the case of the EIS, the potential φ_{eo} is obtained from the solution of the EBEIS system defined by the above equations.

The model proposed above can be "enlarged" by modeling the buffer capacity. It is known (see Sec. 4.2.2) that buffer capacity of a solution becomes extremely important when acidification induced by chemical or biological phenomena must be detected. This is particularly true when we consider an enzyme reaction which produces H^+ ions in a closed volume.

The pH *buffer capacity* of a volume is defined by

$$\frac{dn}{d\text{pH}} = V_b \beta_v \quad (12.44)$$

where dn is the number of moles of OH^- or H^+ ions added to the solution, dpH is the resulting pH change, V_b is the solution volume, and β_v is the volumetric buffer capacity of the solution. The sign of Eq. (12.44) is correct if OH^- ions are considered, otherwise the sign must be changed.

When very small volumes of solution are taken into account, it is necessary to consider the surface buffer capacity. However, we assume that Eq. (12.44) also holds in the case of a small volume, and we will use it as an approximate relationship.

An enzyme reaction producing $[H^+]$ moles per second, i.e.,

$$\frac{dn}{dt} = [H^+] \quad (12.45)$$

will cause a pH change, according to Eq. (12.44), of

$$\frac{d\text{pH}}{dt} = -\frac{[H^+]}{V_b \beta_v} \quad (12.46)$$

A low buffering capacity is useful in increasing the measured chemical change. In general, β_v depends on many features such as the pH of the solution, the number of chemical species, and time.

If we consider a single buffer species (e.g., $NaHCO_3$) in solution and we neglect the surface buffering capacity (as stated before), the volumetric buffer capacity is given by

$$\beta_v = 2.303 \, c_b \frac{e^x}{1 + e^x} \quad (12.47)$$

where $x = 2.303 \, (\text{pK} - \text{pH})$ and c_b is the total buffer concentration.

12.3.6 BIOFET Model in SPICE

The subroutines within SPICE that we have modified for the EIS structure are also valid to model the EBEIS behavior. However, we have introduced in SPICE new subroutines to solve the EBEIS system; in particular, they are ZPHMEM (it returns the value of φ_{eo}); Z0MEM (it computes the solution of the nonlinear equation resulting from the EBEIS system); Z0ENZ (as Z0MEM, but in the presence of enzyme in non-steady-state condition). Ten new parameters are now available to the user of SPICE to be specified in the .MODEL statement: S (zso); E (zenz); c_b (ztamp); K_d (zkd); K_1, K_{-1}, K_2 (zkp1, zkm1, zkp2); d_m (zdm); pK (zpk); V_v (zvbulk). The parentheses contain the .MODEL keywords for the new parameters of the BIOFET model, as listed in Table 12.5. This table also shows that the user of SPICE must set the LEVEL parameter to 5 in order to select the BIOFET structure.

12.3.7 BIOFET Simulation with SPICE

Figure 12.11 shows the results obtained from SPICE simulation under steady-state conditions as a function of substrate concentrations, by letting the reaction products diffuse outside the local compartment (assuming leakage from the membrane edges). As expected, a linear response for a wide range of substrate concentrations is obtained.

BIOFET behavior changes under the assumption of no leakage from the local compartment, i.e., if a tight adhesion of the membrane to the lateral edges of the BIOFET is considered. Under this condition, solving Eq. (12.32), we obtain the substrate concentration as a function of time as

$$[S] + \frac{K_{-1} + K_2}{K_1} \ln \frac{[S]}{[S_0]} = [S_0] - K_2[E_0]t \quad (12.48)$$

where $[S_0]$ is the substrate concentration in the bulk.

If we assume that the product concentration is equal to variations in the substrate concentration, from Eq. (12.48) we can also obtain the product concentration as a function of time. The transient output of the BIOFET under "sealed membrane conditions"

TABLE 12.5 .MODEL Statement for an n-Channel Al_2O_3-Gate BIOFET*

.MODEL isf NMOS LEVEL=5 TOX=100N LD=0.8U XJ=2U NSUB=2e15 TPG=1
+   UO=520 RD=85 RS=85 LAMBDA=0.02 NSS=3e10 XQC=1
+   IS=1F CGSO=40P CGDO=40P CGBO=200P CJ=200U
+   MJ=0.5 CJSW=80N MJSW=0.33
+   ZTOXN=90N ZEPSOX=28.7 ZEPSIL=0.1
+   ZPH=7.0 ZDEFF=1E-18 ZEREL=0.205 ZCHIEO=3M
+   ZPHILJ=1M ZSCON=100M ZNKAP=0.0
+   ZAKAP=12.6N ZBKAP=0.79N ZNOX=8E14
+   ZNNIT=0.0
+   zso=1M zenz=1U ztamp=1M zkd=0.002N
+   zkp1=100 zkm1=.1U zkp2=1K

MISPD 22 1 0 0 isf L=20U W=400U AD=100N AS=100N PS=900U PD=900U

*The lowercase parameters concern the BIOFET model; the uppercase parameters are common to the CHEMFET.

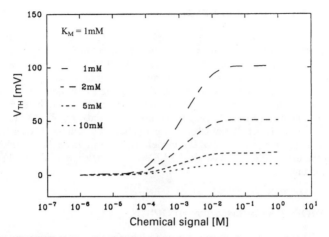

FIGURE 12.11 Chemical signal detection under steady-state conditions.

(i.e., no leakage through membrane), as a function of the membrane-insulator distances, is shown in Fig. 12.12a. Poisoning effects, induced on the enzyme by the pH decrease in the local (sealed) volume, are simulated by making the constant K_2 of the enzyme-product reaction dependent on the pH in the inner compartment via the exponential relationship

$$K_2^* = K_2 \exp\left[\frac{-|\Delta\text{pH}|}{\alpha}\right] \tag{12.49}$$

where α is a damping constant.

Figure 12.12b shows the transient output of the BIOFET, obtained from SPICE simulation, under the above-mentioned conditions. Such effects should be taken into account when considering volumes surrounded by materials which allow no (or very slow) leakage through the membrane.

Finally, as a highly speculative example of a "chemical logical processor," we can consider a situation with two chemical signals S_1 and S_2 and two types of enzyme reactions which produce H^+ and OH^- ions when subjected to S_1 and S_2 signals, respectively. The chemical signal S_1 (H^+ production) generates a low pH value in the local volume. If a sealed condition is assumed (i.e., no or very slow leakage through membrane), then this value remains constant in time, even in the absence of S_1 (i.e., the final value of pH is maintained). It can then be changed by the chemical signal S_2 (OH^- production), which sets the pH to the original value, making the device ready for new measurement. The system output can then be switched in time between two values by acting on S_1 and S_2. Figure 12.13 shows the simulated switching of the system.

12.3.8 Chemical Signal Functions

We have defined and implemented in SPICE chemical signal voltage/current sources to simulate chemical signals that can be used in conjunction with the CHEMFET and BIOFET models described in the previous sections.

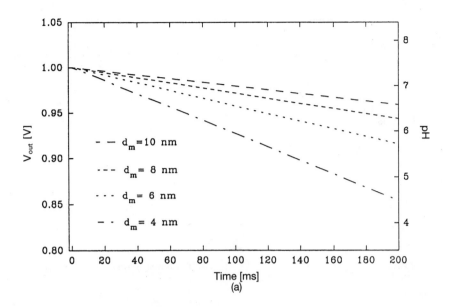

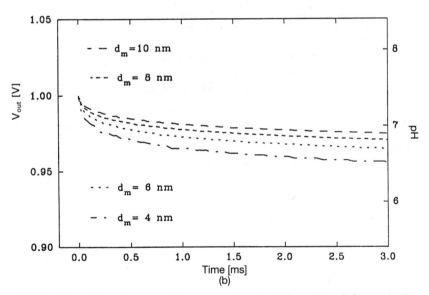

FIGURE 12.12 (*a*) Transient output of the system under the "sealed" condition. (*b*) Same as in *a* but with an acidification-induced reduction in the efficiency of the enzyme reaction.

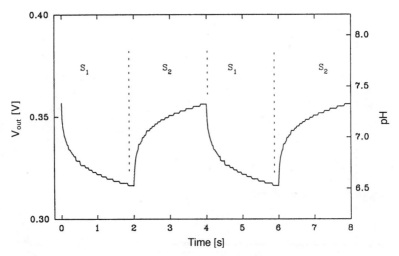

FIGURE 12.13 Simulated switching output of the BIOFET driven by two alternating chemical signals.

In particular, we have introduced in SPICE the following sources.

1. *pH pulse* (or step) and *pH piecewise linear* sources. The general form and the parameters maintain the meaning of the PULSE and PWL sources of SPICE when referred to pH values.
2. *Linear triangular function* source for substrate changes. This source simulates transient changes of the substrate.
3. *Steady-state H^+* source. This source simulates H^+ production at a certain distance from the sensor and for a fixed value of the solution buffer.

To implement these sources in SPICE, we have rearranged some subroutines of the code. In particular, in SORUPD, the new source definitions and their dependence on time have been added; in ELPRNT, the source parameters to be printed in the output list have been inserted; in READIN, the source names (existence) via the ANAM variable have been defined; in SPICE, parameter initialization and flags (TIME, ZPH) have been introduced to control time and pH for the model subroutines; in ERRCHK, error diagnostics have been added. The dimensions of the vectors ASTYPE and ITYPE have been updated to take into account the new sources; the subroutines involved in these modifications are DMPMAT, ERRCHK, FOURAN, NTRPL8, OVTPVT, PLOT, READIN, RUNCON, SETPLT, SETPRN, SORUPD, and SPICE.

Finally, to interface the model subroutines ZPHISF and ZPHMEM to the SORUPD subroutine, a further subroutine (ZSORPH) has been implemented in SPICE.

In Table 12.6 an example of SPICE input for the definition of the implemented chemical signals function sources is shown.

12.3.9 LAPS Model Theory and Its Implementation in SPICE

An alternative semiconductor structure to the CHEMFET, called *light-addressable potentiometric sensor* (LAPS), has been proposed as a pH sensor.[11] The principle of operation of

TABLE 12.6 Definition of Chemical Signal Sources Implemented in SPICE

---

pH pulse source

PHPLS (pH1 pH2 TD TR TF PW PER)
pH1 = initial value of pH
pH2 = pulsed value of pH
TD = delay time
TR = rise time
TF = fall time
PW = pulse width
PER = period

---

pH piecewise linear source

PHPWL (T1 pH1 < T2 pH2 T3 pH3 ... >)
Each pair of values (Ti, pHi) specifies that the value of the source is pHi at time Ti.
The intermediate values are determined by using linear interpolation on the input values.

Chemical signal triangular wave source

SUBST (SO1 SO2 T1 T2 T3)
SO1 = initial value
SO2 = pulsed value
T1 = initial time value for S = SO1
T2 = initial time value for S = SO2
T3 = time value for S = SO1 (end of triangular wave)

H^+ steady-state production

SOURPH (BUFFER RATE DIST PK AGATE TUNIT)
BUFFER = buffer concentration
RATE = H^+ production per s
DIST = source-sensor distance
PK = pK value of the buffer
AGATE = sensor gate area
TUNIT = time scale (0 = s, 1 = min)

---

the LAPS (sketched in Fig. 12.14a) is similar, to a certain extent, to that of the EIS (electrolyte-insulator-semiconductor) capacitor, as both of them are simplified passive versions of the CHEMFET. More specifically, the LAPS uses photoexcitation of the semiconductor to probe the surface potential at the insulator-electrolyte interface. The semiconductor is addressed by a modulated flux of (infrared) photons; this flux results in the generation of hole-electron pairs in the semiconductor. The photogenerated pairs lead to an alternating current that can be measured by means of an external circuit (see Sec. 10.5).

An equivalent electronic circuit[12] for the LAPS (p-substrate and Si_3N_4 as exposed insulator) is shown in Fig. 12.14b, where V_{GB} is the applied bias voltage, C_{GC} is the Gouy-Chapman layer capacitance, C_h the Helmholtz layer capacitance, C_{ox} the insulator capacitance, C_d the depletion region capacitance, C_i the capacitance related to the minority carriers in the inversion layer, C_{ss} the capacitance related to the surface states at the semiconductor-insulator interface, R_e the electrolyte resistance, and R_g the resistance that models the delay effects originated by the generation-recombination processes. All the above capacitances are considered per unit area. The current density source J_p models the effect of the photogeneration. In order to derive an expression for J_p, we make further assumptions

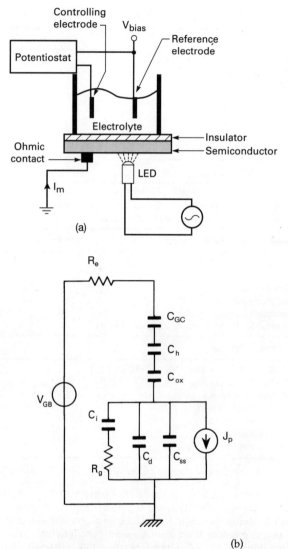

FIGURE 12.14 (a) Sketch of the LAPS. (b) LAPS equivalent circuit.

in addition to those associated with the study of standard MOS theory. In particular, we assume that

1. The structure operates at low-level injection.
2. The depletion region width depends only on V_{GB}.
3. The field in the neutral region is negligible.
4. The illumination is perpendicular to the surface of the insulator, which is assumed to be transparent to radiation.

5. The photogenerated hole-electron pairs are generated in the depletion region only.

Under these assumptions, the current density flowing through the LAPS is

$$J_p = J_{\text{diff}} + J_{\text{phgen}} \tag{12.50}$$

where J_{diff} is the diffusion component and J_{phgen} is the photogenerated component.

In order to derive an expression for J_{diff}, we refer to the continuity equation (see Sec. 3.7) for the excess electrons $n_p'(x, t)$ in the neutral region. Thus, under assumption (3), we have

$$\frac{\partial n_p'(x, t)}{\partial t} = D_n \frac{\partial^2 n_p'(x, t)}{\partial x^2} - \frac{n_p'(x, t)}{\tau_n} + G(x, t) \tag{12.51}$$

where D_n is the electron diffusion constant, τ_n is the electron lifetime, and $G(x, t)$ is the photogeneration rate, which can be expressed as

$$G(x, t) = \alpha \Phi_1 e^{-\alpha x} e^{j\omega t} \tag{12.52}$$

where α is the absorption coefficient and Φ_1 is the flux density of photons.

In our case, the excess of electrons can also be expressed as

$$n_p'(x, t) = \hat{n}_p(x) e^{j\omega t} \tag{12.53}$$

As a result, the diffusion current density is obtained as

$$J_{\text{diff}}(W, t) = qD_n \left[\frac{dn_p'(x, t)}{dx} \right]_{x=W} = q \frac{\alpha \Phi_1 L_n^*}{1 + \alpha L_n^*} e^{-\alpha W} e^{j\omega t} \tag{12.54}$$

where W is the depletion region width and

$$(L_n^*)^2 = \frac{D_n \tau_n}{1 + j\omega \tau_n} \tag{12.55}$$

Neglecting the generation-recombination term due to thermal processes, we can derive the photogeneration component J_{phgen} as

$$J_{\text{phgen}} = q \int_0^W G(x, t)\, dx = q\Phi_1 [1 - e^{-\alpha W}] e^{j\omega t} \tag{12.56}$$

Finally, from Eqs. (12.54) and (12.56), we obtain

$$J_p = q\Phi_1 \left[1 - \frac{e^{-\alpha W}}{1 + \alpha L_n^*} \right] e^{j\omega t} \tag{12.57}$$

Equation (12.57) defines the current source J_p in the circuit in Fig. 12.14b. Equation (12.57), through its amplitude $|J_p|$, has been implemented in SPICE.[12]

The pH sensitivity of the LAPS is modeled by using a generalized site-binding approach to describe the H$^+$ kinetics at the insulator-electrolyte interface. This approach allows us to consider two different binding sites (i.e., silanol and amine groups) as in the case of an Si_3N_4 insulator. The fundamental relationship between the charge at the electrolyte-insulator interface, σ_o, the dissociation constants K_+, K_-, K_{N+}, and the proton concentration $[H^+]_s$ on the insulator surface is again given, by Eq. (12.6).

12.3.10 LAPS Simulation with SPICE

LAPS Response to pH Variations (cell metabolism). One key feature shared by integrated silicon devices like CHEMFETs and LAPSs lies in a fast (millisecond range) response to chemical variations. This feature makes such devices optimal candiates for on-line pH monitoring in chemical or biological microenvironments in flow-through cells.

The model described in the previous section can be used to characterize the response of the LAPS to acidification rates induced in a weakly buffered volume by a cell population layer.[13]

Simulated detections of acidification rates are given in Fig. 12.15 as a function of geometric and chemical parameters, respectively. A complete experiment, with a periodic injection of unacidified fresh medium, is simulated in Fig. 12.16, where a production of 10^8 H^+/s per cell and a buffer concentration of 1 mM are assumed.

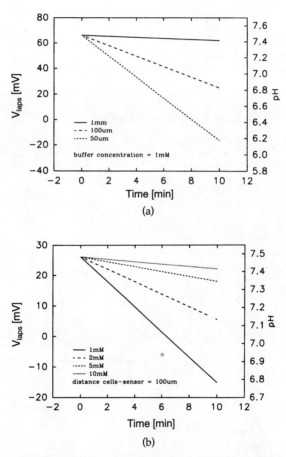

FIGURE 12.15 (*a*) Acidification rate detection as a function of cell layer-sensor distance. (*b*) Acidification rate detection as a function of solution buffer capacity.

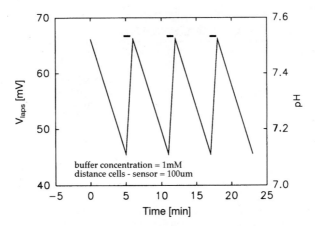

FIGURE 12.16 Simulated LAPS measurement of cell metabolism in a flow-through system. The cell-induced acidification is temporarily counterbalanced by periodic injections of fresh medium (bars).

LAPS Frequency Dependence. The equivalent-circuit representation of the LAPS (see Fig. 12.14b) is well suited to being introduced into circuit simulation programs such as SPICE. In this way, the performance of the device can be easily analyzed as a function of the input parameters. As a specific example, the effects of changing the modulation frequency of the addressing light through variations in the parameters α and L_n are shown in Fig. 12.17. This result was obtained for fixed pH and bias values.

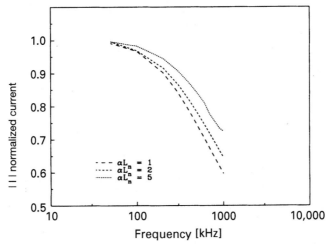

FIGURE 12.17 Output current amplitude versus frequency, with αL_n as a parameter. Current I flows through the voltage source V_{GB}.

12.4 USE OF SPICE FOR MODELING NEURONS (EXCITABLE MEMBRANE)*

In this section we will describe how SPICE can be used to model the behavior of neurons of Hodgkin-Huxley type (excitable membrane) and of postsynaptic membranes.[14]

This type of modeling approach can be considered as a *software algorithm approach* to distinguish it from the *hardware architecture approach* we will describe in Sec. 12.5.

12.4.1 Neuron Circuit Model

We will provide here an equivalent circuit description of the simulation building blocks; in the next subsections the corresponding SPICE implementations will be described.

Excitable Membrane Circuit. A patch of the membrane of a neuron can be modeled by using the *Hodgkin-Huxley* model,[15] which describes the dynamic properties of the sodium and potassium channels and of other leakage ion channels located on the excitable membrane.

These properties are described by the Hodgkin-Huxley (H-H) equations which we summarize for readers' convenience as follows (see Sec. 11.2.) In the following the membrane potential is denoted by V_{mem}.

$$\left(\frac{a}{2R_i}\right)\left(\frac{\delta^2 V_{mem}}{\delta x^2}\right) = \frac{C_m \delta V_{mem}}{\delta t} + g_K n^4 (V_{mem} - V_K) + g_{Na} m^3 h (V_{mem} - V_{Na})$$

$$+ g_l(V_{mem} - V_l) \quad \text{(membrane potential equation)} \quad (12.58)$$

$$\frac{dn}{dt} = \alpha_n(1-n) - \beta_n n \quad \text{(activation of K}^+ \text{ channels)} \quad (12.59)$$

$$\frac{dm}{dt} = \alpha_m(1-m) - \beta_m m \quad \text{(activation of Na}^+ \text{ channels)} \quad (12.60)$$

$$\frac{dh}{dt} = \alpha_h(1-h) - \beta_h h \quad \text{(inactivation of Na}^+ \text{ channels)} \quad (12.61)$$

where
- a = radius of the neuron membrane
- R_i = resistance of the axoplasm
- V_{mem} = membrane potential
- C_m = membrane capacitance
- g_K, g_{Na}, g_l = potassium, sodium, and leakage conductances through the membrane, respectively
- V_K, V_{Na}, V_l = potassium, sodium and leakage equilibrium potentials, respectively
- m, h = degrees of activation and inactivation of the sodium channels, respectively
- n = degree of activation of the potassium channels
- $\alpha_n, \alpha_h, \alpha_m$ = rates by which the channels switch from a closed to an open state
- $\beta_n, \beta_h, \beta_m$ = rates for the reverse

---

*From M. Bove, G. Massobrio, S. Martinoia, and M. Grattarola.[14] Used by permission of Springer-Verlag GmbH & Co. KG, Heidelberg, Germany.

This model formulation leads to the equivalent circuit shown in Fig. 12.18a. The sodium and potassium currents flow through specific channels whose conductance values are time- and voltage-dependent, whereas the value of the leakage ion conductance is considered constant.

Figure 12.18b shows the circuit model used to describe the membrane patch. The nonlinear voltage-dependent current source I_{NaK} represents the sum of the potassium and sodium currents and is expressed as

$$I_{NaK} = [g_K n^4 (V_{mem} - V_K) + g_{Na} m^3 h (V_{mem} - V_{Na})] \qquad (12.62)$$

The single isopotential compartment of Fig. 12.18b constitutes the basic building block, which can be extended to more complex systems.

Compartmental Model Circuit. When the morphology of a neuron is considered and phenomena such as the propagation of the action potential along axonal branches are to be studied, the model of Fig. 12.18b can be used as the basis for the so-called *compartmental model* approach.[16,17] Figure 12.19 shows an equivalent circuit modeling a nerve axon divided into compartments; each compartment is modeled by the H-H kinetics described by Eqs. (12.58) to (12.61). The values of the circuit elements depend on the size and the area of each compartment. Adjacent compartments are connected in series through an axoplasm resistance, denoted by R_i in the circuit.

By using the compartmental model approach, it is possible to switch from the solution of the nonlinear partial differential Eq. (12.58) (which describes a traveling wave in a one-dimensional symmetric axon) to the solution (obtained by the model implemented in SPICE) of a set of ordinary differential equations, each representing an isopotential membrane patch of length Δx. Moreover, it is possible to take into account the anatomical aspects of the neuron through the electrical properties of each compartment.

Simulation results of the action potential propagation in neuronal arborizations are shown in Sec. 12.4.3.

Postsynaptic Circuit. The compartmental model is a useful tool to study the effects of excitatory and inhibitory synapses placed at various distances from the soma on the dendritic arborizations of a postsynaptic cell and activated according to different timings.[18] The current I_{syn} in the postsynaptic region of the nerve cell can be modeled as a time-

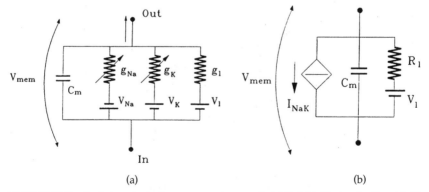

FIGURE 12.18 Equivalent circuit of a neuron membrane patch. (*a*) Hodgkin-Huxley circuit model. (*b*) Equivalent circuit used in the software algorithm approach. The value of I_{NaK} is given by Eq. (12.62).

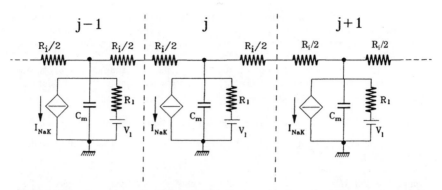

FIGURE 12.19 Compartmental model of a continuous axon region. Example of three compartments connected in series: each compartment is bounded by dashed vertical lines.

varying alpha-function transmembrane conductance G_{syn} connected in series with a synaptic source V_{syn},[19] as shown in Fig. 12.20a. Thus, I_{syn} can be expressed as

$$I_{syn} = G_{syn}(V_{mem} - V_{syn}) \tag{12.63}$$

where

$$G_{syn} = Cte^{-t/t_{peak}} \tag{12.64}$$

and C is a constant. Equation (12.64) reaches its maximum value, $Ct_{peak} \exp(-1)$, at $t = t_{peak}$, which is the time to peak of the conductance transient. This model formulation leads to the equivalent voltage-dependent nonlinear current source shown in Fig. 12.20b, which constitutes the basic building block used in the software algorithm approach model.

Equivalent Circuit of Extracellular Electrical Coupling between Neuron and Microtransducer. A technique for recording the electrical activity of neurons cultured in vitro is based on extracellular recording by means of planar microtransducers physically coupled to the nerve cells. We then show a circuit model which can simulate a patch of

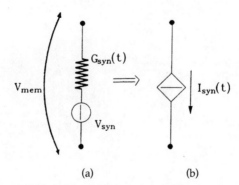

FIGURE 12.20 (a) Equivalent circuit of a postsynaptic current input. (b) Equivalent circuit of a postsynaptic current input used in the software algorithm approach model.

nerve cell membrane coupled to a passive (i.e., noble metal) microtransducer. Figure 12.21 shows the equivalent circuit of a junction between a neuron membrane patch (modeled by the compartmental approach) and a passive microtransducer.[20] The meanings of the symbols of the electrical parameters of the neuron are the same as those in Fig. 12.19. For the coupling of the patch of nerve cell membrane to the microtransducer (see Sec. 11.6), R_{seal} is the sealing resistance between the nerve cell and microtransducer; R_{spread} is the spreading resistance; R_e and C_e are the equivalent elements of the microtransducer-electrolyte interface; and R_{me} is the metallic resistance of the connection path of the microtransducer.

12.4.2 Neuron Model in SPICE

The model previously described has been implemented in SPICE, allowing us to simulate the biophysical behavior of Hodgkin-Huxley model-based membranes. In other words, Eqs. (12.58) to (12.61), (12.63), and (12.64) have been implemented in SPICE by adding five new subroutines (ZCHHMAIN, ZCHHEXTR, ZCHHINIZ, ZCHHRUNGE, and ZCHHRATE) to the code and by modifying four existing subroutines (READIN, RUN-CON, NLCSRC, and EVPOLY).

In particular, the READIN subrountine has been modified to initialize the variable "zncomp," which indicates how many compartments have to be considered during the simulation.

The RUNCON subroutine has been modified to introduce the number of compartments as a variable that the user can specify in the .OPTIONS statement.

The NLCSRC subroutine has been modified to define nonlinear voltage-controlled current sources that depend on the number of compartments, on the H-H parameters coming from the ZCHHMAIN subroutine and on the number of postsynaptic circuits in order to define the sodium, potassium, and synaptic currents.

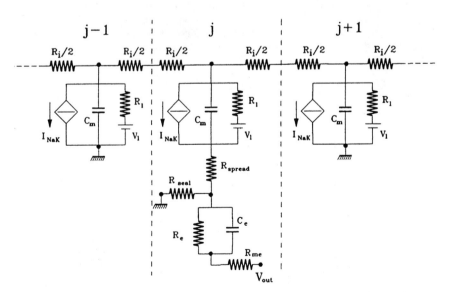

FIGURE 12.21 Equivalent circuit of the junction between a neuron membrane patch (described by the compartmental model) and a passive (i.e., noble metal) microtransducer.

The EVPOLY subroutine has been modified to introduce, as polynomial coefficients, the actual values of the H-H conductances or, in the case of postsynaptic circuits, the values of the synaptic currents that flow through the membrane.

The ZCHHMAIN subroutine contains the main control loop of the H-H model. It begins by checking whether one or more compartments are required: once the number of compartments has been established, an algorithm is activated to synchronize the "time" variable used (at a fixed time step) in the numerical integration subroutine ZCHHRUNGE with the "time" variable of SPICE (using multistep methods). Then the ZCHHEXTR subroutine is called.

The ZCHHEXTR subroutine assigns initial values to the Hodgkin-Huxley parameters; then it invokes the ZCHHINIZ subroutine, when the simulation is at time zero (initial conditions), or the ZCHHRATE subroutine during the simulation.

The ZCHHINIZ subroutine assigns initial conditions to the resting potential of the membrane and to the alpha and beta parameters and defines the alpha and beta rates.

The ZCHHRUNGE subroutine performs the integration of Eqs. (12.58) to (12.61) by the Runge-Kutta fourth-order method.

The ZCHHRATE subroutine updates the alpha and beta rates.

12.4.3 Neuron Simulation with SPICE

Using the just-described model implemented in SPICE, we show the simulation results concerning some typical configurations.

Space Clamp Condition. Figure 12.22 shows the simulation of a membrane action potential in the space clamp condition, when a current stimulus of 10 µA is applied. From now on, the resting potential is conventionally set to zero

Simulations of the Action Potential Propagation in Neuronal Arborizations. A program in C language was developed which automatically generates an input for SPICE from a list of the geometrical parameters (e.g., diameter, length) of neuronal arborizations. The inputs are the geometrical parameters of the soma or of the dendritic/axon

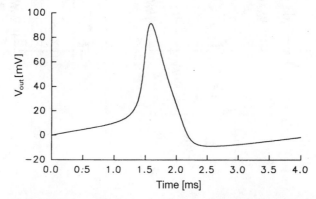

FIGURE 12.22 SPICE output representing a membrane action potential in the space clamp condition, when a current stimulus of 10 µA is applied.

branches or postsynaptic inputs, and the output consists of a set of isopotential compartments. Each compartment is characterized by an electrical equivalent circuit which depends on the morphological and functional properties of the compartment. The basic equivalent circuits and their translation into the input for SPICE are summarized in Fig. 12.23.

In the simulations, the values of the membrane were $R_i = 35.4$ Ω-cm, $C_m = 1.0$ μF/cm^2, $g_{Na} = 120$ mS/cm^2, $g_K = 36$ mS/cm^2, $g_1 = 0.3$ mS/cm^2, $V_{Na} = 115$ mV, $V_K = -12$ mV, and $V_1 = 10.613$ mV.

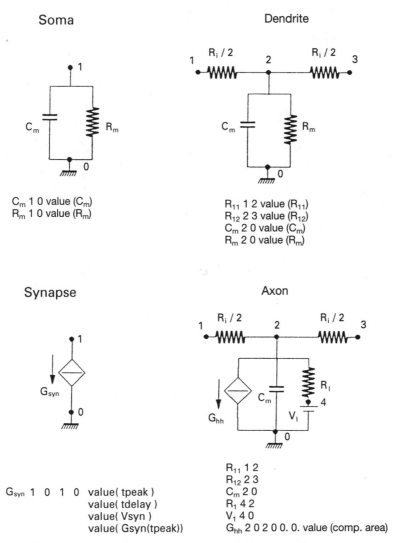

FIGURE 12.23 Translation of functional and morphological properties of a neuron (soma, dendrite, synapse, axon), represented by their equivalent electrical circuits, into an input for SPICE.

The space constant for the excitable membrane was obtained by the relation

$$\lambda = \left\{ \frac{a}{2R_i[g_{Na}m^3(0)h(0) + g_K n^4(0) + g_1]} \right\}^{1/2} \quad (12.65)$$

For each compartment, the segment length, Δx, was set to 0.2λ.

Critical cases have also been simulated in which the behavior of the action potential, while propagating along an axon with a branch point, is characterized by a geometrical ratio GR higher than unity, GR being defined as

$$GR = \frac{\sum_i d_i^{3/2}}{d_p^{3/2}} \quad (12.66)$$

where d_i is the daughter branch diameter and d_p is the parent branch diameter.

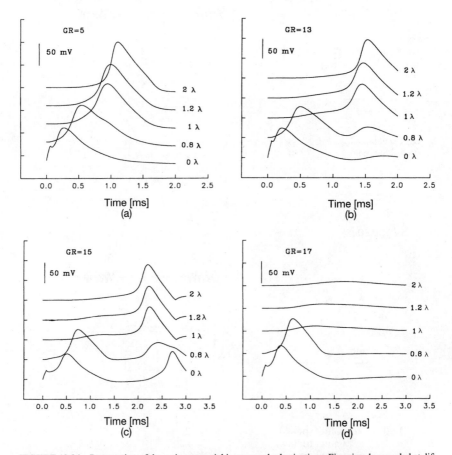

FIGURE 12.24 Propagation of the action potential in neuronal arborizations. Five signals recorded at different points (0λ, 0.8λ, 1λ, 1.2λ, 2λ) of an axon arborization are shown. The numbers near the signals are the length constants from the origin. All the daughter axons are equal in length (1λ) and diameter, and the parent axon is 1λ long from the origin. Thus, GR is equal to the number of daughter axons. The dependence of the back reflection upon GR is shown in (a) GR = 5, (b) GR = 13, (c) = GR = 15, (d) GR = 17.

Figure 12.24 shows how the propagation of the action potential changes as a function of the GRs of the neuron arborizations (the action potential is shown for five points along the axon: 0λ, 0.8λ, 1λ, 1.2λ, 2λ, where λ is the length constant).

Simulation of Action Potential Generation Induced by Synaptic Inputs. Here we present the simulation results, obtained with the software algorithm approach model implemented in SPICE, concerning the action potential generation and propagation induced by synaptic inputs placed along the dendritic arborizations of an idealized neuron (Fig. 12.25). For the passive membrane, the values of the membrane resistance and capacitance and of the cytoplasmic resistivity were chosen according to typical data reported in the literature.[17] For the excitable membrane, the values are those given above in the section on simulations in neuronal arborizations.

Figure 12.26 shows the response of the idealized neuron to 10 excitatory inputs distributed on the soma and on compartments 8, 13, 26, and 29. The synaptic inputs were activated simultaneously at $t_{start} = 0$ and reached a peak at $t_{peak} = 0.2$ ms with a peak conductance equal to 40 nS and a synaptic reversal potential $V_{syn} = 80$ mV.

The generation of the action potential does not occur when six excitatory synaptic inputs distributed on compartments 18, 22 and 28 are activated simultaneously at $t_{start} = 0.5$ ms [$t_{peak} = 0.7$ ms, $G_{syn}(t_{peak}) = 4$ nS, $V_{syn} = 80$ mV] and three inhibitory synaptic inputs placed on compartments 15, 20, and 25 are activated at $t_{start} = 0$ [$t_{peak} = 0.8$ ms, $G_{syn}(t_{peak}) = 40$ nS, $V_{syn} = 0$]. The simulation results are shown in Fig. 12.27. In this case, the inhibitory synaptic inputs activated ionic conductances characterized by a reversal potential equal to the resting membrane potential (shunting or silent inhibition). They were distributed on the direct path from the locations of the excitatory synaptic inputs to the soma. Under these conditions, they could veto the excitatory input effects as described in Ref. 18.

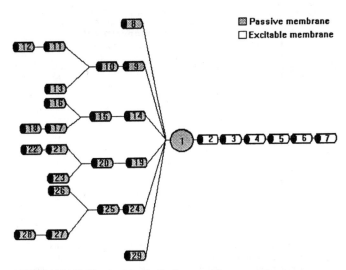

FIGURE 12.25 Sketch of the idealized neuron. The open compartments are related to the excitable membrane and the shaded compartments to the passive membrane.

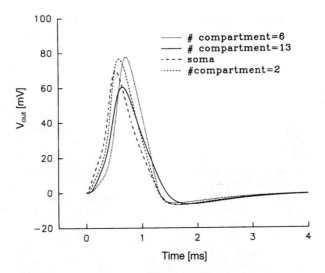

FIGURE 12.26 Response of the idealized neuron to 10 excitatory synaptic inputs distributed along the dendritic branches (compartments 8, 13, 26, 29) and on the soma. The synaptic inputs are activated simultaneously at $t_{start} = 0$ and reach peak conductance at $t_{peak} = 0.2$ ms with a peak conductance $G_{syn}(t_{peak}) = 40$ nS and an equilibrium potential $V_{syn} = 80$ mV.

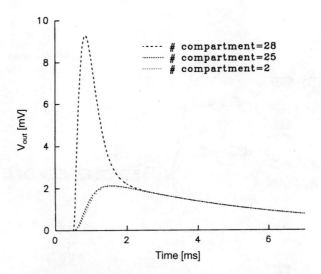

FIGURE 12.27 Response of the idealized neuron to six excitatory synaptic inputs distributed on dendritic tree terminations and three inhibitory synaptic inputs distributed on the direct path from the locations of the excitatory inputs to the soma.

Simulations of Extracellular Recording via Planar Microtransducers. The circuit used for the simulations of extracellular recording via planar microtransducer is that of Fig. 12.21. Figure 12.28 shows the translation of the electrical equivalent circuit into the input for SPICE. The area of the microtransducer considered in these simulations is 10×10 μm^2, and a constraint $\omega R_e C_e = 1$ at a frequency of 1 kHz is assumed.

In the simulations, a patch of excitable membrane 10 μm in diameter is considered. In order to model the coupling between the microtransducer and the excitable membrane, we have used a specific Hodgkin-Huxley compartment 10 μm long. Figure 12.29 shows the results of two simulations of electrical activity recording which differ only in the sealing impedance (5 and 50 MΩ, respectively). The impedance value 5 MΩ corresponds to a "loose patch" condition. The shape of the signal for a sealing impedance of 5 MΩ is similar to the second time derivative of the action potential. A larger and slower signal is obtained with a sealing impedance of 50 MΩ.

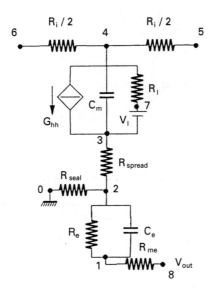

R_{i1} 6 4 value (R_{i1})
R_{i2} 5 4 value (R_{i2})
C_m 4 3 value (C_m)
R_l 4 7 value (R_l)
G_{hh} 4 3 4 0 0. 0. value (compart. area)
V_l 7 3 dc value (V_l)
R_{seal} 2 0 value (R_{seal})
R_{spread} 3 2 value (R_{spread})
C_e 2 1 value (C_e)
R_e 2 1 value (R_e)
R_{me} 1 8 value (R_{me})

FIGURE 12.28 Translation of the electrical equivalent circuit of the junction between a membrane patch and a passive microtransducer into an input for SPICE.

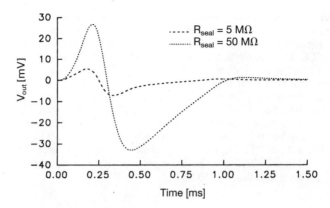

FIGURE 12.29 SPICE outputs of two passive microtransducer recordings of axon activity for R_{seal} equal to 5 and 50 MΩ, respectively (V_{out} amplification stage gain = 20).

12.5 USE OF SPICE FOR MODELING SILICON NEURONS*

In Sec. 12.4, we have dealt with the so-called software algorithm approach to model the neuron. Another approach proposed by C. Mead[21] is the so-called *hardware architecture approach*.

Following this approach, Mahowald and Douglas[22] proposed a device named the *silicon neuron*, i.e., a silicon device whose performance simulates that of its biological counterpart (i.e., a neuron) at the molecular level. In other words, a silicon neuron is an assembly of MOSFET circuits that generates the equivalents of the ionic currents and of the action potentials of a real (biological) neuron. Therefore, it can be described as a realistic, silicon technology–based, physical analog of a neuron.

In this section we show that SPICE can be used as a powerful tool to simulate the behavior of silicon neurons and to design small networks of such neurons.[23]

12.5.1 Silicon Neuron Model

The silicon neuron (SN) model follows directly the arrangement of the MOSFET circuits described in Ref. 22. We have seen in Sec. 12.4.1 that the nerve cell membrane is represented by a fixed-value capacitor C_m and by variable conductances that represent the membrane conductances in series with a battery, which is the equilibrium potential of the corresponding ion.[15] Equation (12.58), which describes the membrane voltage as a function of time, is here rewritten for reader's convenience in the form

$$C_m \frac{dV_{mem}}{dt} = \sum_j I_{ion\,j} \qquad (12.67)$$

---

*From M. Grattarola, M. Bove, S. Martinoia, and G. Massobrio.[23] Used by permission of Peter Peregrinus Limited, Stevenage Herts, United Kingdom.

Each ionic current $I_{ion\,j}$ can be expressed as

$$I_{ion\,j} = g_j m^M n^N (V_{mem} - V_j) \qquad (12.68)$$

where g_j = maximum conductance of the current $I_{ion\,j}$
   m, n = activation and inactivation degrees, respectively
   M, N = respective exponents
   V_{mem} = membrane potential
   V_j = equilibrium potential

Let us consider a generic Hodgkin-Huxley neuron; then we deal only with potassium and sodium conductances. The behaviors of these conductances g_j (activation and inactivation states) can be characterized by the time- and voltage-dependent output of circuits based on low-pass filters, differential pairs, and current mirrors (see Sec. 9.7) connected together to form a specific configuration to model the potassium and sodium conductances. In particular, for example, sodium current can be modeled by implementing in SPICE the circuit of Fig. 12.30.

The sodium current I_{Na} depends on the interaction between the blocks of the activation (modeled by the drain current of MOS4) and inactivation (modeled by the drain current of MOS2) processes. These two drain currents are added at node 30 of the circuit of Fig. 12.30. Circuits analogous to that shown in Fig. 12.30 for the sodium current, have been implemented in SPICE to simulate the potassium current. With such circuits connected in the appropriate configuration, a circuit model (*silicon neuron*) can be obtained to simulate the electrophysiological behavior of a neuron.

12.5.2 Synaptic Circuits

Usually in the nerve cell membrane there are other channels (besides sodium and potassium) that are influenced by the presence of external chemical agents such as neurotransmitters. In this case, synaptic channel kinetics need to be added to Eq. (12.67). Usually, a synaptic current is modeled as a time-dependent conductive pathway in series with a constant voltage source that represents the synaptic equilibrium potential of the corresponding ionic species; the expression for the synaptic conductance is based on the well-known alpha function.[19]

Here we present a different approach to model the synaptic conductance.[24] The synaptic currents are modeled by using Eqs. (12.68), after setting the M and N exponents to 1 and 0, respectively. The activation degree m is defined as the fraction of postsynaptic receptors in the open state. Mathematically, it can be expressed as

$$\frac{dm}{dt} = \alpha[T](1-m) - \beta m \qquad (12.69)$$

where [T] is the concentration of the neurotransmitters in the synapse, and α and β are the forward and backward binding rates, respectively. A pulse of neurotransmitters is released whenever a presynaptic spike occurs.

Starting from the above expression, we have designed three circuits to simulate the behaviors of an inhibitory synapse, an excitatory synapse and the neurotransmitter release, respectively. In the following we describe each circuit.

Inhibitory Synapse Circuit.   The inhibitory synapse circuit is shown in Fig. 12.31. The voltage V_{rel} is low-pass-filtered by a follower-integrator connected to the activation stage.

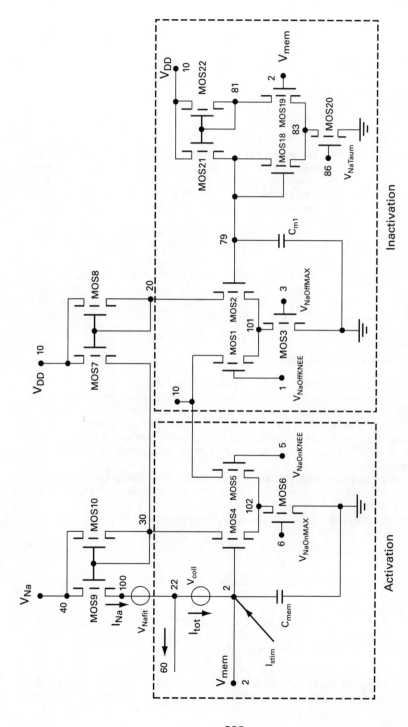

FIGURE 12.30 Circuit which models the sodium current of the action potential in a silicon neuron. MOS3, MOS7, MOS9, MOS10, MOS21, MOS22 are *p*-channel MOSFETs; the other ones are *n*-channel.

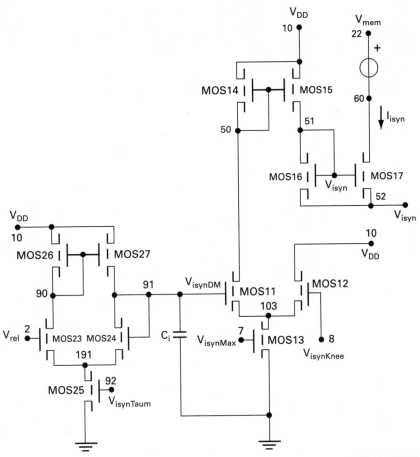

FIGURE 12.31 MOSFET circuit modeling inhibitory inputs in a postsynaptic neuron. MOS14, MOS15, MOS26, MOS27 are *p*-channel MOSFETs; the other ones are *n*-channel.

The voltage V_{rel} represents the pulse of the neurotransmitter released when a presynaptic spike occurs. Both the voltage amplitude (i.e., the concentration of the released neurotransmitter) and the duration of the pulse can be modulated.

The neurotransmitter release will be considered below. The activation stage is controlled by the $V_{isynTaum}$, $V_{isynMax}$, and $V_{isynKnee}$ signals, which are the activation time constant, the maximum-activation voltage, and the half-activation voltage, respectively. A current mirror stage, reflecting the activation signal, allows us to control the synaptic-conductance transistor MOS17. The current that drains in this MOSFET depends on the voltage V_{isyn} applied to the source and on the membrane voltage V_{mem} applied to the drain. The current flows from the drain to the source; thus MOS17 drains the current from the membrane capacitance to drive the membrane voltage to the inhibitory synaptic equilibrium potential V_{isyn} and to hyperpolarize the nerve cell membrane.

The conductance value of the synaptic channel can be modulated by changing the voltage threshold for MOS17.

Excitatory Synapse Circuit. The circuit simulating the behavior of an excitatory synapse is shown in Fig. 12.32. In this circuit, the same activation circuit as used for the inhibitory synapse circuit (Fig. 12.31) connected to the neurotransmitter-release circuit (see below) is implemented. The activation signal is transformed into a voltage control signal and applied to the gate of the synaptic-conductance transistor MOS9. The current flows from the source to the drain and depends on the voltage V_{esyn} (excitatory synaptic equilibrium potential) applied to the source and on the V_{mem} applied to the drain. Thus the current flows in the membrane capacitance to depolarize the nerve cell membrane to the V_{esyn} value.

Neurotransmitter Release Circuit. A circuit that models the neurotransmitter release in the presence of a presynaptic spike is shown in Fig. 12.33. This circuit is represented by a transconductance amplifier and it works as a comparator. The inverting input is connected to a reference voltage which represents the minimum membrane voltage value to detect the presence of a presynaptic spike. The noninverting input is connected to the membrane voltage of the presynaptic neuron, and, as soon as it becomes higher (lower) than the reference voltage, the output voltage switches high (low) and clips against the dc voltage V_h (V_l).

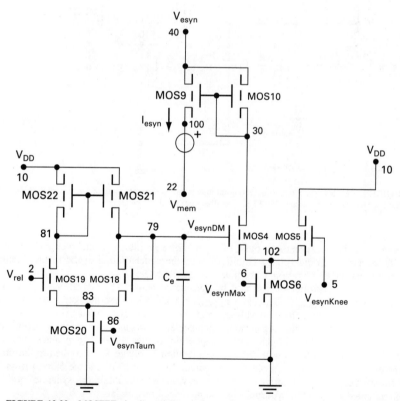

FIGURE 12.32 MOSFET circuit modeling excitatory inputs in a postsynaptic neuron. MOS9, MOS10, MOS21, MOS22 are *p*-channel MOSFETs; the other ones are *n*-channel.

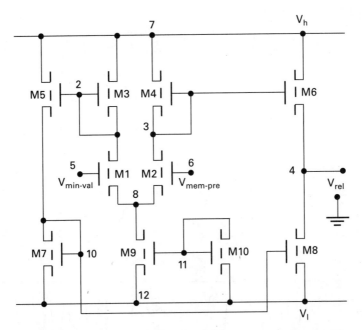

FIGURE 12.33 Comparator circuit represented by a transconductance amplifier modeling the neurotransmitter release in the presence of presynaptic spikes. M3, M4, M5, M6 are *p*-channel MOSFETs; the other ones are *n*-channel.

By changing the value of the reference voltage and the dc voltage (i.e., V_h and V_l), it is possible to modulate the amplitude and the duration of the pulse of the neurotransmitter release.

Structural and functional details of a silicon neuron can be modified by changing the electrical and physical parameters[3] of the MOSFETs that describe the silicon neuron and silicon synapses.

12.5.3 Silicon Neuron Simulation with SPICE

Figure 12.34a shows the connection between two neurons through an excitatory synapse. Figure 12.34b shows how a train of presynaptic spikes (V_{N1}) evokes excitatory currents (I_{esyn}) and a subsequent depolarization phase of the postsynaptic neuron (V_{N2}), by releasing a series of transmitter pulses. As in biological neurons, the firing rate changes as a function of the amplitude of the current stimulus.

If the strength of the postsynaptic conductances is increased, i.e. if the neurotransmitter release (the pulsed value of V_{rel}) is increased, the firing rate of the postsynaptic neuron (V_{N2}) increases as a function of the amplitude of the excitatory currents (Fig. 12.34c).

Figure 12.34d refers to a network of three neurons: two presynaptic neurons (V_{exc} and V_{in}) and one postsynaptic neuron (V_{post}). The effects of excitatory and inhibitory synaptic currents activated at different times by V_{exc} and V_{in}, on the response of V_{post}, are shown. Interestingly enough, when the inhibitory input placed on the postsynaptic neuron is activated, a veto effect occurs on spike generation. This effect is biologically meaningful and

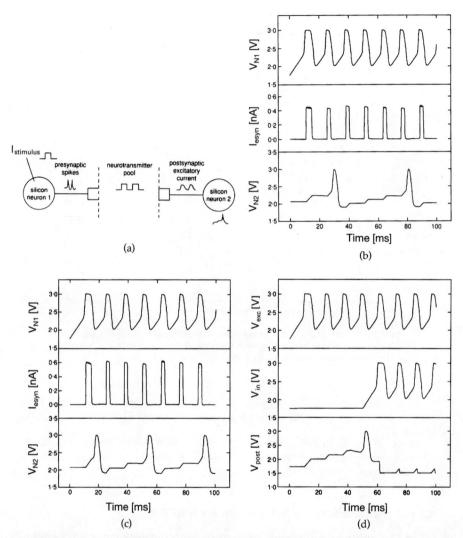

FIGURE 12.34 Models of simple two- and three-neuron networks: (*a*) connection between two neurons through an excitatory synapse; (*b*, *c*) responses of the postsynaptic neuron V_{N2} (lower trace) to excitatory synaptic currents I_{esyn} (middle trace), resulting from a train of spikes in the presynaptic neuron V_{N1} (upper trace), as a function of different values of the excitatory synaptic conductance; (d) responses of the postsynaptic neuron V_{post} (lower trace) to excitatory (V_{exc}, $t_{start} = 0$ ms) (upper trace) and inhibitory (V_{in}, $t_{start} = 50$ ms) (middle trace) presynaptic neurons; in the last condition, a veto effect occurs on spike generation.

has previously been observed in dendritic trees. Its role in information processing has been investigated in the literature as a function of several parameters such as timing and localization.[18] It should be stressed that the present method allows one to introduce synaptic conductances at low computation cost by translating a kinetic model into simple MOSFET circuits.

REFERENCES

1. L. W. Nagel and D. O. Pederson, "Simulation program with integrated circuit emphasis (SPICE)," Electronics Research Laboratory Report, ERL-M382, University of California, Berkeley, 1973.
2. L. W. Nagel, "SPICE2: A computer program to simulate semiconductor circuits," Electronics Research Laboratory Report, ERL-M520, University of California, Berkeley, 1975.
3. G. Massobrio and P. Antognetti, *Semiconductor device modeling with SPICE,* New York: McGraw-Hill, 2d ed., 1993.
4. A. Vladimirescu, K. Zhang, A. R. Newton, D. O. Pederson, and A. Sangiovanni-Vincentelli, *SPICE version 2G user's guide,* Dept. EECS, University of California, Berkeley, 1981.
5. M. Grattarola, G. Massobrio, and S. Martinoia, "Modeling H^+ sensitive FET's with SPICE," *IEEE Trans. Electron Devices,* ED-39(4): 1992.
6. G. Massobrio, S. Martinoia, and M. Grattarola, "Use of SPICE for modeling silicon-based chemical sensors," *Sensors and Materials,* 6(2), 1994.
7. D. E. Yates, S. Levine, and T. W. Healy, "Site-binding model of the electrical double layer at the oxide/water interface," *J. Chem. Soc. Faraday Trans.,* 70(11), 1974.
8. D. L. Harame, "Integrated circuit chemical sensors," Integrated Circuit Laboratory Report G558-12, Stanford University, Stanford, Calif., 1984.
9. S. Martinoia, M. Grattarola, and G. Massobrio, "Modeling non-ideal behaviors in H^+-sensitive FETs with SPICE," *Sensors and Actuators,* B(7), 1992.
10. G. Massobrio and S. Martinoia, "Modeling the ISFET behavior under temperature variations using BIOSPICE," *Electronics Letters,* 32: 10, 1996.
11. D. G. Hafeman, J. W. Parce, and H. M. McConnell, "Light-addressable potentiometric sensor for biochemical systems," *Science,* 240: 1182–1185, 1988.
12. G. Massobrio, S. Martinoia, and M. Grattarola, "Light-addressable chemical sensors: modeling and computer simulations," *Sensors and Actuators,* B(7), 1992.
13. M. Grattarola, S. Martinoia, G. Massobrio, A. Cambiaso, R. Rosichini, and M. Tetti, "Computer simulations of the responses of passive and active integrated microbiosensors to cell activity," *Sensors and Actuators* B(4): 261–265, 1991.
14. M. Bove, G. Massobrio, S. Martinoia, and M. Grattarola, "Realistic simulations of neurons by means of an ad hoc modified version of SPICE," *Biol. Cybern.,* 71: 137–145, 1994.
15. L. Hodgkin and A. F. Huxley, "A quantitative description of membrane current and its applications to conduction and excitation in nerve," *J. Physiol.,* 117: 500–544, 1952.
16. B. Bunow, I. Segev, and J. W. Fleshman, "Modeling the electrical behavior of anatomically complex neurons using a network analysis program: excitable membrane," *Biol. Cybern,* 53: 41–56, 1985.
17. I. Segev, J. W. Fleshman, and R. E. Burke, "Compartmental models of complex neurons," in C. Koch and I. Segev (eds.), *Methods in neuronal modeling: From synapses to network,* Cambridge, Mass.: 63–96, Bradford Book, MIT Press, 1989.
18. C. Koch, T. Poggio, and V. Torre, "Nonlinear interactions in a dendritic tree: localization, timing, and role in information processing," *Proc. Natl. Acad. Sci.,* U.S.A., 85: 4075–4078, 1983.

19. W. Rall, "Distinguishable theoretical synaptic potentials computed for different soma-dendritic distributions of synaptic input," *J. Neurophysiol.,* 30: 1138–1168, 1967.
20. M. Grattarola and S. Martinoia, "Modeling the neuron-microtransducer junction from extracellular to patch recording," *IEEE Trans. Biomed. Eng.,* 40: 35–41, 1993.
21. C. Mead, "Neuromorphic electronic systems," *Proc. IEEE,* 78(10): 1629–1636, 1990.
22. M. Mahowald and R. Douglas, "A silicon neuron," *Nature,* 354: 515–518, 1991.
23. M. Grattarola, M. Bove, S. Martinoia, and G. Massobrio, "Silicon neuron simulation with SPICE: tool for neurobiology and neural networks," *Med. Biol. Eng. & Comput.,* 33: 533–536, 1995.
24. A. Destexhe, Z. F. Mainen, and T. J. Sejnowski, "An efficient method for computing synaptic conductances based on a kinetic model of receptor binding," *Neural Comput.,* 6: 14–18, 1994.

APPENDIX A
PHYSICAL CONSTANTS AND MATERIAL PROPERTIES

A.1 PHYSICAL CONSTANTS

Quantity	Symbol	Value	Units
Avogadro number	N_{AV}	6.02212×10^{23}	mole^{-1}
Boltzmann constant	k	1.38066×10^{-23}	J/K
($= R/N_{AV}$)		8.61738×10^{-5}	eV/K
Electronic charge	q	1.60218×10^{-19}	C
Electron volt	eV	1.60218×10^{-19}	J
		23.053	kcal/mole
Faraday constant	$\mathscr{F}$	9.6485×10^{4}	C/mole
Gas constant	R	8.31451	J/(mole-K)
		1.98719	cal/(mole-K)
Mass of electron at rest	m_0	9.10939×10^{-31}	kg
		0.568×10^{-15}	eV-s^2/cm^2
Mass of neutron at rest	m_n	1.67493×10^{-27}	kg
Mass of proton at rest	m_p	1.67265×10^{-27}	kg
Permeability in vacuum	μ_0	1.25663×10^{-8}	H/cm
($= 4\pi \times 10^{-9}$)			
Permittivity in vacuum	ε_0	8.85418×10^{-14}	F/cm
($= 1/(\mu_0 c^2)$)			
Planck constant	h	6.62608×10^{-34}	J-s
		4.13566×10^{-15}	eV-s
Reduced Planck constant	$\hbar = h/2\pi$	1.05457×10^{-34}	J-s
		0.659×10^{-15}	eV-s
Speed of light in vacuum	c	2.99793×10^{10}	cm/s
Thermal energy at 300 K	kT	25.86×10^{-3}	eV
		4.14×10^{-21}	J
Thermal voltage at 300 K	kT/q	25.86×10^{-3}	V
Joule	J	0.2389	cal

A.2 PROPERTIES OF Si, GaAs, SiO$_2$, Si$_3$N$_4$, Al$_2$O$_3$ (AT 300 K)

Property	Symbol	Units	Si	GaAs	SiO$_2$	Si$_3$N$_4$	Al$_2$O$_3$
Breakdown field	E_B	V/cm	$\simeq 3 \times 10^5$	$\simeq 4 \times 10^5$			
Diffusion coefficient:							
Electron	D_n	cm^2/s	34.91	219.81			
Hole	D_p	cm^2/s	12.41	10.34			
Effective mass:							
Electron	m^*_n/m_0		1.18^a; 0.26^b	0.068			
Hole	m^*_p/m_0		0.81^a; 0.386^b	0.50			
Electron affinity	$q\chi$	eV	4.05	4.07	1.0		
Energy gap	E_g	eV	1.124	1.424	9	4.7	
Intrinsic carrier concentration	n_i	cm^{-3}	1.45×10^{10}	2.1×10^6			
Intrinsic resistivity	ρ	Ω-cm	3.16×10^5	10^8	10^{14}–10^{16}	10^{14}	
Minority carrier lifetime	τ	s	2.5×10^{-3}	$\simeq 10^{-8}$			
Mobility:							
Electron	μ_n	cm^2/V-s	1350^c	8500^c			
Hole	μ_p	cm^2/V-s	480^c	400^c			
Relative permittivity	ε_r		11.9	13.1	3.9	7.5	7.8
Effective density of states:							
Conduction band	N_c	cm^{-3}	2.8×10^{19}	4.7×10^{17}			
Valence band	N_v	cm^{-3}	1.04×10^{19}	7.0×10^{18}			

^{a}Used in density-of-state calculations.
^{b}Used in conductivity calculations.
^{c}Values for intrinsic (undoped) material.

APPENDIX B
MATHEMATICAL OPERATORS

If in any region of space a function f is defined for every point of the region, then this region is called a *field*. Let us assume that the functions under consideration are single-valued and continuous and have continuous first space derivatives in the regions under consideration.

If the function f is a *scalar point function*—that is, if f is a function of the three variables x, y, z such that to each of the values x, y, z a scalar number f is defined—then the field is called a *scalar field*. Examples of such functions are temperature and electrostatic potential.

If the function f is a *vector point function*—that is, a function that defines a vector at every point of the field—then the field is called a *vector field*. An example of a vector point function is the function representing the force at every point of the field under consideration.

By using certain differential operators, it is possible to associate a vector field with each scalar field and vice versa. This connection is of fundamental importance in many investigations in physics and electronics.

The form of the mathematical operators we consider is restricted to the rectangular coordinate form, as this representation is the one used in the book.

B.1 VECTOR DIFFERENTIAL OPERATOR (∇)

The symbol

$$\nabla = \frac{\partial}{\partial x}\mathbf{i} + \frac{\partial}{\partial y}\mathbf{j} + \frac{\partial}{\partial z}\mathbf{k} \qquad (B.1)$$

(read *nabla* or *del*) represents a *vector differential operator*. In Eq. (B.1), $\mathbf{i, j, k}$ are the base vectors of unit length that are directed along the positive directions of the $x, y,$ and z axes, respectively.

It should be borne in mind that the symbol ∇ does not itself represent a vector quantity. It is only a symbol showing the fact that certain operations of differentiation are to be performed on the scalar function which follows it. It must be taken into account that the result is a vector point function which is obtained from the scalar point function f. Hence, ∇f forms the vector field that is associated with the scalar field f.

B.2 LAPLACIAN OPERATOR (∇^2)

The symbol

$$\nabla^2 = \frac{\partial^2}{\partial x^2} + \frac{\partial^2}{\partial y^2} + \frac{\partial^2}{\partial z^2} \qquad (B.2)$$

(read *nabla squared* or *delta*) is called the *Laplacian operator*.

B.3 GRADIENT

Let us consider a *scalar field* in space given by a function $f(P) = f(x, y, z)$, where x, y, z are cartesian coordinates, and let us assume that f has continuous first partial derivatives. Then, the function

$$\text{grad } f = \nabla f = \frac{\partial f}{\partial x}\mathbf{i} + \frac{\partial f}{\partial y}\mathbf{j} + \frac{\partial f}{\partial z}\mathbf{k} \qquad (B.3)$$

defines a *vector*, and it is called the *gradient of the scalar function f*.

Some of the vector fields encountered in physics and electronics are given by vector functions that can be obtained as the gradients of *ad hoc* scalar functions. Such a scalar function is called a *potential function* of the corresponding vector field. The use of potentials simplifies the investigations of the vector fields. As examples of this approach to vector fields, we can consider Newton's law of gravitation or, in electrostatics, the force of attraction (or repulsion) between two particles of opposite (or like) charge.

If the vector function defining a vector field is the gradient of a scalar function, the field is said to be *conservative*, because, in such a field, the work done in moving a particle from a point P_1 to a point P_2 in the field depends only on P_1 and P_2 but not on the path along which the particle is moved from P_1 to P_2.

B.4 DIVERGENCE OF A VECTOR FIELD

Let $\mathbf{v}(x, y, z)$ be a differentiable vector function, where x, y, z are cartesian coordinates, and let v_1, v_2, v_3 be the components of $\mathbf{v}$. Then the function

$$\text{div } \mathbf{v} = \frac{\partial v_1}{\partial x} + \frac{\partial v_2}{\partial y} + \frac{\partial v_3}{\partial z} \qquad (B.4)$$

is called the *divergence of the vector function v (field)*. Another common notation for the divergence of $\mathbf{v}$ is $\nabla \mathbf{v}$

$$\text{div } \mathbf{v} = \nabla \cdot \mathbf{v} = \left(\frac{\partial}{\partial x}\mathbf{i} + \frac{\partial}{\partial y}\mathbf{j} + \frac{\partial}{\partial z}\mathbf{k}\right) \cdot (v_1\mathbf{i} + v_2\mathbf{j} + v_3\mathbf{k})$$

$$= \frac{\partial v_1}{\partial x} + \frac{\partial v_2}{\partial y} + \frac{\partial v_3}{\partial z} \qquad (B.5)$$

where, for example, $(\partial/\partial x) v_1$ in the dot product means the partial derivative $\partial v_1/\partial x$. Note that $\nabla \cdot \mathbf{v}$ means the scalar *div* $\mathbf{v}$, whereas ∇f means the vector grad f defined in Sec. B.3.

B.5 CURL OF A VECTOR FIELD

Let x, y, z be right-handed cartesian coordinate system, and let

$$\mathbf{v}(x, y, z) = v_1 \mathbf{i} + v_2 \mathbf{j} + v_3 \mathbf{k} \tag{B.6}$$

be a differentiable vector function. Then the function

$$\text{curl } \mathbf{v} = \nabla \times \mathbf{v} = \begin{vmatrix} \mathbf{i} & \mathbf{j} & \mathbf{k} \\ \dfrac{\partial}{\partial x} & \dfrac{\partial}{\partial y} & \dfrac{\partial}{\partial z} \\ v_1 & v_2 & v_3 \end{vmatrix}$$

$$= \left(\frac{\partial v_3}{\partial y} - \frac{\partial v_2}{\partial z} \right) \mathbf{i} + \left(\frac{\partial v_1}{\partial z} - \frac{\partial v_3}{\partial x} \right) \mathbf{j} + \left(\frac{\partial v_2}{\partial x} - \frac{\partial v_1}{\partial y} \right) \mathbf{k} \tag{B.7}$$

is called the *curl of the vector function* **v** *(field)*. For a left-handed cartesian coordinate system, the determinant in Eq. (B-7) is preceded by a minus sign.

Instead of a *curl* **v**, the notation *rot* **v** is also used in the literature.

B.6 BASIC RELATIONS FOR THE MATHEMATICAL OPERATORS

$$\nabla(fg) = f\nabla g + g\nabla f, \qquad \nabla(f/g) = (1/g^2)(g\nabla f - f\nabla g) \tag{B.8}$$

$$\text{div }(f\mathbf{v}) = f \text{ div } \mathbf{v} + \mathbf{v} \cdot \nabla f, \qquad \text{div }(f\nabla g) = f\nabla^2 g + \nabla f \cdot \nabla g \tag{B.9}$$

$$\nabla^2 f = \text{div }(\nabla f), \qquad \nabla^2(fg) = g\nabla^2 f + 2\nabla f \cdot \nabla g + f\nabla^2 g \tag{B.10}$$

$$\text{curl }(f\mathbf{v}) = \nabla f \times \mathbf{v} + f \text{ curl } \mathbf{v} \tag{B.11}$$

$$\text{div }(\mathbf{u} \times \mathbf{v}) = \mathbf{v} \cdot \text{curl } \mathbf{u} - \mathbf{u} \cdot \text{curl } \mathbf{v} \tag{B.12}$$

$$\text{curl }(\nabla f) = 0, \qquad \text{div }(\text{curl } \mathbf{v}) = 0 \tag{B.13}$$

In the above equations, f and g are two scalar functions, while **v** and **u** are two vector functions.

INDEX

Absorption of light, 6, 22, 89
   threshold frequency, 22
Acceptor, 35
Acid:
   Brönsted definition, 87
   strong, 86
   weak, 86
Alkanethiols, 307
Amino acid, 92
   alanine, 93
   aspartic acid, 93–94
   tryptophan, 92
Atom:
   Bohr, 4–6
   electronic configuration, 11
Autocatalysis, 128–129
Avogadro constant, 194

Base:
   Brönsted definition, 87
   strong, 86
   weak, 86
Beer-Lambert law, 23, 89
Behavior, 36
Binding sites, 181
Biological field-effect transistor (BIOFET):
   buffer capacity, 370
   model theory, 365–370
   SPICE models, 371–374
Biosensors:
   applications, 287
   biological component, 283
   cell-based, 302–304
   characteristics, 286
   ENFET, 294–300
   ISFET, 287–293, 300–302
   LAPS, 304–306

Biosensors *(Cont.)*
   microfabrication, 307–310
   transducer, 284–285
Bohr atom, 4–6
Boltzmann:
   constant, 7, 173, 194, 205
   distribution, 125, 140,182, 184, 188, 191, 205, 222, 232, 292, 358, 369
   transport equation, 60–63
Bonds:
   chemical, 3, 4
   covalent, 3, 16
   dipole-dipole, 83
   hydrogen, 83
   ion-dipole, 82
   ion-ion, 81
   metallic, 3
   peptide, 92
   physical, 3, 4, 81
Bottom-up, 92, 99
Breakdown:
   avalanche, 161, 246
   voltage, 162
   Zener, 162
Brönsted definition, 85, 87
Brownian motion, 101, 114–117
Buffer capacity, 87, 88, 291–293, 304, 370
Butler-Volmer equation, 180

Capacitance:
   diffusion, 159
   Helmholtz, 357
   junction, 157, 171
   membrane, 202
   oxide, 213
Cell:
   cycle, 106

Cell *(Cont.)*
  cytoskeleton, 105
  eucaryotic, 105–106
  membrane, 101–104
  organelles, 106
Cell-based biosensors, 302–304
Channel:
  fluctuation, 336
  ion, 103–104
  length modulation, 245
  pinch-off, 243, 245, 248–251
Charge carrier:
  acceptors, 35
  collision, 46
  diffusion length, 152–154
  donors, 35
  electrons, 23
  excess, 150
  holes, 24
  intrinsic, 33
  lifetime, 42, 43, 68, 152, 154, 160
  majority, 39
  minority, 42
  mobility, 47–51, 62
  motion, 45, 54
  photogenerated, 23
Chemical field-effect transistor (CHEMFET):
  buried layer, 357, 359
  Gouy-Chapman layer, 357
  Helmholtz layer, 357
  large-signal model, 360
  parameters, 360
  SPICE model, 361–365
  static model, 356–360
  temperature model, 360–361
  threshold voltage, 359
Chemical reactions:
  autocatalysis, 128, 129
  enzyme kinetics, 126–128
  equilibrium constant, 125, 181
  first-order, 123
  irreversible, 122–123
  kinetic rate constant, 122
  Michaelis-Menten constant, 127
  reversible, 124–125
  second-order, 123
  stochastic approach, 125–126
Coagulation, 186
Colloidal particles, 184–187
Combinatorial chemistry, 307
Continuity equation, 71–75
Convection, 111

Coulomb:
  energy, 81
  force, 82
Current density:
  diffusion, 56–58
  displacement, 73
  drift, 51–52
  electron, 58
  hole, 58
  ion, 52, 58, 194
  saturation, 153, 155, 173, 269
  subthreshold, 268–269

Debye length, 184, 190, 233
Deoxyribonucleic acid (*see* DNA)
Depletion:
  approximation, 144
  region width, 146, 148, 171, 215, 223, 246, 248
Diffusion:
  capacitance, 159
  coefficient, 55–58, 62, 113, 195, 199
  electron current, 56
  hole current, 57, 58
  ion current, 58
  length, 152–154
  solution, 109–113
Diode:
  breakdown, 161–163
  diffusion capacitance, 159
  forward characteristic, 163
  generation-recombination, 163, 164
  high-level injection, 165
  ideal equation, 155
  junction capacitance, 157
  large-signal model, 168
  long-base, 152, 158
  low-level injection, 150
  resistance, 166
  reverse characteristic, 161–163
  reverse saturation current, 153–155
  Schottky, 173
  short-base, 153, 158
  small-signal model, 168
  SPICE models, 167–168
  static model, 168
  transit time, 168
  voltage drop in neutral regions, 165
  Zener, 162
Dirac delta function, 111, 115, 331
Discontinuous flow, 204–207

INDEX

DNA:
  A-form, 96
  B-form, 96
  bases, 96
  computation, 98
  exons, 97
  introns, 97
  transcription, 97
  Z-form, 96
Donnan equilibrium, 202–204, 317
Donor, 35
Drift:
  electron current, 51
  hole current, 52
  ion current, 52
  velocity, 46, 47

Effective density of states, 32–33
Effective mass, 34–35, 46
Einstein relations, 57, 62
Einstein-Smoluchowsky equation, 113, 117
Electrode:
  mercury, 179
  reference, 178, 287, 289, 290, 302, 360
Electrolyte solution, 81, 84–86
Electrophoresis, 117–121
  sodium dodecylsulfate (SDS), 120
Energy :
  bands, 14–19
  binding, 36
  gap, 17, 27
  rotational, 90–91
  vibrational, 90–91
Enzyme:
  autocatalysis, 128–129
  kinetics, 126–128
  Michaelis-Menten constant, 127
  velocity of the reaction, 127
Enzyme field-effect transistor (ENFET):
  diffusion modulus, 297
  enzyme-catalyzed reactions, 295–296
  mode of operation, 297–300

Faraday constant, 193
Fermi-Dirac statistics, 27–31
Fermi energy level, 31, 40
Fick laws, 55, 109, 195
Fitzhugh-Nagumo model, 326
Flash of light, 40–43
Flatband voltage, 228, 234

Flocculation, 186
Fluorescence, 25, 89–90
  lifetime, 90
  quantum yield, 89
Fokker-Planck equation, 117
Frictional coefficient, 114, 118
Frictional force, 114, 117

Gangliosides, 101
Gauss law, 190, 216, 233
Generation-recombination, 33, 40–43, 63–71, 84
Genetic code, 98
Glycocalyx, 102
Glycoconjugates, 99
Glycolipids, 99
Goldman assumption, 196
Goldman-Hodgkin-Katz (GHK) equation, 198, 318
Gouy-Chapman:
  layer, 179, 184, 357
  potential, 185, 186

Hall:
  coefficient, 76
  effect, 75, 76
  voltage, 76
Helmholtz:
  capacitance, 342, 357
  layer, 179, 187, 342, 357
High-level injection, 70
Hodgkin-Huxley (HH) equations, 322–324, 335, 380, 381
Hund multiplicity rule, 8
Hydrogen bond, 83
Hydronium ion, 86

Insulator, 17
Ion:
  Ca, 83
  channels, 101, 103, 104
  current, 52, 58, 194
  hydrated, 83
  interactions, 81, 82
  K, 83
  Li , 83
  Mg , 83
  mobility, 52
  Na , 83

Ion-sensitive field-effect transistor (ISFET):
  buffer capacity, 291–293
Ion-sensitive field-effect transistor *(Cont.)*
  modes of operation, 289–290, 301, 302
  parameters, 290
  pH sensitivity, 290–293
  threshold voltage, 290
  transient responses, 300–301
Ionic strength, 88, 89
Isoelectric point, 88

Kinetic rate constants, 122

Langevin equation, 114
Light-addressable potentiometric sensor (LAPS):
  model, 305
  operation mechanisms, 304–306
  SPICE models, 374–379
Liposome, 99, 100
Low-level injection, 42, 70, 150

Mass action law, 34, 85, 122
Maxwell distribution, 116, 140
Membrane:
  capacitance, 202, 318
  channels, 101
  proteins, 101
Metal-oxide semiconductor (MOS) structure:
  accumulation operating mode, 212–213, 225, 230, 233
  capacitance-voltage (CV) characteristics, 225–231
  contact potential effects, 227
  depletion operating mode, 214–220, 225, 230, 233
  depletion region width, 215, 218, 220, 223
  flatband voltage, 228, 234
  inversion operating mode, 220–225, 233
  ion implantation, 231
  parasitic charge effects, 227
  surface potential, 218, 223
  threshold voltage, 220, 223, 227, 231, 235
Metal-oxide semiconductor transistor (MOSFET):
  breakdown, 246
  channel-length modulation, 245
  current-voltage characteristics, 240–246, 249–251
  depletion-mode operation, 246–251

Metal-oxide semiconductor transistor *(Cont.)*
  depletion region width, 246, 248
  enhancement-mode operation, 237–246
  operating regions, 240–241, 267–269
  pinch-off, 243, 245, 248, 249, 251
  small-signal model, 260–267
  SPICE models, 352–356
  subthreshold mode operation, 267–269
  symbols and conventions, 239
  threshold voltage, 237, 240, 243, 245, 255, 259, 263
  types, 238–239
Metal-semiconductor (MS) junction:
  built-in voltage, 171
  depletion region width, 171
  diode, 169–173
  energy bands, 169
  junction capacitance, 171
  ohmic contact, 174
  saturation current, 173
  Schottky barrier, 169
Michaelis-Menten constant, 127, 296, 369
Microfabrication technologies, 307
Mobility:
  apparent, 118, 120
  electron, 47–51, 62
  hole, 47–49, 51, 62
  ion, 52
Molar extinction coefficient, 89
Morris-Lecar model, 326–328
MOSFET-based circuits:
  amplifier, 251–256
  biasing circuits, 256–260
  current mirror, 270
  differential pair, 270
  diode connected, 269
  exponential, 272
  graphical analysis, 252–253
  high-frequency operation, 265–266
  load-line, 253
  logarithmic, 273
  low-frequency operation, 264–265
  operational amplifier, 266–267
  organic, 274–276
  signal transfer characteristics, 254
  square-root, 273
  transconductance, 271

Nernst-Planck equation, 193–202
Nernst potential, 139, 197, 200, 203, 289, 322, 327, 335

Neural networks, 315, 333, 348, 395–397
Neurobioengineering, 338
Neuroelectronic junction, 338–346
Neurons:
    action potential, 317–324
    axon, 316
    biology, 315–317
    central pattern generators (CPG), 336–338
    channel conductances, 319–321
    dendrites, 316
    depolarization, 322
    equilibrium potential, 319, 322, 327, 335
    excitatory postsynaptic potential, 333
    Fitzhugh-Nagumo model, 326
    Hodgkin-Huxley (HH) model, 322–324, 335
    hyperpolarization, 322
    inhibitory postsynaptic potential, 334
    integrate-and-fire model, 328, 331
    models, 380–383
    Morris-Lecar model, 326–328
    networks, 336–338
    nodes of Ranvier, 316
    potential fluctuations, 334–336
    refractory period, 322
    resting potential, 318–319
    reversal potential, 331, 333
    silicon, 346–348
    sodium-potassium pump, 318
    soma, 315
    SPICE model, 383–389, 390–397
    synapses, 328–333
    threshold potential, 321

Ohm law, 52, 197, 249
Orbitals:
    atomic, 8–12
    bonding, 13, 16
    molecular, 12–13

Partition coefficient, 195, 204, 206
Pauli exclusion principle, 8
Permeability coefficient, 195, 204, 206
pH, 85, 182, 291
Phospholipids:
    liposome, 99
    micelle, 99
    monolayer, 99
Photoconductivity, 24
Photoemission, 23
Photogeneration, 23, 41
Photolithography, 307

Photons:
    absorption, 6
    emission, 6
    energy, 6, 23
Planck constant, 6, 173, 205
pn junction, 135–168
    abrupt or step, 135
    bias, 148–149, 161, 163
    Boltzmann relations, 140
    boundary values, 150
    built-in potential, 139
    charge-control equation, 160
    continuity equation, 152
    current-voltage characteristic, 149–156
    depletion approximation, 144
    depletion-region width, 138, 146, 148, 163
    diffusion capacitance, 157, 159
    electric field, 144, 145
    energy bands, 140, 141
    junction capacitance, 157
    law of the junction, 150
    linearly graded, 146
    long-base analysis, 152–153
    Poisson equation, 142
    reverse saturation current, 153, 155
    short-base analysis, 153–155, 158
    transient behavior, 160
Point of zero charge, 182, 291
Poisson-Boltzmann equation, 187–192
Poisson equation, 73, 142, 188
Potential:
    action, 317–324, 328
    built-in, 139, 171
    contact, 227
    electrochemical, 194
    excitatory postsynaptic, 333
    fluctuations, 334–336
    Gouy-Chapman, 185, 186
    Hall, 76
    Nernst, 139, 197, 200, 203, 289, 322, 327, 335
    resting, 318–319
    reversal, 331, 333
    surface, 218, 223
    threshold, 321
Proteins:
    α helix, 93
    β sheet, 95
    G-proteins, 103
    membrane, 101–104
    structure, 92–93, 95
    synthesis, 97

Quantum numbers, 8

Random walk, 111–113
Recombination process:
  nonradiative, 25
  radiative, 25
Reference electrode, 178, 287, 289, 290, 302, 360
Resistivity:
  silicon, 52, 53
Ribonucleic acid (*see* RNA)
Ribosomes, 98
Richardson constant, 173
RNA, 96–98
  messenger, 97
  polymerase, 97
  structure, 96
  transfer, 96, 98

Schottky diode, 169–173
  current-voltage characteristic, 172–173
  depletion region width, 171
  diffusion theory, 172
  junction capacitance, 171
  saturation current, 173
  thermionic theory, 172
Self-assembled monolayers, 307
Semiconductor:
  compound, 17
  doped, 35–40
  electronic configuration, 11
  intrinsic, 33
  n-type, 35, 36
  optical properties, 21–26
  properties, 18–19
  p-type, 39
Shockley equation, 155
Shockley-Hall-Read function, 69
Silicon:
  bond structure 16

Silicon *(Cont.)*
  dopants, 18–19
  element, 10
  intrinsic, 33
  microcantilevers, 307, 309–310
Sodium-potassium pump, 318
Solid-electrolyte interface:
  blocked, 178
  electrified, 179
  nonpolarizable, 178–179
  polarizable, 178–179
  semiconductor-insulator-electrolyte, 180
SPICE, 351
Stern layer, 191
Surface stress, 310

Thermal generation, 27, 37, 41
Thermal velocity, 45, 46, 56, 65, 66
Thermal voltage, 57
Threshold:
  frequency, 22, 25
  potential, 321
  voltage, 220, 223, 224, 225, 227, 228, 231, 235, 240, 243, 255, 267, 290, 359, 360
Top-down, 92
Transport equation, 58

Velocity:
  drift, 46, 47
  thermal, 45, 46, 56, 65, 66

Water:
  dipole, 83
  ionized, 84–86
Wave equation, 7

Zener breakdown, 162
Zener diode, 162

ABOUT THE AUTHORS

MASSIMO GRATTAROLA received the Laurea degree in Physics from the University of Genova, Italy, in 1975. He is Professor of Bioelectronics in the Department of Biophysical and Electronic Engineering (DIBE) at the University of Genova, where he teaches courses on bioelectronics, biochemistry, and biomedical technologies, offered mainly to students of the biomedical engineering curricula. His research interests include interfacing networks of neurons to arrays of microelectronic devices (neurobioengineering); silicon-based biosensors and their application to monitor the cell microenvironment (cellular engineering); and development of bioelectrochemical techniques (scanning probe microscopies). On the above topics he has contributed more than 60 papers to international journals.

GIUSEPPE MASSOBRIO received the Laurea degree in Electronic Engineering from the University of Genova, Italy, in 1976. He is Research Associate in the Department of Biophysical and Electronic Engineering (DIBE) at the University of Genova. Since 1976, he has worked on semiconductor power device modeling and circuit design and simulation. Since 1987, he has been working on modeling semiconductor-based biosensors and neuronal structures. His extensive background in microelectronic device modeling includes teaching and research activities. He has contributed several papers in the fields of bioelectronics and semiconductor device modeling. He is coauthor of the book *Semiconductor Device Modeling with SPICE* published by McGraw-Hill.